AF845023

Implement AS 9100 D/EN9100:2018 for Business Excellence

Quality Management Systems - Requirements for Aviation, Space and Defence Organisations (Same as EN 9100: 2016; Includes ISO 9001:2015)

RIVER PUBLISHERS SERIES IN TRANSPORT TECHNOLOGY

Series Editors

HAIM ABRAMOVICH
Technion - Israel Institute of Technology, Israel

THILO BEIN
Fraunhofer LBF, Germany

The "River Publishers Series in Transport Technology" is a series of comprehensive academic and professional books which focus on theory and applications in the various disciplines within Transport Technology, namely Automotive and Aerospace. The series will serve as a multi-disciplinary resource linking Transport Technology with society. The book series fulfils the rapidly growing worldwide interest in these areas.

Books published in the series include research monographs, edited volumes, handbooks and textbooks. The books provide professionals, researchers, educators, and advanced students in the field with an invaluable insight into the latest research and developments.

Topics covered in the series include, but are by no means restricted to the following:

- Automotive
- Aerodynamics
- Aerospace Engineering
- Aeronautics
- Multifunctional Materials
- Structural Mechanics

For a list of other books in this series, visit www.riverpublishers.com

Implement AS 9100 D/EN9100:2018 for Business Excellence

Quality Management Systems - Requirements for Aviation, Space and Defence Organisations (Same as EN 9100: 2016; Includes ISO 9001:2015)

Thomas Kannattumadom

Aerospace Industry Experienced Auditor, India

River Publishers

Routledge
Taylor & Francis Group
NEW YORK AND LONDON

Published 2025 by River Publishers
River Publishers
Broagervej 10, 9260 Gistrup, Denmark
www.riverpublishers.com

Distributed exclusively by Routledge
605 Third Avenue, New York, NY 10017, USA
4 Park Square, Milton Park, Abingdon, Oxon OX14 4RN

Implement AS 9100 D/EN9100:2018 for Business Excellence – Quality Management Systems - Requirements for Aviation, Space and Defence Organisations (Same as EN 9100: 2016; Includes ISO 9001:2015) / Thomas Kannattumadom.

©2025 River Publishers. All rights reserved. No part of this publication may be reproduced, stored in a retrieval systems, or transmitted in any form or by any means, mechanical, photocopying, recording or otherwise, without prior written permission of the publishers.

Routledge is an imprint of the Taylor & Francis Group, an informa business

ISBN 978-87-4380-068-2 (hardback)
ISBN 978-87-4380-182-5 (paperback)
ISBN 978-87-4380-070-5 (online)
ISBN 978-8-743-80069-9 (ebook master)

While every effort is made to provide dependable information, the publisher, authors, and editors cannot be held responsible for any errors or omissions.

Contents

Foreword

Mr Thomas Kannattumadom (K. T. Thomas) served in the quality assurance department of Indian military aviation for nearly 30 years, making significant contributions to our defence industries and the military services. I am delighted that he continued his passion for improving Quality Management Systems in the aviation, space and defence industries after his superannuation from the Government assignment. Over the last 10 years, he has interacted with more than 200 organisations to guide world-class aviation quality systems and processes. This book (following the earlier edition, which I also had read delightfully) will help Mr Thomas contribute to many industries dealing with the quality system of aviation products.

Quality Management Systems have continuously evolved over the last 25 years, culminating in the latest iteration, AS 9100 D/EN 9100:2018. More and more organisations have adopted these quality systems for their business excellence. Business in aviation, space and defence organisations demands the continuous growth of safety and quality standards in the aerospace industry.

This book is an ideal reference guide for many organisations that wish to achieve excellence in quality by adopting world-class practices in upgrading their quality management processes. Throughout the text, Mr Thomas has demonstrated his depth of experience in quality assurance, both as a regulatory authority official in the Directorate of Aviation Quality Assurance in the Ministry of Defence, where I had the opportunity to work as the Director General, and as an expert in aviation quality systems. Mr Thomas is now an independent auditor for many organisations in India and abroad. I am sure the deliberations in the book will benefit many quality system practitioners in establishing the culture of right first time and every time across multiple industries.

This book is a rich contribution by a practising professional passionate about spreading the message of quality standards. The style is easy to follow.

The author has made painstaking efforts to explain specific quality system elements, such as addressing operational risks, managing regulatory requirements within the organisation's context, aligning with design specifics, etc. Interleaved with engaging anecdotes, explained in the book, it will be an asset for any aviation, space and defence professional.

I am sure this book will provide valuable practical input to all budding professionals who are driving quality in aviation and related sectors. My best wishes to Mr Thomas for this great stride!

Pulak Sengupta
Emeritus Professor, National Institute of Technology Patna
Former Director General, Aeronautical Quality Assurance,
Ministry of Defence, and
Former Executive Director (Quality),
Hindustan Aeronautics Limited, Bangalore

Preface

This book is a continuation of the first guidebook titled 'A Complete Guide to Implement Aerospace Standard AS 9100 (Including ISO 9001:2015, AS 9110 REV C and AS 9120B) Quality Management System Requirements for Aviation, Space and Defence Organisations, MRO Organisations and Distributors' published in 2012. ISO 9001 and AS 9100 standards have been revised and updated. This book explains, clause by clause, all the changes and improvements made to both Standards. Organisations and professionals will use the revised Standard to establish, implement, maintain and continually improve the Quality Management System and to get the organisation audited and certified. While primarily developed for the aviation, space and defence industries, this Standard can also be used in all other industry sectors where a Quality Management System with additional requirements over an ISO 9001 system is needed.

Obtaining certification for the QMS in the organisation should not be a goal for any organisation. Instead, it is the best practice in the world regarding Quality Management Systems. The systematic and effective implementation of the AQMS throughout the organisation can significantly benefit the organisation, leading to improved processes, customer satisfaction and business excellence. It is the starting point in pursuing continual improvement for business excellence.

The intention of the guidance material given in this book is to assist organisations and personnel in understanding the various topics' requirements and view them through the interpretations of a practising auditor who has a rich and wide variety of experience in the design, development, production and maintenance of aircraft and aero-engines for civil and military applications. Therefore, using the guidance material helps comply with the referenced QMS standards. The tools, templates, sample documents, procedures and formats for necessary records are all available in the first guidebook: 'A Complete Guide to Implement Aerospace Standard AS 9100 C'. The same documents, procedures and formats can be used/referred to implement the updated Standard AS 9100:2016 with customised changes.

It is advised to purchase copies of the Standards AS 9100 D/EN 9100:2018 from www.sae.org and use them while reading this book. All chapters and paragraph numbers in this book align with the Standard clauses for easy reference.

Thomas Kannattumadom
Bangalore 560077

Acknowledgements

All the information in this self-help guide are the lessons learned, practices followed and seen during the four decades of service/experience in the industry, as well as my opinions and views formed during the assessment service. It became possible to share the subject information in this manner only through the involvement, interaction, encouragement and support of so many personnel. Though unable to name them all, my wholehearted and sincere thanks go to each.

My special thanks go to:

- My colleagues in DGAQA and peers in certification bodies.
- My friends in various industries and ASD organisations.
- Shri. Pulak Sengupta, Emeritus Professor, National Institute of Technology, Patna, who wrote the foreword for this work.
- Shri Sreelal Sreedhar, Associate Director (Reliability and Quality Assurance), GTRE, DRDO, Bangalore.
- Shri Sastry Veluri, who is the senior industry principal of aerospace engineering at Infosys Bangalore for the industry expert reviews.
- My family for the continued support they provide.

This work is dedicated to:

- My parents who taught me to love, to share values, to improve performance and to achieve objectives.
- Two of my mentors and guides throughout my education and career: Shri K. T. Antony, Kannattumadom and Lt. Gen. (Retd.) T. C. Joseph, AVSM, respectively.

The figures and text taken from ISO 9000:2015 Quality Management Systems – Fundamentals and Vocabulary and ISO 9001:2015 (E) Quality Management Systems – Requirements, respectively, are reproduced with permission to reproduce extracts from British Standards granted by BSI Standards Limited (BSI) via email on Monday, 11 July 2016, 19:22:15 IST. No other use of this material is permitted. These standards can be obtained from any ISO member or the website of the ISO Central Secretariat at the following address: www.iso.org. Copyright remains with ISO. It is learned that the standards' revision status will remain unchanged until 2030. The original publisher-authorised revision status of specification, SAE and national or international standard numbers must be referred to for correct knowledge.

It is acknowledged that information has been gathered on the 'Requirements' from the latest AS/EN/JISQ 9100:2016 standard and various other SAE published standards (Bibliography), such as AS 9101 Rev G (November 2022), etc. Copyright © SAE International. Any revision of the requirements may happen after the next revision of ISO 9001:2015, which will be likely after 2030.

Requirements of the applicable QMS standards shall be followed for certification purposes. Commentaries described in this book are for guidance and compliance of personnel and organisations interested in the Aerospace Quality Management System Standards described herein. Even though all the discussions/opinions expressed agree with the Standards requirements, the views and opinions expressed in this book are those of the author in his capacity. They are not intended to replace the criteria specified in the international standards.

About the Book

This is the 2024 updated/revised edition of 'Implement AS 9100 D/EN9100:2018 for Business Excellence'. The first edition hard copy was sold only in India, and the soft copy on Kindle was sold worldwide in 2017. Revision updating has been done over 2023 and 2024. The next revision of the Standard will likely be in 2030, and the book is relevant until then.

SAE Aerospace Standard AS 9100 Rev D (Including ISO 9001:2015) Quality Management System Requirements for Aviation, Space and Defence Organisations, ISO 9001 and SAE AS 9100 standards have been revised and updated. This book indicates and explains clause by clause all the changes and improvements made to both the Standards. Organisations and professionals will use the revised Standard to establish, implement, maintain and continually improve the Quality Management System and to get the organisation audited and certified. While primarily developed for the aviation, space and defence industries, this Standard can also be used in all other industry sectors where a Quality Management System with additional requirements over an ISO 9001 system is needed.

Obtaining certification for the QMS in the organisation should not be a goal for any organisation. Instead, it is the best practice in the world regarding Quality Management Systems. The systematic and effective implementation of the AQMS throughout the organisation can benefit the organisation. It is the starting point in pursuing continual improvement for business excellence. Experience of various problems and solutions during the production and maintenance of ASD products are juxtaposed within the implementation guidance for a clear understanding.

The intention of the guidance material given in this book is to assist organisations and personnel in understanding the various topics' requirements and view them through the interpretations of a practising auditor who has a rich and wide variety of experience in the design, development, production and maintenance of aircraft and aero-engines for civil and military applications for about 30 years and 16 years' experience in the certification audit during post-retirement career.

List of Figures and Table

List of Abbreviations

AAB	Auditor Authentication Body
AAQG	Americas Aerospace Quality Group
AAQSC	Americas Aerospace Quality Standards Committee
AATT	Aerospace Auditor's Transition Training
AB	Accreditation bodies
ABC	Activity-based costing
AHSP	Authority Holding Sealed Particulars
AIA	Aerospace Industries Association
AMI	Advance modification information
ANAB	ANSI National Accreditation Body
ANSI	American National Standards Institute
APAQG	Asia-Pacific Aerospace Quality Group
ASD	Aviation, space and defence
ASIP	Aircraft Structural Integrity Programme
ATA	Air Transport Association
ATE	Automated test equipment
C of C	Certificate of competency
CAB	Conformity Assessment Body
CAD	Computer-aided design
CAE	Computer-aided engineering
CAM	Computer aided-manufacture
CATIA	Computer-aided three-dimensional interactive approach
CB	Certification body
CDR	Critical design reviews
CEMILAC	Centre for Military Airworthiness and Certification
CFD	Computational fluid dynamics
CMM	Coordinate measuring machine
CMP	Compression moulding
CNC	Computerized numerical control
COP	Customer-oriented processes
COPQ	Cost of poor quality
COTS	Commercial off-the-shelf

CP	Core processing
CSI	Critical safety item
CSM	Characteristic selection matrix
CTQ	Critical to Quality
DDA	Digital radiographic testing
DGCA	Director General of Civil Aviation
DMSMS	Diminishing manufacturing sources and material shortages
DOE	Design of experiments
EAQG	European Aerospace Quality Group
EAR	Export Administration Regulations
EASA	European Aviation Safety Agency
ECG	Electrochemical grinding
ECM	Electrochemical machining
ECU	Engine control unit
EDM	Electrical discharge machining
EEC	Electronic engine controller
ERM	Enterprise risk management
ERP	Enterprise resource planning
ESD	Electrostatic discharge
EVM	Earned Value Management
FAA	Federal Aviation Agency
FADEC	Full authority digital engine control
FAI	First Article Inspection
FAIR	First Article Inspection Report
FEA	Finite element analysis
FEM	Finite element method
FIFO	First in first out
FLIMSS	Function, life, interchangeability, maintainability, safety and strength
FMEA	Failure mode and effects analysis
FMECA	Failure mode, effects and criticality analysis
FMS	Flight management system
FOD	Foreign object damage/ Foreign object debris
FTA	Fault-tree analysis
HLS	High-level structure
IAF	International Accreditation Forum
IAQG	International Aerospace Quality Group
ICT	Information and communication technology
ICY	Interchangeability

IDTO	Inter-divisional transfer order
IPC	Illustrated parts catalogues
ISIC	International Standard Industrial Classification of All Economic Activities
IT	Information technology
ITAR	International Traffic in Arms Regulations
IV&V	Independent verification and validation
KPI	Key performance indicator
LBM	Laser beam machining
LRP	Liquid resin processing
LRU	Line replaceable unit
MandTE	Measuring and test equipment
MB	Metal bonding
MBD	Model-based definition
MCU	Machine control unit
MEDA	Maintenance Error Data Analysis and Reporting to Customer
MOE	Maintenance organisation expositions
MOP	Management-oriented processes
MOU	Memorandum of understandings
MPS	Master process sheet
MR	Management representative
MRO	Maintenance repair and overhaul (sometimes Operations)
MSDS	Material safety data sheet
MTBF	Mean time between failures
MTTR	Mean time to repair
NACE	Statistical Classification of Economic Activities in the European Community
NADCAP	National Aerospace and Defence Contractors Accreditation Program
NAS	National Aviation, Defence, and Space Standard
NCR	Nonconformity report
NCSL	National Conference of Standards Laboratories
NDE	Non-destructive evaluation
NDI	Non-destructive inspection
NDT	Non-destructive testing
OASIS	Online Aerospace Supplier Information System
OCM	Original component manufacturer
OEM	Original equipment manufacturer

OFI	Opportunities for improvement
OIN	OASIS identification number
OPMT	Other Party Management Team
OTACE	On Time and Above Customer Expectations
OTD	On-time delivery
PAR	Prepreg/adhesive bonding/resin film infusion
PDCA	Plan-Do-Check-Act
PDR	Preliminary design reviews
PEAR	Process Effectiveness Assessment Report
PEARL	Process Enhancement Framework and Repository of Best Practices and Learning
PFMEA	Process failure mode effect analysis
PMA	Part manufacturer approval
PO	Purchase order
POE	Production organisation exposition
PPC	Production planning and control
PPE	Personnel protective equipment
PPV	Production process verification (Previously known as First Article Inspection (FAI)
QFD	Quality function deployment
QPL	Qualified products list
REACH	Registration, Evaluation, Authorisation and Restriction of Chemicals
RFP	Request for purchase
RFQ	Request for quotation
RPN	Risk exposure number/Risk priority number
SI	System international
SIPOC	Supplier-Input-Process-Output-Customer
SJAC	Society of Japanese Aerospace Companies
SLA	Service-level agreement
SMART	Specific, measurable, achievable, realistic, and time-bound
SME	Small- and medium-sized enterprise
SOP	Standard of preparation
SOP	Standard operating procedures
SOP	Support-oriented processes
SOR	Statement of requirements
SOW	Statement of works
SPC	Statistical process control
SUP	Suspected unused part

SWOT	Strength, weakness, opportunities and threats
TBO	Time between overhauls
TC	Technical Committee
TSO	Technical Standard Order
UFO	Unidentified flying object
USML	United States Munitions List
VSM	Value stream mapping
VTOL	Vertical take-off and landing

Introduction

0.1 General Comments

The quality of an object is defined as the degree to which a set of inherent characteristics fulfils requirements (ISO 9000–2015). Two keywords in the definition are 'Characteristics' and 'Requirements'. Characteristics are:

1. Physical (dimensions, size, weight, etc.)
2. Sensory (sight, smell, taste, touch and hearing)
3. Behavioural (activities and interactions, courtesy, honesty)
4. Temporal (relating to time, earthly, punctuality, reliability and availability)
5. Ergonomic (the science of designing the job, equipment and workplace to suit the operator. Proper ergonomic design is necessary to prevent injuries/effects (affecting health and productivity), which can develop over some time and can lead to long-term disability, physiological characteristics or related to human safety)
6. Functional (able to fulfil its purpose or function, measurable characteristics; these may be either the characteristics of the physical means used to deliver the service, e.g., the maximum speed of an aircraft, or of the environment in which the service is provided, e.g., the interior temperature or facilities in an aircraft).

Requirements are the needs or expectations, which are stated (through indicated requirements, expressed desires), implied (suggested without being directly or explicitly stated; tacitly understood) or obligatory (compulsory, morally or legally binding). A specific type of requirement may be a product requirement, Quality Management System requirement or customer requirement. Different interested parties can generate requirements. The requirement is an expression in the content of a document conveying criteria to be fulfilled if compliance with the document is to be claimed and from which no deviation is permitted.

'Generally implied' means that it is custom or common practice for the organisation, its customers and other interested parties that the need or expectation under consideration is implied.

Quality control is part of quality management focused on fulfilling quality requirements. Quality assurance is that part of quality management focused on providing confidence that quality requirements will be fulfilled. Therefore, quality control and assurance have different and distinct/separate functions. Quality control is a direct function, whereas quality assurance, generally carried out by a third party, ensures that the organisation does quality control effectively.

Conformity, Effectiveness and Efficiency

Organisations generally benefit from having an effective Quality Management System (QMS). A Quality Management System is typically defined as: 'A set of coordinated activities to direct and control all the processes in an organisation. QMS enables the organisation to carry out processes conforming to the best practices in the world, continually improve the performance and efficiency of its processes and, most importantly, to be effective'. Customer expectations inevitably drive and define 'performance' criteria and standards. Therefore, Quality Management Systems focus on customer expectations and continually review and improve its processes. The main thrust of a QMS is defining and implementing the methods that will produce quality products and services. A Quality Management System defines how an organisation can meet the requirements of its customers and other stakeholders affected by its work, as defined by the Standard. It helps businesses and organisations to be more efficient and improve customer satisfaction.

Implementing a Quality Management System helps your organisation to:

a. Assess the overall context of the organisation to define who is affected by the work and what they expect from your organisation. This will enable your organisation to state the objectives and identify new business opportunities.

b. Customers are the focus of the business, ensuring you consistently meet their needs and enhance their satisfaction. This can lead to repeat orders from existing customers, getting new clients and increased business for your organisation.

c. Work more efficiently as all the processes in the organisation will be aligned and understood by everyone in the organisation. This increases productivity and efficiency and brings down internal operating costs.

d. Meet the necessary statutory and regulatory requirements.

e. Expand into new market segments.

f. Identify and address the type of risks associated with your organisation.

International Organisation for Standardization (ISO) and QMS Principles

ISO is a worldwide federation of national standards bodies and a network of National Standards Institutions from 165 countries. Through its members, it brings together experts to share knowledge and develop voluntary, consensus-based, market-relevant international standards that support innovation and provide solutions to global challenges. ISO started operation in February 1947. It is a non-governmental organisation with its headquarters located in Geneva, Switzerland. The organisation's objective is 'facilitating international cooperation and unification of industrial standards'. Preparing international standards is carried out through ISO technical committees in liaison with international organisations and governmental and non-governmental bodies. The current sets of standards issued by ISO for Quality Management Systems are:

- ISO 9000:2015, Quality Management Systems – fundamentals and vocabulary
- ISO 9001:2015 Quality Management Systems – requirements
- ISO 9004:2009 Managing for the sustained success of an organisation – a quality management approach

ISO standards provide a model for setting up and operating a management system. They result from international, expert consensus and offer the benefit of global management experience and good practice. These standards are generic and can be applied to any organisation, large or small, whatever the product or service, regardless of the sector of activity.

The benefits of an effective management system include the following:

- more efficient use of resources,
- improved risk management, and

- increased customer satisfaction as services and products are delivered consistently, as they promise.

Quality management standards based on the ISO now number more than 21,000. QMS Standards are periodically updated to remain relevant by considering the different challenges and needs that businesses face occasionally. ISO has a scheduled revision of each Standard every five to seven years. They are built around business processes, with a strong emphasis on improvement and a focus on meeting customers' requirements. The updated standards originated from a regular periodic review and are intended to be generic and adaptable to various organisations.

Every organisation is unique. Therefore, strategic planning and organisational decisions are matters of pragmatic choice regarding what helps the organisation best. Adopting a QMS needs to be an organisation's strategic decision. It is influenced by varying needs, objectives, the products/services provided, the processes employed and the size and structure of the organisation. A QMS must ensure that the products/services offered meet customer requirements and expectations and the organisation's objectives.

A new version of the Standard, ISO 9001:2015, has just been released, replacing the previous version (ISO 9001:2008). The significant changes/improvements made over the earlier versions of the ISO 9001 are detailed below in subsequent paragraphs.

- AS/EN 9100 is a quality management standard requirement for the aviation, space and defence (ASD) organisations/industry, published by the International Aerospace Quality Group (IAQG). All previous versions of the AS/EN 9100 quality standards were based on ISO 9001, adding requirements specific to the ASD segment while striving to meet the needs of stakeholders. The latest version of the Standard AS/EN 9100 Rev D (2016) was released in September 2016 after aligning with the ISO 9001:2015. The latest AS/EN 9100 Rev D (2016) version incorporates ISO TC (Technical Committee) 176 changes to the ISO 9001:2015 requirements. Further, IAQG clarifications to the requirements were added as notes. In addition, aviation, space and defence stakeholders' needs since the last revision were also identified and incorporated. Some of the improvements made in the requirements of ISO 9001:2015 were already there in the AS/EN 9100:2009 as additional requirements for aviation, space and defence organisations (ASD). These sector-specific requirements have been reiterated, re-emphasised and re-worded when the AS/EN 9100 Rev D aligned with ISO 9001:2015.

- International Aerospace Quality Group (IAQG) comprises three sectors, divisible geographically: the Americas, Asia-Pacific and Europe. Each sector manages its own rules of membership and is responsible for deploying IAQG initiatives within its sector.
- Americas Aerospace Quality Group publishes AS 9100 D standards through SAE International.
- European Aerospace Quality Group publishes EN 9100:2018 standards through the ASD-STAN.
- Asia Pacific Aerospace Quality Group publishes EN 9100: 2018 standards through the Society of Japanese Aerospace Companies (SJAC).

All three versions of the standards are precisely similar. It standardises Quality Management System requirements to the greatest extent possible. It can be used at all levels of the supply chain by organisations around the world.

The change in the title of the Standard is made accordingly. Further improvements have been made in the description of the requirements to bring more clarity to the understanding. More details have been included for compliance with the requirements, such as:

a. Context of the organisation

b. Product safety

c. Prevention of counterfeit parts

d. FOD control and visual checks

e. Risk register

f. Human factors

Some of the templates (Annexures) are also reviewed and updated.

The significant changes/improvements made in ISO 9001:2015 over the previous versions of ISO 9001 are in terminology and high-level structure, and in the focus on risk-based thinking.

Terminology and high-level structure (HLS): ISO 9001:2015 now follows the same overall structure/format, having 10 clauses. This has been done in accordance with 'Annex SL of ISO/IEC Directives Part 1 (the rules for developing ISO standards)'. All ISO management system standards will hereafter look the same, with the same high-level structure. It will be easier for any organisation to integrate multiple management systems.

The changes in terminology, in brief, are as follows:

Previous version (ISO 9001:2008)	**Current updated version (ISO 9001:2015)**
Products	Products and services
Exclusions	The scope of the QMS is to be formally defined, and all requirements are applicable if they are in the scope.
Documentation, documented procedures, records	Documented information: Documents or procedures are maintained. Records are retained for defined periods (specify).
Purchased product	Externally provided products and services
Supplier	External provider

The structure follows the PDCA approach. As ISO 9001 is the basis for 9100, the new clause structure in AS/EN 9100 Rev D (2016) is derived from ISO 9001.

Introductory clauses	1	Scope
	2	Normative reference
	3	Terms and definitions
Plan	4	Context of the organisation
	5	Leadership
	6	Planning for the QMS
	7	Support
Do	8	Operation
Check	9	Performance evaluation
Act	10	Improvement

From the point of implementation considerations, it is always desirable to review the current QMS structure and adapt and integrate the above QMS structure to the business processes in the organisation.

The structure of clauses is intended to provide a coherent presentation of requirements rather than a model for documenting an organisation's policies, objectives and processes. Therefore, it is unnecessary to reflect the changes in the structure and terminology in the documentation of an organisation's Quality Management System.

Certified organisations can consider re-arranging their QMS documentation around the value stream being followed. A value-stream-based QMS allows the organisation to customise their documentation to unique business needs. The documentation then describes precisely what the organisation does. It integrates the business processes with the QMS requirements. It

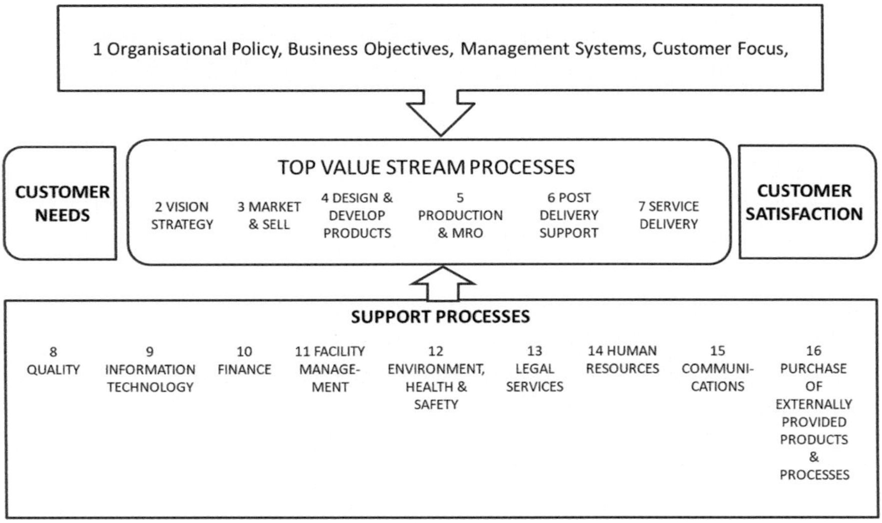

Figure 0.1 Process-based QMS and value stream.

supports the process's compliance with the QMS requirements specified in the Standard. It is a better foundation for future improvements and performance excellence.

An example of a process QMS and value stream is indicated in Figure 0.1:

Focus on risk-based thinking: The new version of the Standard gives increased prominence to risk-based thinking and risk management. Risk-based thinking is something we do, such as the difficulties that we must overcome while planning to do or achieve something. We do that often subconsciously to get the desirable result. The concept of risk has always been implicit in previous versions of ISO 9001. But in the latest version, it is made more explicit. In the revised Standard AS 9100 D/EN 9100:2018, risk-based thinking and its management are reiterated and made an essential part of the whole management system. Risk is the effect of uncertainty on achieving these objectives and intended results.

Risk-based thinking ensures risk is considered right from the beginning, identified, assessed and managed periodically throughout the product realisation. Risk management, otherwise preventive action, becomes part of strategic and operational planning. Successful organisations have taken a risk-based approach as a strategy for their business decisions because, ultimately, it has brought benefits to them. They understand the impact of risk on their operational processes. They have been able to improve customer confidence and satisfaction. Identifying and managing the risks within their capability limits has helped them ensure consistency of goods and services. Preventive actions achieve real improvement in quality. Establishing a proactive culture of preventing errors and mistakes and eliminating potential defects and deficiencies helps improve performance excellence.

Risk-based thinking is a philosophy or culture that the top management must develop across all organisational levels. Risk is the effect of uncertainty on achieving these objectives and intended results. Risk determination and controls must be proactive rather than reactive in preventing or reducing undesired effects through early identification and action. The Standard has deleted reference to preventive actions as a separate clause and has replaced it with a holistic risk management process. Preventive action is the mitigation actions taken while managing the risks.

0.2 Quality Management Principles

ISO 9001:2015/AS 9100D is based on the quality management principles described in ISO 9000. The descriptions include a statement of each principle, a rationale of why the principle is important for the organisation, some examples of benefits associated with the principle and examples of typical actions to improve the organisation's performance when applying the principle. Quality Management System standards developed by the ISO have the following principles embedded in their structure and requirements. The quality management principles are:

1. **Customer focus:** Focus on your customers. Organisations rely on customers. Therefore, organisations must understand customer needs. Organisations must meet customer requirements. Organisations must exceed customer expectations. Time delivery of the products/services and delivering the products or services fully meeting the defined quality requirements, as agreed with the customer, are the two most important aspects of the customer focus.

2. **Leadership:** Provide leadership. Organisations rely on leaders and management. Therefore, leaders must establish a unity of purpose and set the organisation's direction. Leaders must create an environment that encourages people to achieve the organisation's objectives and customer goals.

3. **Engagement of people:** Involve all the people in the organisation. Organisations rely on people. Therefore, organisations must encourage

people's involvement at all levels to achieve their objectives and performance goals. Organisations must encourage people to develop and use their abilities, competency and skills to benefit the organisation and customers.

4. **Process approach:** Use a process approach for all the activities (processes) in the organisation. See the 'Schematic Representation of the Elements of a Single Process'. Organisations can be more efficient and effective when they use a process approach and treat every process with identified input, steps/stages and output. Control the inputs and actions to get the desired output. Identify the potential problems/ nonconformities that can happen in the process, assess their impact on achieving the objectives and manage the high risks to keep them within acceptable levels. Measure the output/product/results against planned product requirements and process objectives. The 'Plan-Do-Check-Act' (PDCA cycle) methodology can be applied to all processes. Therefore, organisations must use a process approach to manage all activities. Organisations are more efficient and effective when they use a process approach. Therefore, organisations must identify the sequence and interactions of all interrelated processes and treat them as a system (see the business process map/Sun and Planets Diagram as an example, not a requirement). Organisations must use a systems approach to manage all the interrelated processes. The system approach earlier indicated as one of the eight quality management principles has been merged with the process approach for managing the processes.

5. **Improvement:** Encourage improvements. Organisations are more efficient and effective when they continually try to improve every process, bit by bit or innovating. Therefore, organisations must make a permanent commitment to continually (step-by-step process) improve their overall performance process objectives, improve quality and on-time delivery performance, and improve product/service features and characteristics for the benefit of both the customers and the organisation.

6. **Evidence-based decision-making:** Get the facts before making any decisions. Organisations perform better when their findings are based on facts and figures. Therefore, organisations must base their findings on analysing information and data collected from every process.

7. **Relationship management:** Work with the suppliers (external products and service providers), guide and control them. Organisations depend on their suppliers to create value for their products and services

to benefit the customers and the organisation. Therefore, organisations must maintain a mutually beneficial relationship with their suppliers. Suppliers are long-time business partners. All communications must be practical and closed loop. Requirements must be communicated effectively and without any ambiguity. If you are not clear and specific in indicating or defining the requirements in writing, what the external providers provide may not meet those requirements. Any discrepancy can lead to (the opposite of beneficial = detrimental, damaging, harmful) a mutually unhelpful relationship! In a buyer–seller relationship, the buyer is responsible for the quality of what is bought! 'The organisation shall be responsible for the conformity of all externally provided processes, products, and services, including from sources defined by the customer' (8.4.1).

Quality is achieved through planning and appropriate communication to prevent problems. It is cheaper and more effective to run preventive systems than inspection-based ones. Preventive measures include planning activities and processes, analysing inputs, steps and outputs, training, writing and implementing procedures or exact plans and auditing the activities. The concept of preventive action is now addressed throughout the Standard by risk identification and mitigation.

It is necessary to deal with the 'process approach' in more detail since it is the most important principle that must be followed for establishing, documenting, implementing and maintaining a Quality Management System and continually improving its effectiveness per the requirements of the international standards ISO 9001 and AS 9100 series, to enhance customer satisfaction by meeting customer requirements. The process approach is now strengthened by integrating the QMS into the organisation's business processes.

0.3 Process Approach and QMS

'The application of a system of processes within an organisation, together with the identification and interactions of these processes, and their management, can be referred to as the process approach'. Identification of various processes in the organisation is the first step. The next step is identifying the sequence and interactions of all the interrelated processes. If the organisation intends to function effectively, that is, carry out the business by ensuring conformity and effectiveness, it must manage all the interrelated or linked processes as a system. Hence, effective Quality Management Systems and process approaches are inseparable.

A process needs to be determined with inherent controls and checks on performance, and it promotes improvement. The process approach enables an organisation to plan its operations and their interactions.

The PDCA cycle within each process enables an organisation to ensure that its operations are adequately planned, resourced and managed and that opportunities for improvement are determined and acted upon.

Consistent and predictable results are achieved more effectively and efficiently when activities are understood and managed as interrelated processes that function as a coherent system.

Process effectiveness is the extent to which planned activities are realised and planned results achieved.

0.3.1 General

All activities and resources can be managed more effectively if considered a process. An activity using resources and managing to enable the transformation of inputs into outputs can be considered a process. Product is the result or output of a process. There are four different types of products: hardware, software, processed materials and services. Some organisations do not have a tangible product but provide services like education, training, heat treatment, special processing and other value-added activities on materials provided by customers. Therefore, whenever one refers to a product, it could mean any product, including services. A procedure is a specified way to carry out a process. Suppose the procedure is written or documented as a text for uniform understanding and practice. In that case, it may be called a documented procedure or documented information.

The output of one process may be the input of another process. Any activity is a chain of processes. Consistent inputs produce consistent outputs. We should control the process inputs and monitor the characteristics for variations and possible improvement. The process approach is pictorially depicted in a simple schematic sketch (Figure 0.2).

It is necessary to periodically analyse and review the processes for adequacy and continual improvement. Adequate control for the inputs and steps ensures desired outputs that meet requirements.

Controls of all processes are achieved through the control of the '7 M' words:

1. Men – by training or qualification, competency
2. Materials – by control of suppliers and specification
3. Machines – by maintenance systems

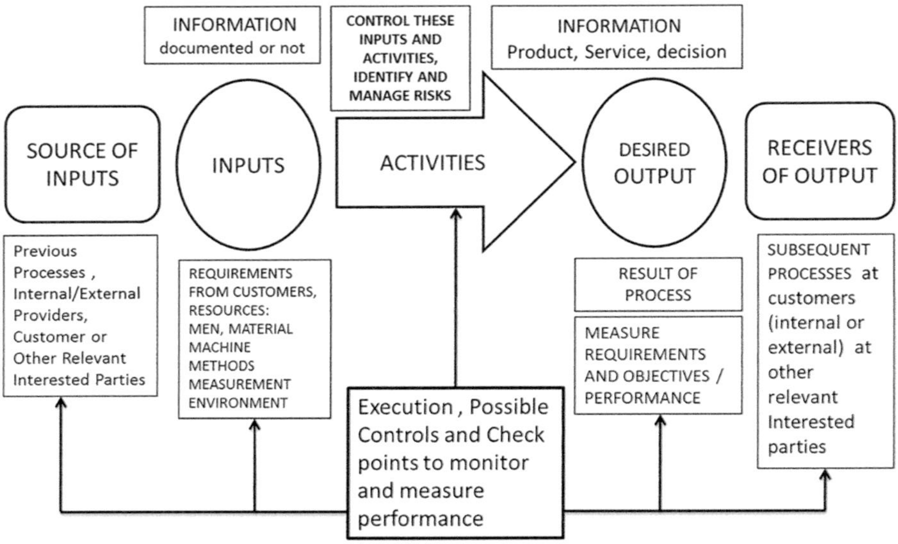

Figure 0.2 Schematic representation of the elements of a single process.

4. Methods – by procedures and standards
5. Measurement – by traceability, accuracy, etc.
6. Environment (mother earth) – by organising and caring
7. We mitigate or reduce the risk level by identifying risks directly or indirectly related to the process and taking appropriate preventive actions.

Process or product nonconformities occur due to inadequate control of these 7M letter words. Some examples are as follows:

a. Manpower: Inadequate capability/competency, lack of knowledge, lack of skill, stress or improper motivation
b. Materials: Defective raw material, wrong type for job, lack of raw material
c. Machine/equipment: Incorrect tool selection, poor maintenance or design, poor equipment or tool placement, defective equipment or tool
d. Methods: Lack of clarity or poor procedures, practices are not the same as written procedures, poor communication
e. Measurement of Quality Management System for effectiveness (management system and measuring devices): Inadequate training or

education, poor employee involvement, poor recognition of hazards and previously identified risks/hazards were not eliminated. Lack of management/supervisory involvement, inadequate attention to task, risk/hazards not dealt with properly, stress demands, lack of process, lack of communication, etc., could also be considered nonconformities arising from the management system.

f. Environment (mother earth): Disorderly workplace, poor job design and layout of work, poorly maintained surfaces, inability to meet physical demands of the task, forces of nature, etc.

g. Risk management is detailed in para 6.1, which addresses the risks and opportunities when planning for the organisation's Quality Management System. The scope of clause 8.1.1 is limited to the risks associated with the operational processes needed to provide products and services.

All the ISO standards and those Quality Management System standards evolved after adopting these principles, which are also reflected in the sector-specific standards, including Aviation, Space and Defence (ASD) Standards 9100/9110/9120. Conformity and effectiveness of the processes are the essence of Quality Management System practices! Conformity is the extent to which planned activities are realised, and effectiveness is the extent to which planned results are achieved.

The efficiency of the process is the ratio between the planned objectives or performance results achieved and the resources used, measured in terms of cost or similar units. Process efficiency is outputs divided by inputs as a percentage, both output and input being expressed in the same units of measurement.

QMS processes are generally classified into three categories as follows:

- Customer-oriented processes (COP): Processes directly determine/influence the quality and delivery of products by the organisation to the customer. These processes are directly linked to external customers by input and output. Examples: Market analysis, submission of quotation, bidding/tendering, product design/validation, product manufacturing, after-sales service, customer complaint handling, etc.

- Management-oriented processes (MOP): Processes primarily driven by the top management and focused on planning, monitoring, evaluation and improvements. Examples: Business development and planning, strategic planning, quality policy and setting the objectives/goal, risk management, resource planning, internal audits, management reviews, etc.

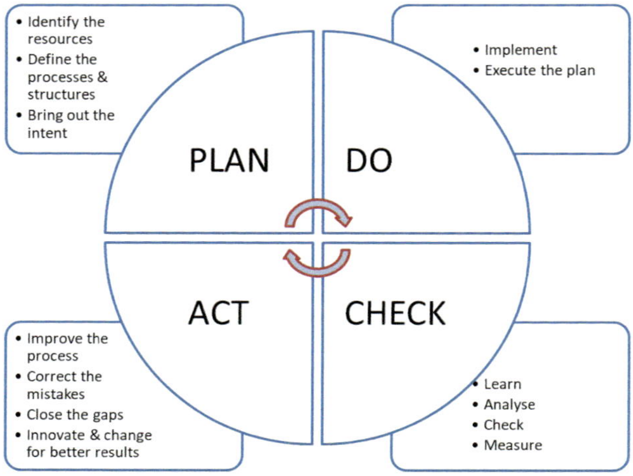

Figure 0.3 Plan-Do-Check-Act cycle (Deming's wheel).

- Support-oriented processes (SOP): These processes directly or indirectly support the MOPs and COPs. Examples: Human resources, training, tooling, maintenance, supplier development, procurement, administration, process and facility planning, finance, etc.

0.3.2 Plan-Do-Check-Act cycle

Every human activity can be perceived as a process. The activity can be completed efficiently and improved by following the 'Plan-Do-Check-Act' cycle, known as 'Deming's wheel'. Refer to Figure 0.3 Plan-Do-Check-Act cycle (Deming's wheel). It is interrelated with the process approach. 'PDCA' cycle can be deployed within each of the processes and to the entire system of processes in an organisation. Deming's PDCA cycle can be illustrated as follows:

Plan: Plan what you want to do. Establish the objectives and processes necessary to deliver results in accordance with customer requirements and the organisation's policies. The organisation's plan and strategy should contain objectives, key performance indicators (KPIs), customer inputs, management inputs, benchmarks, requirements or complaints from other processes, audit results, environmental requirements, previous analyses' results, clearly defining the actions required, resources, etc. Plans may be improved occasionally based on the inputs and outcomes of checking and analysing.

Do: Do what you have planned. Implement the process strictly following the planned activities, stages, sequence, monitoring and measurement. Operations should be managed utilising the required resources.

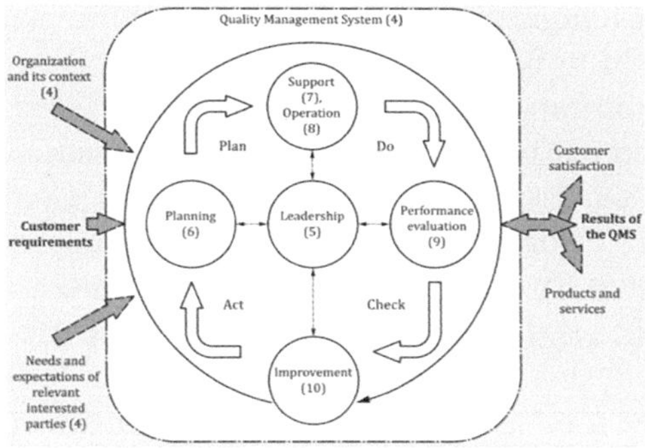

Figure 0.4 Representation of the structure of this international standard in the PDCA cycle. (Note: Numbers in brackets refer to the clauses in ISO 9001:2015/AS 9100D.)

Check: Check whether what you have done is strictly as per plan or not. Did you get the results as planned? Monitor and measure processes and products against policies, organisational or process objectives and requirements for the product. Report the results. Monitoring and measuring the output is to see that the pre-determined requirements for the product and planned objectives are achieved. Analysis would help in arriving at the correct decision – to accept, reject or where to act for improvement.

Act: Write down what is achieved against what was planned. Act on the difference. Take actions to improve process performance and objectives continually. The results give a better understanding of the process. Establish improvements if the planned and desired outcomes are not achieved. Use best practices known or innovative to get the desired improvements and results.

Taken together, the process approach, the PDCA cycle, the focus on customer satisfaction, quality and on-time delivery of product/service and continual improvement of the process by identifying and managing the risks and integrating the business processes with the Quality Management System provide an overall framework for managing the business. Refer to Figure 0.4 Representation of the structure of this international standard in the PDCA cycle. For an organisation to carry out its functions effectively, it must identify and manage several connected activities and processes. Activities using resources, controlling and managing the resources in a series of actions and converting the inputs into outputs, products or services can be considered a process. It may be reiterated here that a process's output often becomes another input. Improving the process and objectives is by analysing the results of all processes. Analysis brings out the factors that directly impact customer satisfaction. If it is beneficial to enhance customer satisfaction or to meet additional or unique customer requirements, the plan and inputs/steps for the process need to be changed or improved for the results of analyses. Evaluating and confirming whether the changes introduced enhance customer satisfaction and improve organisational objectives is necessary. Then, only the processes can be stated as conforming and effective. Conformity and effectiveness are the two cardinal requirements for any process: conformity to the Quality Management System and effectiveness of product quality, on-time delivery performance and meeting the organisation's performance objectives.

In case of analysing an ongoing business process for improvements, the PDCA cycle can be used as a tool in the following manner:

- Plan: Design or revise business process components to improve results
- Do: Implement the plan and measure its performance
- Check: Assess the measurements and report the results to decision-makers
- Act: Decide on changes needed to improve the process

When used with a Quality Management System, the advantage of the process approach is that it provides an unbroken chain of controls over all the processes, individually and over the complete system of interrelated processes within the organisation. Together it proclaims the importance of the following:

- Need to consider processes in terms of adding value to the product/customer
- Understanding and meeting requirements of the output/product/service/customer (conformity)
- Achieving planned results of processes and organisational performance goals (effectiveness)
- Continual improvement based on measurement of requirements and objectives of each of the processes.

A process tool can quickly identify critical components in a business process/customer-oriented process. Using any of the process tools to identify the processes in an organisation is not a mandatory requirement of the Standard. However, adequate descriptions of the process, Supplier-Input-Process-Output-Customer (SIPOC) format, flow chart, process map, turtle diagram, planet diagram, etc., are all examples of process tools. Do not confuse the organisation chart for a collective list of all processes in an organisation – an example of a process map we may call the planet diagram is shown below. Refer to Figure 0.5 Typical process map/planet diagram for a product realisation

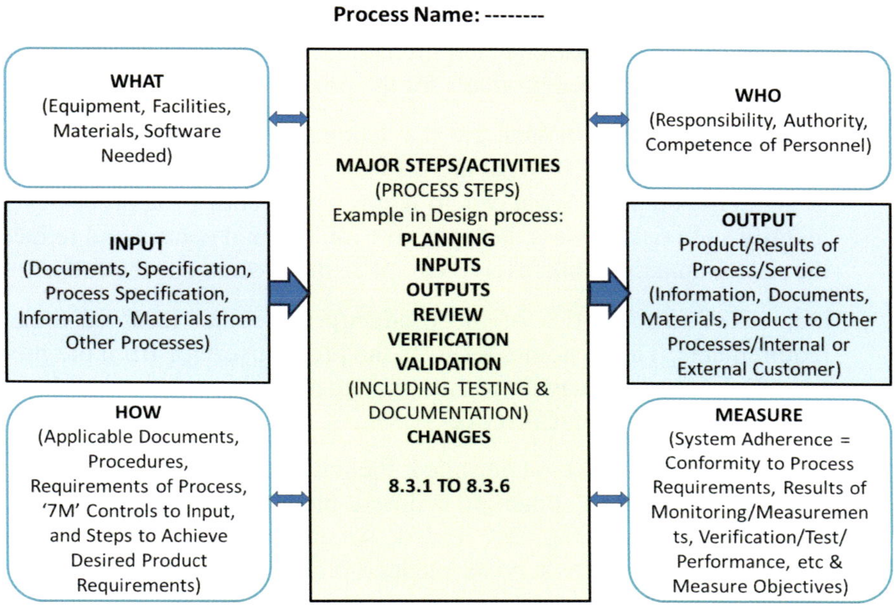

Figure 0.5 Typical process map/planet diagram for a product realisation process .

process. Whichever process tool is used should contain the minimum information for effective management. They are the name of the process and its owner, activities or significant stages/steps, inputs, outputs, controls established (monitoring and measurement) for the process, document/procedure reference, resources necessary for the process, customer and legal requirements and applicable standard clauses (requirements). A typical process map/planet diagram for a customer-oriented process is indicated below as an example.

The following explains further the process elements:

a. Process identification: Process name and process steps as documented or otherwise determined.

b. What: Computer, plant and machinery, process equipment, building, hardware, software, measuring equipment, maintenance, storage/packing, customer property, transportation equipment, raw materials needed for the process, etc.

c. Who: Who is responsible for the process? What are the competency requirements? Who does the process? Is the person authorised to carry out the procedure? Any logistics/assistance required, etc.

d. Inputs: What are the customer requirements for this process, what is the equipment needed, what inputs are required for the conversion process to get the desired output, statutory and regulatory requirements to be fulfilled, and organisational requirements/objectives for the process? What are the purchased products for the procedure?

e. Major steps/activities: What are the actions to reach the output using the inputs and controls? What are the steps in the appropriate sequence to get to the output? ‘What can go wrong’ in executing the processes – identify such risks, assess their possible impact, and control and reduce the risk to an acceptable level to complete the process successfully.

f. Outputs: Characteristics of the product/process that meet customer requirements. The customer receives the product/service from the process. The customer may be within the organisation or the external one who receives the product/service.

g. How: How is the process controlled, including documented procedure, process specification, other procedures, methods, work instructions, requirements of records, etc.? How do we apply the controls necessary to produce/provide the service, manage production process/document changes, any associated sub-processes, support processes, etc.?

h. Measure the results: Measurement/analysis of the product characteristics that should meet the customer inputs, measurement of organisational objectives and performance requirements, conformity to the quality system requirements within the organisation and customer quality system requirements, process output or results achieved and effectiveness of the process, customer satisfaction, including addressing any process/product nonconformities encountered, etc.

Any organisation typically will have several processes depending on the size and type, complexity of the processes and their interactions. A quality manual is required to describe the sequence and interaction between the processes covered by the Quality Management System. Using a business process map or a ‘Sun and Planets Diagram’ or any other tools to indicate the sequence and interactions of the processes in an organisation is not a mandatory requirement of the Standard. However, a pictorial depiction conveys the same purpose as a business process map if it identifies the various processes correctly, especially if it indicates the higher-level processes and their interactions. Below is an example or illustration of identifying the multiple processes that focus on the customer needs linking to the organisation at the centre. Let us call it the ‘Sun and Planets Diagram’, as the Sun provides the energy for all planets, including earthly processes. The organisation is represented here by the Sun, and various top-level core processes, which revolve around the Sun, are defined by

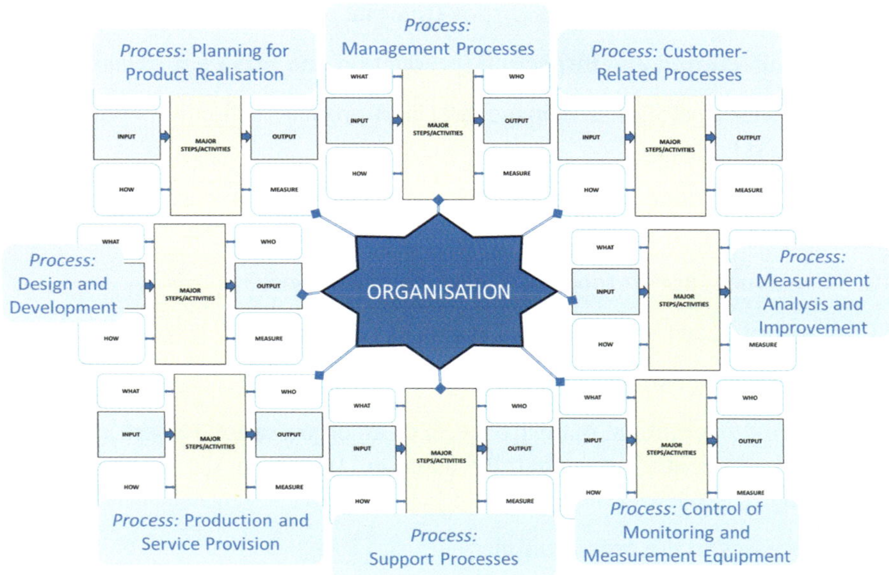

Figure 0.6 Sun and Planets Diagram for typical processes in an organisation.

Planets. The top management (Sun) provides the energy to establish, implement and monitor each process, previously identified as process maps/planet diagrams. A Sun and Planets Diagram for typical processes in an organisation, are depicted in Figure 0.6. Every process may have several sub-processes, which we may call satellite processes. The diagram below does not indicate the satellite processes.

Project management, risk management, configuration management and control of work transfers are all planned activities that bring out a desired result; therefore, processes must be carried out throughout the product realisation stages, as applicable. The exclusion of these requirements is not expected from an organisation's scope of the Quality Management System.

It is cited here as an example of an organisation that has identified the following 12 processes within its quality management system. They are:

1. Sales, master scheduling and sequencing
2. Contract requirements – flow down (customer process)
3. Design and development
4. Supplier sourcing, selection and approval (purchasing process)
5. Planning for product realisation (plant, material, skills, capacity planning and scheduling)

6. Order management and logistics (internal and external)
7. Manufacturing and inspection (production and service provision)
8. Supplier operational management and product validation (purchasing process)
9. Control of nonconformities, corrective and preventive actions (on time, on quality)
10. Customer support (control of service operations)
11. Performance management of business processes
12. Customer satisfaction monitoring

Strategic and operational maturity levels of an organisation to perform effectively and efficiently are generally assessed based on the following five parameters:

a. Processes in the organisation
b. People in the organisation
c. Resources used for the processes and their relevant data
d. Performance metrics of each of the processes
e. Performance metrics and quality objectives of the organisation.

Organisations can assess their maturity level on a scale of 1–5 and find themselves at levels 1, 2, 3 or 4+. Scoring 5 should be considered as not having the desire to improve further. Here is a pen picture of typical organisations when they achieve the levels, say:

- **Level 1:** Beginner organisation. Management is interested in completing production or giving the service. Quality Management System compliance and PDCA approach missing. Results are by chance and not by choice. Improvements are seldom planned and only as a reaction to requests or complaints from the customer.
- **Level 2:** Proactive organisation. Customer focus is visible. Quality Management System implemented. Some results are planned, but planned results are not achieved. Corrective and preventive actions are taken, but they are not effective.
- **Level 3:** Flexible organisation. Focus on the customers and stakeholder needs. A Quality Management System is implemented and moderately

effective. Organisation and processes have defined performance goals and objectives. Planned results are achieved in most cases. Actions are taken when intended results are not fulfilled.

- **Level 4:** Innovative organisation. There is a balanced focus and strategies for enhancing customer satisfaction and stakeholder needs. Conforming to QMS processes and is effective in meeting customer expectations. Effective implementation achieved planned results and organisational performance goals consistently using higher learning and experience for productivity and continual Improvement.
- **Level 5:** Sustainable organisation. Has the ability to maintain high performance consistently. Industry leadership and performance excellence are used for continued growth and sustainability in the field/domain.

Detailed guidelines for strategic assessment by the management and operational review by various process owners, using the PDCA cycle, are given in the guidelines standard ISO 9004: 2009 Managing for the sustained success of an organisation – a quality management approach (see reference).

The effectiveness of processes in an aviation, space and defence organisation is determined by assessing process effectiveness levels (refer AS 9101E).

Aerospace Standard AS 9100

Sector-specific requirements emerged for the aeronautical and aerospace industry and aviation business. The nature of the aerospace business demands certain concepts, issues and industry practices to be more critical than others. While most of the industrial world accepted the ISO 9001 Quality Management System standard as applicable requirements for any product or service, the following were the additional requirements explicitly required for the aeronautical aviation/space/defence sector:

- Safety: Safety for the products and their users. While using a vehicle or equipment on the ground, in case of any defect or damage, it could be stopped, waylaid and the problem could be attended to without any human safety issues affecting the user or society. The matter is entirely different if one is flying an aircraft as no signposts in the sky indicate 'repairs undertaken'!
- Reliability: Reliability is the probability that a product will perform its intended function under stated conditions without failure for a given period. Reliability of any equipment is that 'when required the

equipment should work'. Many efforts go behind the phrase to make it possible that failures seldom occur, and the equipment is available for use as required for the given period and environmental conditions. Failures happen, but a systematic and dedicated effort is made during design and manufacture to ensure that the equipment under use does not fail, which is more important than attending to and rectifying failures.

- Maintainability: Maintainability is the probability that a given maintenance action for an item under given usage conditions can be performed within a stated time interval when the maintenance is performed under stated conditions using stated procedures and resources. Maintainability has two categories: serviceability (the ease of conducting scheduled inspections and servicing) and repairability (the ease of restoring service after a failure). Modern aviation and military aircraft missions demand that the flying machine is ready for use after necessary servicing in the shortest amount of time. To facilitate easy and faster maintainability, the flying machines and associated equipment must be available, accessible and easily replaceable when required. For this design, manufacturing and maintenance activities are to be planned and executed effectively within the parameters of a quality management system.
- Statutory and regulatory requirements and their compliance by industry: Statutory requirements are the laws of the land passed by parliament or legislative assemblies. Regulatory requirements are rules or directives passed by an authority, generally the concerned ministry. The government must see that the products or services of the industry or business operations are in the public's interest and do not adversely affect society. The industry must follow these statutory and regulatory requirements while undertaking production or service functions related to aviation, space or defence equipment. Public awareness and involvement while using the products make it essential that the government imposes these requirements. Regulatory oversight is a traditional function of all modern governments.
- Aviation, space and defence sectors use leading-edge technologies that, if not proven for their safety and reliability, may invite risks while using the products or services.
- Industries or business operations/services in aviation, space and defence sectors are highly competitive with greater liability and visibility. Therefore, safeguards ensure the safety, reliability and fulfilment of the requirements to the satisfaction of all.

The Standard applies to all aerospace organisations/suppliers, from the most significant prime organisations to the smallest suppliers/external providers. A two-year change over time ensured that the Aerospace Standard AS 9100 D/EN 9100:2018 would become operational in September 2018.

The OASIS database, an online resource, contains a list of suppliers certified/registered under the IAQG rules to comply with the aerospace Quality Management System requirements, 18585, as of 5 July 2016.

IAQG maintains an Online Aerospace Supplier Information System (OASIS), which contains information on all Certificated/Registered Suppliers since July 2003. The online site indicates National Accreditation Boards, Accredited Certification Bodies, Approved Auditors, Industry Associations, etc. Detailed assessments of all organisations against the Aerospace Standard are available to IAQG members.

SAE (IAQG) has also published the following standards specifically for MRO and Stockist Distributor organisations, respectively, and they also have been revised.

- AS 9110 C – Requirements for aviation maintenance organisations
- AS 9120B – Aerospace requirements for (stockist) distributors

Apart from the certification standards, there are several guideline standards issued by ISO and SAE, which complement the Standards and give guidelines for implementation. Some of them are listed below:

- ISO 9000: 2015: Quality Management Systems Fundamentals and Vocabulary
- AS 9102 Rev C: Aerospace First Article Inspection Requirement (FAI)
- AS 9103A:2012: Variation Management of Key Characteristics
- ISO 10005:2005: Quality Management Systems – Guidelines for quality plans
- ISO 10007:2003: Quality Management Systems – Guidelines for configuration management
- ISO 10012:2003: Measurement Management Systems – Requirements for measurement processes and measuring equipment
- ISO 19011:2018: Auditing Management Systems
- ISO 10013:2001: Guidelines for developing quality manuals

- AS 9131 Rev C: 'Aerospace Series – Quality Management Systems – Non-conformance Data Definition and Documentation', Quality Systems Non-Conformance Documentation
- AS 9132A: Data Matrix Quality Requirements for Parts Marking
- ISO 9004:2009: Managing for the sustained success of an organisation quality management approach.

The oversight system and the database have been helpful to IAQG members for confidently using supplier assessments from recognised certification bodies around the globe as one of the tools in their supplier quality management activities. Oversight at all levels is the key to building and maintaining confidence. The aerospace industry prioritizes quality management to ensure that products are safe, reliable, and meet customer expectations. This includes compliance with regulations, efficient processes, and continual improvement. They always derive customer satisfaction.

When AC Kermode defined aircraft as 'a heavier than air flying machine supported by aerofoils, designed to obtain when driven through the air at an angle inclined to the direction of motion, a reaction from the air approximately at right angles to their surfaces', it appeared to be a simple flying machine. As you know, the first flying machine had production problems and customer complaints. Modern aircraft systems provide everything necessary for a sustained flight, depending on the requirements of such a flying machine, whether the customers state the requirements. When you know more about modern aircraft, you will realise that it is not just an object heavier than air, driven through the air, but consists of several systems and subsystems that enable it to have a sustained flight and satisfy the varied requirements of various types of customers. From the inspection and quality control systems, which were followed during the Wright brothers' time, to the Quality Management System that currently fulfils the requirements in aviation, space and defence industries/organisations, which design, produce, maintain or operate modern types of aircraft, it has been a long and continuous journey for improvements. It has been a long and constant journey for enhancements. Improvements in safety, reliability and maintainability have been achieved for aerospace and defence products, though more needs to be completed, at the same time meeting the statutory and regulatory requirements, using state-of-the-art technologies, all under the leadership of a Quality Management System, which eventually evolved, and continually improving, through the pioneering efforts of many organisations and personalities. Aviation, Space and Defence Standard 9100 Rev D is now the brightest star in the sky, which shows the light for organisations to follow the right Quality Management System for business excellence, ensuring the safety and reliability of the equipment they produce for the customer's delight.

Processes within an Organisation

Details of the processes within an organisation are listed in Table 0.1 Requirements of the Standard (Business Processes); Clauses 4–10.

0.3.3 Risk-based thinking

The products created by and used by the aviation, space and defence organisations (aircraft, spacecraft and weapon systems) have inherently more significant risks than other products. Therefore, Quality Management System – requirements for aviation, space and defence organisations – Standard AS 9100 D/EN 9100:2018 – not only identifies risk as a concept but also identifies risk management as an essential ingredient in the healthy functioning of the organisation's QMS. Preventive action carried out as a procedure for improvement earlier has been dropped out and merged with the implemented actions to address risks and opportunities. Preventive measures to eliminate potential nonconformities, analysing any nonconformities that may occur, and taking action to prevent recurrence appropriate to the effects of the nonconformity have become part of the organisation-wide risk management.

By considering risk-based thinking throughout the organisation, the probability of achieving stated objectives is improved, output is more consistent, and customers gain confidence that they will receive the desired product or service. Therefore, risk-based thinking helps organisations build a solid knowledge base, establish a proactive culture of improvement, ensure consistency of quality of goods or services to customers and derive improved customer confidence and satisfaction.

Specific risk management requirements are indicated in the Standard under the following clauses:

Risk management is explicit or implied in the Standard throughout and is applicable for decision-making concerning the controls and acceptability of the processes and product characteristics. Quality Management Systems requirements and processes such as planning, design, development, production and continual improvement imply risk management. Risk management is desired, and it is indirectly indicated in the following clauses of the Standards:

- Clause 0.4 of ISO 9001:2015: This international standard enables an organisation to use the process approach, the PDCA cycle, and

Table 0.1 Requirements of the Standard (Business Processes); Clauses 4–10.

Plan				Do	Check	Act
4	**5**	**6**	**7**	**8**	**9**	**10**
Context of the organisation	Leadership	Planning for the QMS	Support	Operation	Performance Evaluation	Improvement
4.1 Understanding Context	5.1 Leadership and commitment	6.1 Action to address risk and opportunity	7.1 Resources	8.1 Operational planning and control	9.1.1 Monitoring, measurement, analysis and evaluation	10.1 General
4.2 Interested parties	5.2 Policy	6.2 Objectives and planning	7.2 Competence	8.2 Determination of requirements for products and services	9.1.2 Customer satisfaction	10.2 Nonconformity and corrective action
4.3 Scope	5.3 Organisational goals, responsibilities, and authorities	6,3 Planning of changes	7.3 Awareness	8.3 Design and development of processes and services	9.1.3 Analysis and evaluation	10.3 Continual improvement
4.4 Processes			7.4 Communications	8.4 Control of externally provided processes and services	9.2 Internal audit	
			7.5 Documented information	8.5 Production and service provision	9.3 Management review	
				8.6 Release of products and services		
				8.7 Control of nonconforming outputs		

risk-based thinking to align or integrate its Quality Management System with the requirements of other management system standards.

- Para 0.1 of AS 9100 D/EN 9100:2018: The potential benefits to an organisation of implementing a Quality Management System based on this international standard include: c. addressing risks and opportunities associated with its context and objectives.
- Clause 0. 3.2 of AS 9100 D/EN 9100:2018 PDCA cycle- Planning includes 'actions to identify and address risks and opportunities'.
- Clause 3.2 Critical Items: Controlling the production and availability of items or parts that significantly affect the product realisation process is a risk management activity.
- Clause 3.3 Key Characteristics: Requirements of identifying the key characteristics are indicated while evaluating the requirements related to the product, determined by the customer or the organisation concerning performance, safety and delivery of the product, including customer satisfaction, review of requirements related to the product and while making the design output documents such as drawings and specifications. Variation control of key characteristics to ensure form, fit and function, performance or service life during design development and production is a risk management activity (refer AS 9103:2012).
- Clause 3.4 Product Safety: The state in which a product can perform to its designed or intended purpose without causing unacceptable risk of harm to persons or property damage.
- Clause 3.5 Special Requirements: The organisation must determine the unique requirements of the customer, which have high risks of being achieved. Hence, these are to be included in the risk management process.
- Clause 6.1: Actions to address risks and opportunities.
- Clause 7.1.3 of AS 9100 D/EN 9100:2018 Infrastructure: The provision and maintenance of infrastructure, equipment and environment needed to achieve conformity to product requirements should be suitable and adequate to manage the associated risks.
- Clause 7.2 of AS 9100 D/EN 9100:2018 Competence, training and awareness: By directly or indirectly meeting the requirements of personnel performing work affecting product quality. That is, managing the risks associated with human resources is possible by having the necessary competence.

- Clause 8.1 Process capability measurements. The process capability index or ratio, which is the ability of a process to produce output within the specification limits (e.g., minimum process capability index Cpk of 1.33 for two-sided specifications and 2.00 for six sigma quality processes) after process capability studies and maintaining the process capability at that level is a risk management process.
- Clause 8.1 Work Transfer: The organisation shall establish, implement and maintain a process to plan and control the temporary or permanent work transfer to ensure the work's continuing conformity to requirements. The procedure shall ensure that work transfer impacts and risks are managed.
- Clause 8.1.1 Operational Risk Management: The organisation shall plan, implement and control a process for managing operational risks to the achievement of applicable requirements.
- Clause 8.1.3 Product Safety: The organisation shall plan, implement and control the processes needed to assure product safety during the entire product lifecycle, as appropriate to the organisation and the product. Note: Examples of these processes include assessing hazards and managing associated risks.
- Clause 8.2.2 Review of requirements related to product and services: Operational risks associated with new technology, i.e., new or unproven technology, short delivery schedules required by the customer, etc., must be identified. Availability or acquisition of the latest technology, either an improvement on the existing one or an entirely new process, needs to be assessed by the organisation if there is an identified risk while using the current/available technological approach. It is not likely to give the desired results. The ability and capacity to provide the product and services must also be identified and assessed.
- Clause 8.3.6 Control of design and development changes: It is necessary to evaluate the effect of the modifications on the constituent parts and the equipment already delivered while effecting the changes in design.
- Clause 8.4.1 Purchasing process: Determine and manage the risks when selecting and using suppliers. Criteria for evaluation of suppliers need to be reviewed by the organisation and ensured that the selected supplier would be able to deliver a product or service with the desired result when chosen and used. The supplier's capability to meet the requirements of the product/service and supply conditions to the

satisfaction of customers is to be evaluated. Risks in dealing with the selected supplier concerning quality and on-time delivery of the product (product, process and personnel competency at the supplier's organisation) need to be examined to identify the potential risks and mitigation plans. Suppliers (external providers of products and services) are also to be evaluated for their ability to meet the flow down of corrective action requirements when it is determined that the external provider is responsible for the nonconformity of the supplies. 'Supply Chain Risk Management Guidelines' via document SAE ARP 9134:2004 is a helpful guideline.

- Clause 9.1.2 of AS 9100 D/EN 9100:2018 Customer Satisfaction: While determining customer satisfaction, it is also essential to identify the risks and potential causes associated with customer dissatisfaction, which may risk the organisation's reputation and image.
- Clause 9.3.3 Management Review Outputs shall include risks identified and other matters decided.
- Clause 10.3 Continual Improvement: The organisation shall continually improve the suitability, adequacy and effectiveness of the Quality Management System through risk management reviews. The analysis and evaluation results and the management review outputs are considered to determine if some needs or opportunities shall be addressed as part of continual improvement. Monitoring the implementation of improvement activities and evaluating the effectiveness of the result are actions considered risk management. When FMEA is carried out, it is an opportunity to identify the preventive measures and a risk mitigation/management activity. Design improvements by establishing error proofing, redundancy and taking modification actions for the problems reported by customers and other external sources are all risk management activities. While determining actions to eliminate the causes of potential nonconformities to prevent their occurrence, one must conduct a risk assessment.

In the organisation's planning processes, consideration should be given to possible financial and other risks. The organisation's management should assess the risks accompanying the innovation activities and prepare preventive actions to avoid or minimise the risks, including contingency plans where necessary.

'Risk is the effect of uncertainty. Any such uncertainty can have positive or negative effects. A positive deviation from risk can provide an opportunity, but not all positive effects of risk result in opportunities' (AS 9100 D). Integrate risk management into planning at all levels.

A risk register for identification and management, being maintained for the operational risk, would be adequate for maintaining documented information/records related to actions to address risks and opportunities in the organisation.

All the identified risks, as discussed above for various clauses of the Standard/processes in the organisation, can be included in the risk register.

1

Scope of the Standard

All clauses of ISO 9001:2015 and additional requirements specified are applicable for Aviation, Space and Defence Standard AS 9100 D/EN 9100:2018. Every contract with customers will always have technical requirements, logistics requirements, payment terms, etc. Any applicable laws, requirements or terms and conditions are not indicated in the Quality Management System standards. They are typically specified or implied in the contract between customers and the organisation. In addition, there are always statutory laws like obtaining a license from the government for running the business, eight hours of work for the staff/industrial workers per shift, 'No adult worker shall be required or allowed to work in a factory for more than forty-eight hours in any week', pollution control laws, etc. They are the statutory requirements being followed by any industry or business practice. The industry follows several such requirements, for example, Air consent (for emission/continuation of emission under section 21 of the Air (Prevention and Control of Pollution Act, 1981), Diesel Storage License, ATF Storage License, Testing of Pressure Vessels, Testing of Cranes/Lifts/Weighing machines and Compliance of Electrical Safety for Sub-stations.

Regulatory requirements are instituted by executive orders from the government, such as 'the products and effluents from the organisation shall not cause any harm to the society', 'the respective ministry in the government approves the prices of petroleum products and telecom tariffs', etc. Any such applicable regulatory requirements related to the product or service offered by the organisation are not indicated in the Quality Management System standard. Therefore, it must be borne in mind that the Quality Management System requirements are a method to be followed or a tool to execute the contractual and applicable statutory and regulatory requirements. Therefore, QMS requirements cannot be considered a substitute or alternative for the contract with customers or for not following the relevant statutory or regulatory (legal) requirements.

QMS is a self-made discipline that helps the production, business or service activity to be carried out efficiently, effectively and economically, meeting the contract/customer/statutory and regulatory requirements. Yes, actions are to be carried out efficiently and effectively. Efficiently means that it is to the organisation's satisfaction and that the results achieved are at a cost less than that of the cost of using resources. Effectively means that it is to the satisfaction of the organisation or the customers and that all the targeted/planned results are achieved.

It is also to be understood that the statutory and regulatory requirements of the country are more important and shall be followed. If any clause of the Standard conflicts with or violates the rules of the government, the law of the land prevails. That means if there is a conflict between the requirements of Standard AS 9100 and applicable statutory or regulatory requirements, the latter takes precedence. This is important for aviation, space and defence industries because governments regulate such sectors.

An organisation accepts and adopts any Quality Management System standards when it is necessary to demonstrate to the customers and other stakeholders that it can consistently provide products that meet customer and applicable regulatory requirements. The organisation should enhance or increase customer satisfaction by providing improved products and carrying out the business operation efficiently, conforming and effectively by implementing the Quality Management System. It should establish methods of improving the product and business operations so that it always helps the customer and complies with or obeys the legal requirements. The Quality Management Systems standard, therefore, indicates the methods to convert customer requirements into products/services that conform to the requirements and ways and means to achieve business excellence, higher levels of customer satisfaction and other key performance indicators (KPIs) or objectives of the organisation and customers. The standard can be used by organisations other than aviation, space and defence if they wish to demonstrate their ability to produce and supply their products to customers.

Notes added in the Standard guide the organisation in understanding or clarifying the associated requirement. The term 'product' means a product meant for a customer or required by any customer. The by-products of a manufacturing activity, such as waste or effluents (solid liquid or gaseous industrial waste), are also considered as products. In other words, there are four different types of products – hardware, software, processed materials and service.

2

Normative Reference

Two documents are indispensable to understanding and implementing the Standard AS 9100 D requirements in daily business operations. It is necessary to refer to the latest editions of these standards for the correct and meaningful understanding:

a. ISO 9000:2015 Quality Management Systems – Fundamentals and vocabulary

b. ISO 9001:2015 Quality Management Systems – Requirements

The vocabulary helps to choose the right word for effective communication. The language defines all the relevant terms used in the Standard AS 9100D.

(The complete ISO 9000 and 9001 standards can be obtained from any ISO member or the ISO Central Secretariat: 1, Ch. de la Voie-Creuse, Case post ale 56, CH-1211 Geneva 20, SWITZERLAND, or visit www.iso.org. Copyright remains with ISO.)

Buying, using and maintaining these external standards and reference documents are a good practice. Regularly verify the current validity status for continued use or updating.

Some of the essential terms and their definitions are indicated in Chapter 3. Aerospace Standards use Quality Management Systems – Fundamentals and vocabulary as per ISO 9000:2015. Therefore, when Quality Management System documentation, implementation and practices are established, we must use terms and words in such a way that they imply the meanings and definitions given in the ISO: 9000-2015.

3

Terms and Definitions

Appropriate use of words is quality care. It is good that you want to *be right the first time and every time*. Doing things precisely the first time may be suitable. But how many of us can do that? People achieve things through hard work. Where is room for hard work if things are done right the first time? If things are completed quickly, life becomes monotonous and stagnant. So, you can do things better the second time by analysing what went wrong. Plan better, correct the mistakes, close the gaps and take corrective action so that the errors and gaps do not recur. So, change the motto or slogan to 'Doing things right the first time may be good, but doing things exactly right the second time is better'.

It also encourages one to improve continually. Improvement is better than being right the first time. QMS standards emphasise continual improvement in that context. There is no limit for improvements, or say, improvements are those for which no limit has been established.

While describing the methods of achieving or meeting the requirements of the Standard, wherever the specific terms of risk, special requirements, critical items and key characteristics are used, the significance of these terms is to be clearly understood. Substituting words should not dilute the meanings or intention of the Standard. As mentioned above, care, consideration and action are required while documenting and implementing the standard requirements regarding the definition of terms used.

The supply chain is defined as the 'Supplier to Organisation to Customer'. The supplier may be an organisation, or an organisation may be a supplier to the customer; for a supplier, the organisation is the customer, depending on the link in the supply chain regarding a contract. Standard terms we use, such as contractor/sub-contractor/vendor/supplier, etc., for external providers and company for organisation, are not at all used in the Standard. Therefore, it is desirable to refrain from using such words in the documentation of the Quality Management System or day-to-day practice. Several guideline standards are referred to within the Standard and in its bibliography. These complement the understanding and help implement and practise the Aerospace Standards. It is a good practice to indicate these guidelines and standards as references in the quality manual/documented information. The latest version is available in the organisation.

We have already seen what the term product means. The word product, wherever it appears, can also mean 'service' the organisation provides. In addition to the terms and definitions given in ISO 9000:2015, additional terms such as risk, special requirements, counterfeit parts, critical items and key characteristics are defined in the aerospace standard. These terms describe the requirements related to the product or the business operations. Some of the Quality Management System terms, their definitions and explanations are given below (permission to reproduce extracts from British Standards is granted by BSI Standards Limited (BSI)):

Audit: 'A systematic, independent and documented process for obtaining objective evidence and evaluating it objectively to determine the extent to which the audit criteria are fulfilled' . An audit is a systematic evidence-gathering process. Audits must be independent. Evidence must be evaluated objectively to determine how well audit criteria are being met. There are three types of audits: first-party, second-party and third-party. First-party audits are internal audits, while second and third-party audits are external audits.

Conformity: 'conformity fulfilment of a requirement'. Conformity is the 'fulfilment of a requirement'. To conform means to meet or comply with the requirement. The requirement is a need, expectation or obligation. There are many requirements: customer, quality, Quality Management Systems, management, product and service, contractual, statutory and regulatory requirements.

Correction, containment action and corrective action: Defined in Chapter 'Clause 8 Operations', Section 8.7.1

Customer satisfaction: 'customer's perception of the degree to which the customer's expectations have been fulfilled'. Customer satisfaction is a perception. It is also a question of degree/level. It can vary from high satisfaction to low satisfaction. Customers who believe you have met their requirements experience high satisfaction. If they think that you have not met their requirements, they experience low satisfaction. Since satisfaction is a perception, customers may not be satisfied even though you have completed all the contractual requirements. Primarily when customers assess or record in a survey sheet sent by you. Therefore, it is always better to measure the customer's perception of satisfaction, level or degree of satisfaction when you have met their expectations of quality, on-time delivery and similar performance. Just because you have not received any complaints does not mean that customers are satisfied. There are many ways to monitor and measure customer satisfaction. This measurement must be done by you/your organisation. Customers may also assess this perception, but this assessment may not always be accurate to the level of confidence with which you measure a customer's perception of satisfaction. You can use customer satisfaction surveys/scorecards and user opinion surveys to collect product quality data (post-delivery), track warranty claims, examine dealer reports, study customer compliments and criticisms and analyse lost business opportunities.

Documented information: 'information required to be controlled and maintained by an organisation and the medium on which it is contained'. Recorded information includes information about the management system and related processes. It also includes all the information organisations need to operate and all the information they use to document the results they achieve (archived records). Information created for the organisation to operate is documentation, the evidence of outcomes achieved records.

Words like 'document' and 'record' commonly require correct understanding to avoid confusing usage. ISO 9000:2015 defined the word 'document' more precisely than in a dictionary. 'Document is information (meaningful data) in its supporting medium'. Documents can be reviewed, amended, updated, re-approved or revised. None of the above is appropriate in the case of a record since the record is the history of past events. A record is a document stating results achieved or providing evidence of activities performed. As you are aware, the Indian Constitution is a document which has been amended/updated by Parliament 100 times till September 2015 to become Amendment acts. A drawing or a specification issued by an organisation is a document. Only the originator has the responsibility and authority to amend it. The issue/revision status is updated every time the document is amended. Your birth certificate, degree certificate or inspection record sheet are all examples of records. They are all recordings of past events that cannot be altered, amended or revised. Any unauthorised amendment to a record constitutes a criminal act.

A set of documents, such as a drawing, quality plan and inspection record, may be frequently referred to as 'documentation'. Quality manuals, procedures, work instructions and formats are called QMS documentation. People often use the words document and record without caring for their meaning. Interview call letters ask you to bring the relevant documents; in such cases, all you carry to the board are records! Now consider this. Is your marriage certificate a document or a record? As per ISO 9000:2015, documented information and its supporting medium must be controlled and maintained. Written knowledge can be in any format and medium.

Interested party: 'interested party stakeholder person or organisation that can affect, be affected by or perceive itself to be affected by a decision or activity. EXAMPLE: Customers, owners, people in an organisation, providers, bankers, regulators, unions, partners or society that can include competitors or opposing pressure groups'. An interested party is anyone who can affect, be affected by or believe they are affected by a decision or activity. An interested party is a person, group or organisation with an interest or a stake in a decision or action.

Monitoring: 'Monitoring determining the status of a system, a process, a product, a service,\ or an activity'. Monitoring means determining the status of an activity, process or system at different stages or times. To assess status, you must supervise, continually check and critically observe the monitored action, procedure or system.

Nonconformity: Nonconformity is nonfulfillment of a requirement or failure to meet a requirement. A requirement is a need, expectation or obligation. It can be stated or implied by an organisation or interested parties.

Objective: 'objective result to be achieved'. An objective is a result you intend to achieve. Objectives can be strategic, tactical or operational. Objectives can apply to an organisation as a whole or a system, process, project, product or service. Objectives may also be called targets, aims, goals, intended outcomes or key performance indicators. Quality objectives are generally based on or derived from an organisation's quality policy, which must be consistent with the policy. A quality objective is a quality result that you intend to achieve. Quality objectives are based on or derived from an organisation's quality policy and must be consistent. They are usually formulated at all organisational levels and for all relevant functions.

Objective audit evidence: Objective audit evidence is verifiable information and generally consists of records and other statements relevant to the audit criteria.

Objective evidence: 'objective evidence data supporting the existence or verity of something'. Objective evidence is data that shows or proves that something exists or is true. Factual evidence can be collected by performing observations, measurements, tests or other suitable methods, including statements by the process owner.

Product safety: The state in which a product can perform its designed or intended purpose without causing unacceptable risk of harm to persons or property damage.

Provider: 'supplier organisation that provides a product or a service'. A provider is a person or an organisation that supplies or provides products or services. Providers can be either internal or external to the organisation. Internal providers supply products or services to people within their organisation. In contrast, external providers provide products or services to other organisations. The supplier is a provider. Examples of suppliers include organisations and people who produce, distribute or market products, provide services or publish information. While ISO still consists of a definition for this term, supplier, the new ISO 9001 2015 standard no longer uses it. It prefers, instead, to use the term external provider.

Risk: 'An undesirable situation or circumstance with both a likelihood of occurring and a potentially negative consequence'. According to ISO 9000:2015, the risk is the 'effect of uncertainty of an expected result', and an effect is a positive or negative deviation from what is expected. The following two paragraphs will explain what this means. This definition recognises that all of us operate in an uncertain world. Walking on or crossing the road with vehicular traffic is a risk. We need to cross the highway without getting injured. Whatever we do, we do it positively with a desirable expectation. However, there is a possibility that it can go wrong. Whenever we try to achieve something, there is always a chance that things will not go according to plan. Sometimes, we get positive results, sometimes negative results and occasionally both. Because of this, we need to reduce uncertainty as much as possible. Uncertainty (or lack of certainty) is a state or condition that involves a deficiency of information and leads to inadequate or incomplete knowledge or understanding. In the context of risk management, uncertainty exists whenever the knowledge or experience of an event, consequence or likelihood is inadequate or incomplete. While this definition argues that risk can be positive and negative, a note acknowledges that 'risk is sometimes used when there is only the possibility of negative consequences'. Risk management is all we do subconsciously or in a planned manner to ensure that nothing goes wrong till the end of the process/project. The aim is to minimise the probability that a particular risk will occur or reduce the impact/severity of the effects when it does happen.

Risk-based thinking refers to a coordinated set of activities and methods organisations use to manage and control the many risks that affect their ability to achieve objectives. Risk-based thinking replaces what the old standard used to call preventive action. While risk-based thinking is now an essential part of the new standard, it does not expect you to implement a formal risk management process or document your organisation's risk-based approach.

Validation: 'Validation is confirmation, through the provision of objective evidence, that the requirements for a specific intended use or application have been fulfilled'. Validation is a process. It uses objective evidence to confirm that the requirements defining an intended use or application have been met. Whenever all requirements have been met, a validated status is established. Validation can occur under realistic use conditions or within a simulated user environment. There are several ways to confirm that the requirements defining an intended use or application have been met. For example, you could do tests, carry out alternative calculations or examine documents before you issue them. Whatever the validation method, it must be agreed upon between the organisation and the customer.

Verification: 'Verification is confirmation, through the provision of objective evidence, that specified requirements have been fulfilled. Note 1 to entry: The objective evidence needed for verification can result from an inspection or other forms of determination such as performing alternative calculations or reviewing documents'. Verification is again a process. It uses objective evidence to confirm that specified requirements have been met. Whenever specified conditions have been met, a verified status is achieved. There are many ways to ensure that needs have been met. They are inspected per the quality plan, tested per the test schedule, carried out alternative calculations or verified documents before you issue them as acceptable.

In Quality Management Systems, verification and validation are the activities of reviewing, inspecting or testing, putting it to use and ensuring that it meets statutory, regulatory or technical standards of qualification and acceptance. It establishes and documents that a product, service or system is satisfactory.

Special requirements are requirements identified by the customer or determined by the organisation. It may be challenging to achieve these requirements, or there may be potential risks while attempting to achieve them. Therefore, caution, care and additional controls may need to be established before attaining it. Hence, special requirements may be included in the risk management process. These requirements are related to the product in terms of unique technology/processes to be followed during manufacture, accuracy or performance requirements of the product specified by the customer/

drawing/specification, process capability or machine capability that are at the limit of the industry's capability. Sometimes, the requirements may be new or unknown to the organisation. The organisation needs to evaluate their capability while doing the risk analysis to undertake the requirements. If you do not have machines capable of giving an accuracy of ± 5 microns, undertaking the job of machining components to that accuracy/precision is a risk. Mitigation plans may be to acquire high-precision machines or outsource such jobs to an approved source, an external provider, where the high-precision machine is available. Customer requirements may sometimes exceed the capacity of the organisation. It may be at the limit of its technical or process capabilities. Customer-given special requirements are identified when determining and reviewing requirements related to the product (see clauses 8.2.2 and 8.2.3). Special requirements are often identified during the input and output of the design process (8.3.3 and 8.3.5). Special requirements relating to purchased products may be determined while placing the indents or purchase orders (8.4.3). Special requirements for unique processes capability and capacity may be identified while planning such processes in-house or at the external provider/supplier source (8.5.1.2).

These requirements may include critical items, which may further flow to the list of crucial characteristics requiring additional control to prevent variation. Special requirements can originate from customers, regulatory authorities and environmental or market conditions. Internally, special needs may be identified based on the capabilities, personnel competence, equipment availability, etc. and during the design process.

Critical items are hardware or software, processes, characteristics or functions, which, if a deviation happens, may cause catastrophic or mission failures. Any deviation/defect may affect the product's 'Function, Life, Interchangeability, Maintenance, Strength or Safety (FLIMSS)'. Therefore, special care, considerations and actions are necessary to ensure no deviations occur on such critical items during design, manufacture and maintenance. Essential Items, including key characteristics, significantly affect product realisation and use, requiring specific actions to ensure they are adequately managed. Any component or item that triggers the failure of a primary assembly or the final equipment itself could be considered a critical item.

Fracture-critical items, items with limited life or work life and items prone to wear, brittle or degrade over time, which is lower than the designed life, are also critical items. No deviations are permitted on essential items during the production or maintenance of the equipment. Necessary items may be listed and maintained for monitoring throughout the production stages, testing and the equipment's regular use/operation and maintenance phase. Specific essential items may be identified as interchangeability (ICY) items, component life register with life items and replacement due dates, component replacement list, safety-critical items/characteristics, durability critical items, etc. Ejection seats, personal survival packs, parachutes, cartridges for canopy severance/separation systems, oxygen systems, etc., are safety-critical items in a military aircraft.

If any item is not previously proven to give satisfactory performance as per design or reliability is not established, the item could be considered as a critical item. Similarly, items identified as highly sensitive to adverse environments, such as heat, contamination, electromagnetic interference, etc., could be considered necessary. Such items require meticulous handling, storage, preservation and use where the items are safeguarded against adverse environments. Consumables with shelf life and recommended preservation and use conditions are critical items. Any degradation in quality due to exposure to adverse environments would affect the quality and reliability of the final product where these vital items are used.

Designers generally identify critical items based on the studies and product assurance analyses concerning reliability, availability, maintainability and safety (RAMS). Design organisations typically specify such essential items by defining their classification and method of identification to control them. Drawings may determine their classification. Route Cards or Master Process Sheets may specifically identify them for the care taken during manufacturing. Critical items may need periodic monitoring for satisfactory conditions and performance. Even when the design does not indicate criticality, special requirements or essential characteristics of the product or when the customer does not explicitly indicate them in the enquiry or order, the organisation must evaluate the design drawing/specification for the critical items, key characteristics and unique requirements. The organisation undertaking production is responsible for identifying and controlling them to meet the requirements during production and subsequently during use. Production organisations may evaluate it using techniques such as FMEA, PFMEA, reliability analysis or endurance tests. Production constraints, if any, for the control of critical items to meet quality requirements are to be clarified by production engineering with the design and development agency. Before proceeding, actions and controls are necessary to ensure product conformity may be mutually agreed to.

Key characteristics (see definition of quality) or attributes or features whose variation significantly affects product form, fit and function, performance, service life or manufacturability. The most essential characteristics of aviation, space and defence products are function, life, interchangeability, maintenance, strength and safety (FLIMSS). Therefore, such components or products require special care, consideration and actions to control variation or deviation. Critical items and key characteristics must be identified in design output documents such as drawings and specifications. Key characteristics may be determined for a part, sub-assembly or system, such as certain geometrical features, material properties and functional features. For example, in a unique process like nitriding or case-carburising, case depth and surface hardness may be identified as key characteristics in a gear. Variations' key critical characteristics must be controlled within acceptable limits to meet safety, performance, maintainability and reliability requirements. If not contained within proper boundaries, variables in key characteristics may expose deviations sooner or later and significantly impact some of the customer requirements. 'Variation Management of Key Characteristics' is given in AS 9103:2001. The production engineering documents/controller process sheet/quality plans should also define the exceptional care, consideration and actions necessary to achieve the characteristics and control their variations, if any, during the manufacturing/maintenance.

Special requirements, critical items and key characteristics are all interchangeable features and some of the examples are indicated below:

Critical dimensions having a close tolerance, geometric features like run out, perpendicularity, parallelism and concentricity with a tight tolerance, high surface finishes, angular dimensions with a total tolerance range of 1 degree (60 minutes) or less, test methods and acceptance criteria for non-destructive testing (e.g., magnetic particle, liquid penetrant, radiographic inspection, ultrasonic, eddy current), hardness testing (e.g., Rockwell), shot peen requirements, material properties and material certifications, dynamic balancing of rotating units and static balancing of flight control surfaces, flow checks for blades and vanes, spray pattern requirements for fuel nozzles (including afterburner rings), weight checks and compliance to DO 178C for software.

Aerospace Standard AS9110 Rev C Quality Management Systems – Requirements for Aviation Maintenance Organisations define more words such as authority, maintenance, technical data, human factors, release certificate, etc., clearly explained in the Standard. The authority for civil aviation in India is the Director General of Civil Aviation and for military aviation, the Centre for Military Airworthiness and Certification (CEMILAC) (design authority) and the Director General of Aeronautical Quality Assurance, DGAQA (quality assurance authority), both under the ministry of defence. These authorities have clearly defined roles for statutory and regulatory functions in the aviation sector. Maintenance means periodic servicing, repair, overhaul and embodiment of modifications to maintain components, equipment and leading equipment in serviceable condition for continued use and improved use. Explanations are provided below for some of the additional terms mentioned in AS 9110 Rev C:

Human factors: Human factors are human physical and psychological behaviour, considering human capabilities and limitations that impact aircraft operations' safety and efficiency. When applied to aircraft design, certification, training, operation and maintenance, the behaviour should always be emphasised for the safe interface between human beings and aircraft system components. The constraints of personnel performing tasks are physical fitness, psychological characteristics, personality, stress, fatigue, distraction, communication and attitude. It is to be ensured that the constraints and human errors due to such conditions shall not adversely affect the safety, quality and efficiency of the aircraft maintenance operations when the interface occurs between the personnel and the equipment/environment. Personnel are trained to overcome the constraints and develop human factor skills such as communication, teamwork, situational awareness, etc. Encouraging positive attitudes towards safety culture and discouraging unsafe behaviours are also included in the training for all categories of personnel.

Human factors refer to environmental, organisational, piloting and maintenance factors and the human and individual characteristics which influence behaviours at work in a way that can affect aircraft safety and performance. Error and fatigue management are human factors specifically related to the aviation industry for an effective 'Safety Management System'. Human factors identify and evaluate mainly 'human to machine' interactions to better understand these interactions and develop a safety management framework. Human-to-machine interactions focus on human error, fatigue management, organisational culture, procedures (Standard Operating Procedures), communication, workload, decision-making, human reliability, stress and safety management issues. The success of a mission depends on maximising the ability of an individual or team to operate and maintain a system at required levels by eliminating design-induced difficulty and error. The design of large systems, like aircraft, focuses on organisational culture, personnel training and operating procedures. International Civil Aviation Organisation and National Civil Aviation authorities develop human factors training for maintenance personnel. Human factors training include an organisation's safety management system, a coordinated approach to managing safety, emphasising error management and developing a safety culture. Safety management systems have a non-punitive confidential reporting system to encourage disclosure of events that may threaten safety.

The 9100:2018 requires the consideration of human factors when performing causal analysis to ensure an actual root cause is evaluated. Reviewing the 'FAA Dirty Dozen' for examples of human factors may help. The 'Twelve Common Human Error Preconditions' are conditions that can be precursors to accidents or defects that cause the incidences or accidents.

These twelve elements influence people to make mistakes. The Dirty Dozen is a concept developed by Gordon Dupont in 1993 while working for Transport Canada and formed part of an elementary training programme for human performance in maintenance. It has become a cornerstone of human factors in maintenance training courses worldwide. The original list was developed for aircraft maintenance. Errors, when they occur, will likely be found to be caused by one, or even more common, a combination of the following causes. These dirty dozen causes of error are:

Lack of communication, distraction, lack of resources, stress, complacency, lack of teamwork, pressure, lack of awareness, lack of knowledge, fatigue, lack of assertiveness and norms.

Human factors safety training is critical to addressing human factor error preconditions. Airside Safety HANDBOOK, published by Airports Council International World, Geneva, Switzerland, helps airports build strong defences against human errors by enhancing the awareness of factors that can impact performance and lead to mistakes.

Such training courses provide participants with a comprehensive introduction to the primary concepts of human factors and ensure that they can recognise, understand and prevent human errors on the airside. We hope to avoid the dirty dozen and their consequences only by training and constant awareness.

Release certificate: 'Authorised Release Certificate' is an airworthiness approval tag accompanying an aerospace article/part/component/assembly. It means that the part has been manufactured as per approved design data and is in a condition for safe operation when used. Usually, organisations using such a release certificate will have a release certificate or release note signatory authorised by the regulatory authority. (FAA uses a format 8130–3 for such a release certificate. All airworthy, fit-for-use components and equipment shall be accompanied by the release certificate when released from the Original Equipment Manufacturer and signed by the authorised release note signatory.)

The standard AS 9120 Rev C, Quality Management Systems – Requirements for Aviation, Space and Defence Distributors, defines words like airworthiness certificate, certificate of conformity, counterfeit part, distributor, splitting, suspected unapproved part and test report, etc. An airworthiness certificate is a document issued by the civil or military authority that the part has been designed and manufactured according to the applicable airworthiness regulations and is fit for flying. All aircraft parts or parts used are designed and qualification tested to declare that they are airworthy. Only components accepted after qualification tests have been completed to meet the physical and functional requirements are expressed as airworthy.

The standard AS 9120 B also defines the difference between counterfeit and unapproved parts in the supply chain. Counterfeit parts are fraudulent, imitations of existing products and have no authority for use. Suspected unused parts (SUPs) are parts or products with unknown authority or may be unacceptable.

Splitting – batch splitting – the separation of entities, such as sheets, bars, components, parts, fasteners and containers belonging to the same production batch – product splitting – physically dividing a solid entity such as bar, sheet, plate (metallic or non-metallic material) or partial decanting of a gaseous or liquid entity, where the physical and metallurgical properties or chemical characteristics are not altered. Splitting should not affect the conformity of the product as defined by the original product specification.

This standard AS 9120 B and the Standard's requirements only apply to stockist distributor organisations. In this case, the supply chain is 'manufacturer – organisation – customer'. The manufacturer here is the supplier, and the stockist distributor is the organisation that distributes/sells the OEM spares or part manufacturer's approved parts to customers who own the aircraft and aeronautical equipment.

Shall: Wherever the standard uses the auxiliary verb 'shall', the requirement is mandatory and the necessity is imposed upon the user organisation. 'Should' indicates a recommendation; 'may' indicates permission; and 'can' indicates a possibility or a capability.

Information marked as 'NOTE' is for guidance in understanding or clarifying the associated requirement.

Continual: We commonly use the word continuous for an ongoing activity. There are seven places where the aerospace standard uses the word 'continual' instead of 'continuous' as an adjective for improvements. There is a difference between the words continual and continuous, though both indicate duration and are not synonyms. Continuous means duration without interruption. Continual indicates duration that continues over a long period but with intervals of interruption. Suppose you plot the graph of a rocket piercing the sky at a uniform velocity. In that case, the graph of time (x-axis) versus height (y-axis) will look like a right-angled triangle with its hypotenuse going up continuously without interruption. Improvements cannot be designed and implemented consistently. It takes some time to design and implement improvements. After achieving the improvements, sustaining and evaluating it for further improvements is necessary, which also needs time. Plotting a similar graph for time versus improvements will look like a ladder, a step-by-step process of progress upwards. This step-by-step (intervals of interruption) improvement process is termed continual improvement. Continual improvement tools in the Quality Management System include the corrective action analysis, preventive action analysis and customer satisfaction analysis.

Where applicable, where appropriate and as relevant/proper, etc.: 'As appropriate', 'where applicable' and 'where practicable' are words used in the Standard, allowing the organisations to decide on the applicability of the requirements specified in the Standard. If it is acceptable to the organisation to do so, it is appropriate. Appropriateness and relevance are to be judged by the organisation's scope of activities. The critical question to be considered is whether compliance to the particular 'as appropriate', where applicable' and 'where practicable' requirements add value to the process in the organisation. Does the absence of compliance with the requirement increase the risk of the organisation not meeting the customer's requirements, or will there be any negative consequences? For example, planning to determine or consider the reliability, availability and maintainability and recycling or final disposal of the product at the end of its life is 'inappropriate' for an organisation that machines aircraft components as per the customer drawings. Examples of 'where appropriate' are economically viable, increasing satisfaction, a mundane and practical way of looking at things and not just philosophical. Another explanation can be added as follows:

Identifying resources to support the use and maintenance of the product to keep it airworthy is also not in the scope of work undertaken by the manufacturers who machine components using CNC machine software programs. For them, maintenance is not applicable or not appropriate. In several clauses of the Standard and associated notes, many things are requested to be indicated and complied with as appropriate. Therefore, compliance with some of such notes or requirements may not be insisted upon if, in the organisation's view, the determination of that note or requirement is not appropriate to the processes followed by the organisation.

The standard addresses requirements generally applicable to the aviation, space and defence industries. The specifics will be determined by the organisation, which has decided to follow the QMS standard.

Finally, a word of caution: The structure of clauses in the Standard AS 9100 D/EN 9100:2018 is intended to provide a coherent presentation of requirements rather than a model for documenting an organisation's policies, objectives and processes. The changes in the structure and terminology due to the revision of the Standard need not be reflected in the existing documentation of an organisation's Quality Management System. However, it would help if you considered re-arranging your QMS documentation around your organisation's value stream. A value-stream-based QMS allows you to customise your documentation to your unique business needs that make sense to your leadership, customers and other stakeholders since it will describe exactly what you do. It will also support integrating the Quality Management System into your business processes while making efforts to comply with the new requirements of the Standard. Further changes, if any, will be dictated by the business, not by a structure change of the Standard.

4

Context of the Organisation

Executive actions or the aerospace standard's operative parts commence here; the requirements are indicated in Clauses 4 to 10.

The standard requirements/clauses are arranged by the bigger picture 'PLAN-DO-CHECK-ACT' Cycle (Deming's Wheel). Chapter 4 explains the context of the organisation (including the customer requirements, needs and expectations of relevant interested parties), Chapter 5 explains leadership, Chapter 6 explains planning and Chapter 7 explains support (all resources necessary). They are all part of the 'PLAN' section, and the requirement details are indicated in the respective sub-paras of each chapter.

Context means the circumstances that form the setting for an event, statement or idea and what can be fully understood. The organisation's context defines the position, strength and circumstances in which it is carrying out its business. It may explain its scope of economic activities, the type of products it brings out, the processes deployed, resources used and checks and balances it maintains with different methods used in the organisation. The applicable statutory requirements and the organisation's ultimate purpose establish compliance with regulations and business norms. The organisation's context is the statement of mission, vision and values it upholds, the customers it serves, listing other interested parties, etc. Information that becomes the 'Context of an organisation' is synonymous with an individual's resume, broadly indicating his human behaviour. It is also a means to detect risks and opportunities regarding the business context.

4.1 Understanding the Organisation and Its Context

This part is about understanding the organisation's purpose, the management system and who the stakeholders are. It describes how to set up the Quality Management System with a broad understanding of the situation and needs of the business. The organisation's context establishes the QMS and supports the business strategy. The context of the Organisation forms the basis of the rest of the revised standard. It is necessary to define the influences of various elements on the organisation and how they reflect on the QMS, the company's culture, objectives and goals, the complexity of products, the flow of processes and information, the organisation's size, markets, customers, etc.

It allows an organisation to identify and understand the factors and parties in their environment that support the QMS. Firstly, the organisation must determine external and internal issues relevant to its purpose. 'Relevant' is a keyword here. External issues like market competition and internal issues like the attrition of competent human resources are appropriate. To determine the external context, issues arising from its social, technological, environmental, ethical, political, legal and economic environment are to be considered. Other examples of external context may include the organisation's competition, government regulations and changes in the law and economic shifts in the organisation's market, such as variations in foreign exchange rates. These issues impact what the organisation does or would affect its ability to achieve the intended outcomes of its business/QMS. It should be noted that the term 'issues' covers problems that would have been the subject of proactive actions earlier and essential topics for the management system to address. In fulfilling this clause, the organisation should focus mainly on issues affecting customer satisfaction and delivery of quality products and services.

Each organisation will also need to identify their unique set of interested parties relevant to its QMS; over time, these may change in line with the organisation's strategic direction. These groups could include owners, partners, shareholders/investors, employees, bankers, customers, suppliers/external providers, pressure groups, statutory/regulatory authorities, casual workers and local bodies. It could include the whole organisation, or some specifically identified functions, ring-fenced to comply with the requirements. Any outsourced functions or processes must also be considered in the organisation's scope if they are relevant to the QMS. Understanding the organisation and its context indicates the necessity of greater integration between the QMS and organisational business planning activities. The organisation must periodically ascertain, monitor and review internal and external issues relevant to its purpose. The changes in the matters may influence the organisation/business's strategic direction. They can impact the QMS and its intended results.

4.1.1 Understanding the needs and expectations of the interested parties

This requirement is broadening the scope of work beyond just customers. The organisation has already determined the customer's requirements. Therefore, the organisation will know what its external customers expect. The organisation needs to determine the requirements of interested parties that are relevant to the Quality Management System concerning the delivery and use of the product or services, in addition to the needs and expectations of the customer. For example, suppose the product realisation process includes the emission of hazardous pollutants or effluents discharged into a river or the sea and damaging the environment. In that case, the needs or expectations of many interested parties, people or organisations will be affected. That means the organisation's decision or activity can affect the interested parties or others who perceive themselves as involved. The Standard requires that the organisation continues to determine, monitor and review information about these interested parties and their relevant needs, expectations and requirements.

'Understanding the context of an organisation' means determination of the factors influencing the organisation's purpose, objectives and sustainability. It considers internal factors, external factors and the requirements of relevant interested parties. Para 3.2.2 of ISO 9000:2015 defines the 'context of the organisation'.

The new requirement to determine, monitor and review information about the organisation's context is stated in paras 4.1 and 4.2 of the standard ISO 9001:2015. While auditing several organisations to comply with this new requirement, it was generally found that the explanations provided by the senior management or the organisation's documentation do not give adequate evidence for conformity.

The QMS standard requires the organisation to demonstrate – demonstrate is the key word – the compliance. Compliance of what? Organisations can consistently provide products and services that meet customer and applicable statutory and regulatory requirements even as they aim to enhance customer satisfaction. The goal of achieving the intended results of its Quality Management System is demonstrated by the achievement of the following three organisational objectives:

a. fully desired quality,
b. on-time delivery and
c. Improvement in customer satisfaction.

Most senior management explains the company's background history, capability, market reputation, vision to be among the top ten, or efforts to have the lion's market share. However, they seldom demonstrate their efforts to identify new factors such as external and internal issues, changes to the problems from time to time, challenges faced by the organisation or the varying needs of customers and other stakeholders. Their documentation is generally limited to a few statements they 'have determined, periodically monitor and review information about the organisation's context'. At the most, they may indicate their mission, vision, policy and objectives or list external and internal issues and interested parties. Whatever the case, they fail to demonstrate the methodology they should have adopted to identify the external and internal issues, monitor the status of or changes to, the problems identified and review the information to achieve their organisational objectives. Generally, nobody requires them to show the documented information on the context (since it is not a mandatory requirement of the standard). The managers seldom pick up the information from related documentation, which could have demonstrated their method of identifying, reviewing and managing the issues. This may be due to a lack of awareness of 'the context' or fear of failure to convince the auditor. If they possess business acumen and follow a strategy within the organisation, they could demonstrate compliance with the review of external and internal issues and stakeholders' requirements by referring to or showing the related documentation. Some examples of such documentation are periodic management review reports or minutes of meetings with customers or stakeholders and strategic business development plans, which often happen and have been documented. Information about external and internal issues can be found from many more such sources, like internal recorded information and meetings, national and international press, websites, publications, national statistics offices and other government departments, professional and technical publications, conferences, meetings with relevant agencies, meetings with customers and appropriate interested parties and professional associations. Similarly, the documented information required to demonstrate the methods of determining and reviewing the information related to the organisation's context can also be found.

When planning to achieve organisational objectives, the organisation considers relevant external and internal issues. Occasional changes to those factors/issues are also considered since they can influence the administrative processes and achievement of objectives. ('Note 1: Issues can include positive and negative factors or conditions for consideration'.)

'Relevant or not relevant ' assesses the issues based on the sole criteria – whether the problem impacts the organisation's ability to meet quality and on-time delivery performance. The performance of an organisation is its ability to consistently provide products and services that meet customer and applicable statutory and regulatory requirements and thereby achieve the organisation's objective of enhancing customer satisfaction. The organisation is responsible for looking for any additional points regarding the relevance of its identified issues. The combined effect of the relevant internal and external cases can significantly influence the organisation's approach to developing and achieving its objectives.

Sub-clause 4.1 of 9001:2015 intends to 'understand the external and internal issues relevant to the organisation's purpose and strategic direction and can positively or negatively affect the organisation's ability to achieve the intended results of its QMS. Ultimately, the organisation must demonstrate its awareness that all business and quality system planning is based on 'risk-based thinking'. The organisation also needs to demonstrate that every improvement in its business or quality system comes from periodic reviews of its context. The context's determination, monitoring and review are done by identifying and managing the various risks related to the relevant external and internal issues and the needs and expectations of relevant interested parties. 'Risk-based thinking' is translated into specific action points to reduce the risks in business and utilise the opportunities for performance improvement. Identifying and managing various types of risks and opportunities are usually consolidated in a risk register maintained by the organisation. Suppose any relevant issues/requirements are likely to adversely affect the organisation's efforts towards achieving the desired results or meeting the performance targets. In that case, the issue/requirement should be considered a risk. Any adverse risk should be viewed as an opportunity.

The organisation should be aware that external and internal issues can change and should be monitored and reviewed. Periodic reviews of information about the organisation's context lead the organisation to evaluate the changes in the risks and opportunities and take timely mitigation actions for unacceptable threats. The risk register or the management review record reflects the periodic reviews.

4.1.2 Understanding the requirements and implementation

The term 'context of the organisation' is often equated to other terms such as 'business environment', 'organisational environment' or 'ecosystem' of an organisation. Understanding the organisational context is the starting point for developing a sustainable Quality Management System (QMS). It applies equally to profit-seeking, not-for-profit and public service organisations.

Understanding the context of an organisation is an activity which determines the factors that influence the organisation's ability to achieve sustained success. There are three key factors to consider when choosing the context of an organisation, namely:

- External issues

- Internal issues
- Needs and expectations of its customers and other relevant interested parties.

The occasional changes (occurring or appearing irregularly) to the above factors are monitored and reviewed periodically (while checking the risks and opportunities).

The organisation should align the QMS with its strategic direction while implementing an effective QMS practice. The context of the organisation influences the implementation of the QMS. The context is determined during the initial stages of starting a new business or business plan. The plan and strategy are improved every time after every context review. Subsequent changes to the context are also reviewed periodically to ensure that the organisational objectives are being achieved and the business is sustainable and flourishing.

4.1.3 Understanding the organisation and its context

External issues: : PESTLE analysis determines the external issues in a business environment. This ensures control of the problems so they will have the most negligible impact on the business. If the organisation cannot control the external factors, it must learn to adapt to them.

The analysis includes the following:

a. Political factors, such as political stability, public investments, local infrastructure and international trade agreements: Political factors that influence the economy of an industry. (Examples: National/state issues, current legislation, anticipated future legislation, international legislation, regulatory bodies and regulations and regional or local issues, government policy, political stability or instability, corruption, foreign trade policy, tax policy, labour law, environmental law and trade restrictions, national education policy, infrastructure and health regulations.)

b. Economic factors, such as money exchange rate, the general financial situation, inflation forecasts and credit availability. Economic factors directly or indirectly impact business since they can affect consumers' purchasing power or how the companies price their products and services. (Examples: National economies and trends, inflation, interest rates, general taxation issues, taxation to products and services, seasonality or weather issues, market and trade cycles, exchange rates, economic growth, disposable income of consumers and unemployment rates.)

c. Social factors, such as local unemployment rates, safety perceptions, educational levels, public holidays and working days: Social factors include the organisation's population's demographic characteristics, norms, customs and values. Some of these factors are especially important for organisations when targeting specific customers. (Examples: population growth rate, age distribution, income distribution, career attitudes, local unemployment, holidays, safety perceptions, health consciousness, lifestyle trends, cultural barriers, consumer attitudes/opinions, competition in the market, media views, social image of the organisation)

d. Technological factors include new sector technology, materials and equipment, patent expirations and professional codes of ethics: Technological factors that may influence decision-making include whether to enter a market segment, diversify certain products or outsource production activities, invest a large amount of money in developing a technology that would become obsolete very soon, etc. Innovations in technology may affect the operations and sales in an industry favourably or unfavourably. (Example: New technologies and materials, complex processes, unique processes, competing technology development, replacement technology/solutions, maturity of technology, innovation potential, technology incentives, the level of innovation, automation, research and development activity, technological change and the amount of technical awareness that a market possesses.)

e. Legal factors affecting the work environment include trade union regulations and legal and statutory requirements. (Examples: environmental legislations and codes.) Some of the statutory and regulatory factors may overlap with the political factors. The organisation needs to review the appropriate legal position for an ethical business. If a company is global, the organisation needs to review the rules and regulations applicable to each country. Export limitations/controls include International Traffic in Arms Regulations (ITAR) and Export Administration Regulation (EAR). (Example: trade unions, regulations and legal factors, which may include discrimination law, consumer protection law, antitrust law, employment law, health and safety law, copyright and patent laws)

f. Environmental factors, some of which have become prominent and have assumed global proportions recently. (Examples: Pollution targets and carbon footprint targets set by governments, and resource depletion, especially of raw materials, weather, climate, environmental offsets and climate change, which may affect industries such as agriculture, tourism, farming and insurance.)

g. Market factors, such as competition, including the organisation's market share, similar or substitute products or services, market leader trends, customer growth trends, market stability, supply chain relationships and speed of product obsolescence.

Internal issues: The organisation has control over internal issues, some examples of which are mentioned below:

- Performance factors: products and service offerings, financial results, regulatory requirements.

  Key performance indicators (KPIs) include specific organisational objectives, organisational knowledge, core values and business culture.
- Resource factors: including infrastructure, environment for the operation of the processes, organisational knowledge, assets, capabilities, information systems, and monitoring and measuring resources
- Human factors, such as competence of personnel, organisational behaviour and culture, relationships with unions, suppliers and partners, and operational safety
- Operational factors, such as process or production and service provision capabilities, the performance of the Quality Management System, monitoring customer satisfaction, the complexity of the Process, QMS, the products and services and customer satisfaction.
- Factors in the governance of the organisation, such as its rules and procedures for decision-making or the organisation's structure and size

SWOT (Strength, Weakness, Opportunities and Threats) analysis is a possible substitute for identifying the external and internal issues. Strengths (favourable) and weaknesses (unfavourable) are considered internal factors for achieving the objectives over which the organisation has some control. Opportunities (profitable) and threats (adverse) are considered external factors for attaining the goals over which the organisation has no or least control.

4.1.4 Needs and expectations of interested parties

Determining external and internal issues helps the organisation assess its relevant interested parties later.

'Interested Party' (stakeholder) is a person or organisation that can affect, be affected by or perceive itself to be affected by a decision or activity (para 3.2.3, ISO 9001:2015) within the scope of the management system.

The sub-clause intends to ensure that the organisation considers the requirements of relevant interested parties beyond just those of its direct customers. The intention is to focus on only those appropriate interested parties, which can impact the organisation's ability to provide products and services that meet customer requirements.

Examples of relevant interested parties that an organisation can include the organisation's shareholders, parent and subsidiary organisations, owners, shareholders, joint venture partners, customers, statutory and regulatory authorities, owners of intellectual property, employees and others working on behalf of the organisation, trade unions, external providers (suppliers), end users and competitors. While describing the relevant interested parties, the organisation needs to explain or document the relevance of the person/organisation and their part in the business activities and connections/relationships with the business.

Requirements of interested parties may include specifications from raw materials manufacturers/suppliers, Concerns indicated on employee feedback forms, contractual requirements from customers (quality, time delivery), corporate policies and procedures, legal requirements (say pollution controls), etc.

Any interested party which is not relevant to the QMS need not be considered. Similarly, any requirement of the interested party not relevant to the Quality Management System need not be considered. However, there may also be requirements specified by internal interested parties, such as management, staff, shareholders, trade unions, etc.

Several activities and methods can be used to gather information concerning the needs and expectations of relevant interested parties. Some of them are indicated below:

- Contract reviews for enquiries and orders received,
- Compliance status of statutory and regulatory requirements
- Participation in relevant associations/professional institutions
- Benchmarking the best practices
- Market surveillance
- Review of supply chain management and relationship with suppliers
- Conducting customer/user surveys and obtaining feedback.

Business management starts with quality system planning based on 'risk-based thinking'. 'determining the context' means identifying the relevant external and internal issues/factors the organisation must consider managing any risk the identified issue/factor may unveil. Therefore, quality system planning requires an initial assessment of the risk levels for each identified issue and need; the evaluation uses the probability of occurrence and impact on the business objectives.

Several external/internal issues and requirements of the stakeholders can be identified by looking at the examples above. It is necessary to evaluate every identified issue/requirement for the adequacy of controls established. Evaluate whether the issue/requirement has an impact either on sustainable performance or in achieving and improving the objectives of the organisation. This evaluation includes assessing the probability of occurrence of each issue and likely impact/severity on the business on a scale of 1–5, 1 being the lowest and 5 being the highest. Multiply the severity and probability scores to arrive at a risk exposure number/risk priority number (RPN). Pre-determine a threshold level to accept the risk (on a scale of 1–25, say 9). If the risk level assessed is at or below the threshold, take the level of risk as residual risk. The organisation can proceed with the ordinary business as already implemented controls are adequate. (Record the review and results in the risk register for future assessments.) If the risk level (RPN) is above the threshold, i.e., the risk level is unacceptable. The organisation should identify appropriate mitigation actions/preventive actions and implement them with a planned completion date to reduce the risk (impact × probability) to an acceptable level. A subsequent review of the risk and opportunities may confirm the effectiveness of this risk control. Appropriate assessments and actions should be taken if desired results are not obtained.

Audit criteria 4.1: 'The organisation shall determine, monitor and review the information about external and internal issues'. There is no specific requirement that this information is to be documented. Therefore, there is no room to issue a 'nonconformity' if the report is available but not documented. Evaluate the statements in an interview with the organisation's top management members. Assess relevant information from related documentation to 'determine, monitor and review' and manage identified risks likely to arise from the issues. Explore how the review outputs become the inputs into the QMS planning process (risk-based thinking), including the risk register maintained by the organisation. From the above, the auditor should understand the adequacy of the organisation's review of its context.

Audit criteria 4.2: Verify 'determination of the interested parties (with a general description of relevant interested parties necessary as per vide 4.4.2 (b) of AS 9100 D, if complied), determine the requirements of these interested parties, monitor and review information about these interested parties and their relevant requirements'.

Where the organisation has determined that any interested party and its requirements are irrelevant to QMS, it does not have to take any action to address them. The auditor should evaluate the way an organisation has decided on the requirements of interested parties, which are relevant for the QMS, by considering:

- The range of interested parties considered
- Criteria to select relevant interested parties
- Aspects of choosing applicable requirements.

Therefore, the auditor should be able to conclude the appropriateness of these practices and how this information is reviewed and monitored, such as through management review records and the risk register. Verifying the risk register provides information about the management of identified risks likely to arise from the needs and expectations of stakeholders.

The auditor should conduct this review in an interview with top management and follow these issues throughout the audit. The auditor needs to collect objective evidence to substantiate that the outputs of this activity are consistently reflected in the review of risks and opportunities, communication and other relevant areas of their QMS.

NCs such as 'Not all issues have been determined' are unacceptable. An opportunity for improvement (OFI) is good enough. However, auditors can issue a 'nonconformity' if:

- During the interview with top management, they cannot articulate a discussion about the determination of internal and external issues, the expectations of the stakeholders and the review of changes from time to time.
- Results of management reviews do not include changes in internal and external issues. (Refer 9.3.2 (b).)

4.2 Determining the Scope of the Quality Management System

Scope defines the boundaries within which the Quality Management System is established for the business in the organisation. The organisation must consider whether all the requirements of the Standard are relevant to its activities based on the nature of the organisation itself, its products and the processes it uses to meet customer, statutory and regulatory requirements. In addition, the organisation may consider its commitments in its quality policy and objectives. Organisations need to examine how these could affect any production process. All of these are considered for identifying the scope of the organisation's QMS.

Suppose any requirement of the Aerospace Standard cannot be applied in the organisation due to the nature of its product/business model. In that case, this can be considered as not applicable. However, these requirements that are not applicable do not affect the organisation's ability or responsibility to provide products that meet customer and applicable statutory and regulatory requirements. The scope of the organisation's Quality Management System must be maintained as documented information.

The scope statement should include the types of products and services covered in the organisation's Quality Management System.

Examples are shown in the typical 'Documented Information' in Annexure C.

> *'It was seen that some organisations did not consider the relevant activities applicable to their functions and Quality Management Systems in the past. They declared and documented that they did comply with the requirements of 'Monitoring and Measuring Resources' (7.1.5, earlier 7.6) and 'Validation and Control of Special Processes' (8.5.1.2, earlier 7.5.2). While auditing, none of the processes could demonstrate the use of any measuring equipment for their products/services nor show the validation records of any method. It was a software/engineering services organisation! The organisation could have documented that those clauses do not apply to their business/Quality Management System. Note: This and other sidebars include anonymous examples from personal audit and consulting experiences to protect individual/company identities'.*

If any sub-clause requirement is segregated as 'not applicable' the decision must be justified.

Some of the requirements stated as 'not applicable' by some of the certified organisations are indicated below with their justifications, as examples:

a. Clause 8.5.5 (Post-delivery Activities) collection and analysis of in-service data (e.g., performance, reliability, lessons learned); control, update and provide technical documentation for product use, maintenance, repair and overhaul; controls required for work undertaken external to the organisation (e.g., off-site work); product/customer support (e.g., queries, training, warranties, maintenance, replacement parts, resources, obsolescence). The indicated post-delivery activities are not applicable and are not carried out since the customers do not expect the same from the organisation. The organisation has no mechanism to deal with actions such as periodic maintenance or servicing of the items supplied. The organisation has no accountability for these post-delivery activities. The organisation is justified in indicating the indicated post-delivery activities as 'not applicable' as none of its contracts provide follow-up activities related to the delivered product/service.

b. 8.5.1.2 Validation and Control of Special Processes: 'Validation and Control of Special Processes' is not applicable. The processes are not carried out since the customers do not expect the same from the organisation. The products do not require any unique processes during their realisation. The requirements of sub-clause 8.5.1.2 Validation and Control of special Processes are not applicable for production because it is possible to inspect the result of all the processes in the organisation to determine conformance or nonconformance to the specification. Since the resulting output will be verified by subsequent monitoring or measurement and deficiencies, if any, can be detected through in-house monitoring and measurement before the product is delivered to the customer, validation and control of unique processes are not required.

c. 7.1.5 Monitoring and Measuring Resources: The engineering services provided are software-based, and therefore, no physical monitoring and measuring devices/equipment, which require periodic verification/-calibration, are used. The resources are not used in the operation and support processes. Hence, 7.1.5 Monitoring and Measuring Resources is 'not applicable'.

d. 8.3 Design and Development of Products and Services: Not applicable since the parts and assemblies are manufactured according to the customer's design or-supplied drawings and specifications. The organisation has no responsibility or authority to make any changes in the original/customer-given drawings and specifications without a directive from the customer. We do not design any product and design is not a customer requirement.

Since 'design and development' is defined as a set of processes that transforms requirements into specified characteristics or specifications of a product, and if the organisation has no responsibility for design and development activity, it can be excluded from the scope statement for certification/registration. It also means that design and development are neither part of the production process nor the organisation outsources such activity.

> '*One organisation wanted to include the words "Design and Fabrication of Stamping tools and metal stamping for precision components." The tooling design was not a delivered product/service to the customer. Hence, they claimed the Exclusion of Clause 7.3 (AS 9100C) with justification. Since "design" was included in the scope statement, the certification body did not allow the organisation to exclude 7.3 "Design and Development of Products and Services. They were advised to drop the word 'design' from the scope statement if they claimed exclusion/ applicability of 'design and development of products and services." The scope statement must strictly indicate delivered products to avoid confusion*'.

With the revision of the standard, there are very few requirements which will be subject to a 'non-applicability' declaration, such as for a service organisation that has no 'measuring equipment to monitor or measure that needs traceability' in, which case it may declare non-applicability of Clause 7.1.5 (AS 9100D).

It is also necessary for organisations to declare that the requirements determined as not being applicable do not affect the organisation's ability or responsibility to ensure the conformity of its products and services and enhance customer satisfaction.

If you call a single source of documented information kept by your organisation a 'Quality manual', indicate the scope of the quality manual first and then the scope of the certification/registration. Below the scope of certification, list the sub-clause-wise requirements claimed as 'Not Applicable' and write down their justifications as to why it is not applicable. While conducting audits, the certification body shall review the validity and justification for all 'Not Applicable' requirements claimed by the organisation.

Add or delete requirements previously declared as 'Not Applicable' when necessary. It may be done only with appropriate changes in the documented information and justifications.

Information needs to be given to the certification body well in advance so that they may take appropriate action to include or delete the sub-clauses during their next assessment/audit.

It is desirable to begin documenting the information required by the QMS by giving a profile of the organisation, otherwise known as the 'Context of the Organisation'. A foreword/preface/authorisation/approval by the organisation's chief, calling upon the employees to implement and practice the system and improve the same to achieve the organisational goals continually can follow.

When the organisation documents the Quality Management System in a source of recorded information called a 'quality manual', the scope should be written so that it will not mislead customers. The data should facilitate the user to determine which types and categories of products and production processes are included within the scope of certification. Responsibilities for product design, manufacture and after-sale servicing must be identified.

The scope of the QMS is commonly used to describe the organisation's processes, products and services and related locations, departments, divisions, etc., to which the organisation applies a formal QMS. The QMS may not include some of the organisation's processes, products, locations, departments, divisions, etc. Therefore, the scope of the QMS should be based on the type of products, their production process and the service delivery activities.

The overall objective of third-party certification is to give confidence to customers and the public who rely on certification. The main principles for inspiring confidence are competence, impartiality and independence in action and appearance. Suppose the quality manual is prepared to get the quality system assessed and certified by a third-party certification body. In that case, the scope is to be indicated in two parts, (a) and (b) separately, as given below, concerning the organisation's core business activity.

The scope of the QMS manual and the scope of registration/certification may sometimes be understood interchangeably. However, certification bodies require the scope of the 'QMS manual/Documented Information'; the scope of registration/certification is indicated separately within the quality manual. This is to avoid possible confusion among customers and end users. The scope of registration/certification needs to be clearly defined in the organisation's 'quality manual' and in any other publicly available promotional and marketing documents. However, promotional statements like, 'we are the best, we produce world-class products and we are the number one in the field', etc., are not permitted to be included in the scope of registration/certification.

During the assessment, the certification body auditor (person competent to conduct an external audit) verifies that the final statement of the scope of registration/certification is not misleading. The scope only refers to the organisation's processes, products, locations, departments, divisions, etc., assessed during the registration/certification audit. It is also verified that the scope defines any requirement(s) as having been determined as not applicable to the scope of its Quality Management System and that justification for any such decision of non-applicability is provided and is reasonable. Some examples of the scope statements are given below:

- Scope of the manual: 'This manual describes the organisation of (Name M/s...) and defines the Quality Management System established to comply with the requirements of the aviation, space and defence standard AS 9100 D, including the requirements of ISO 9001:2015'.
- Scope of certification: 'Design and Development of Fixed Wing Aircraft, Aerospace Systems and Associated Equipment like.....'. (For Aerospace Design Organisation)
- Scope of certification: 'Manufacture of Precision Engineering Components, Assemblies and Tools for Aerospace Applications'. (For a small-scale industry, machining components and making sub-assemblies.)
- Scope of certification: 'We manufacture, repair and post-delivery service of aircraft, components and assemblies for civil/military aircraft and aerospace applications'. (For an aircraft manufacturing unit.)
- Scope of certification: 'Providing Engineering Design and Technology Services for Aerospace Components and Systems from Concept to Productionisation'. (For a Software Engineering Unit, Engineering Service provider.)

Documented information can be kept in any media, like in print or a folder/word file on a computer. Keep a copy of the 'quality manual' approved by the organisation's chief either in soft form in the computer system (read-only access) or a hard copy in print, with signature of approval. Identify the master copy and issue controlled copies to other process owners/departments when issued as hard copies.

The organisation must ensure that it has a methodology for identifying, maintaining and periodically updating all applicable statutory and regulatory requirements. The requirements are handled and complied with while utilising the process inputs and monitoring the outputs. The organisation shall be able to demonstrate the involvement of the statutory authorities and regulatory authorities in some of their functions. While undergoing the certification/registration process, if the auditors notice any violation or non-compliance, for example, if the machine shop has no license from the state government, the overhead crane is not certified for safe load, or the weighing machine is not authorised by the authority for weights and measures of the state government authorities, it is a violation of the statutory requirements. Compliance with statutory laws on labour, health, safety and environment is also mandatory.

4.3 Quality Management System and Its Processes

4.3.1 Process elements

The requirement of clause 4.4.1 is to establish, implement, maintain and continually improve a Quality Management System, including the processes needed and their interactions. This requires adopting a process approach. Since every organisation is unique, documented information such as process diagrams, planet diagrams, business process maps or written procedures could be used to support this.

Each process can be depicted as including the following:

- Input, outputs
- Sequence and interactions
- Criteria and methods (including monitoring, measurements and related performance indicators) needed to ensure the effective operation and control of these processes
- Resources required for these processes and ensuring their availability
- Responsibilities and authorities for these processes.

While determining its QMS processes, an organisation must evaluate the risks and opportunities that may affect its context. Identifying, assessing and managing such risks in the risk register maintained according to Chapter/ Clause 6.1 is easy. Using the PDCA cycle, the organisation must integrate the actions to address risks and opportunities into its QMS processes. Not all processes of a Quality Management System represent the same level of risk in terms of the organisation's ability to meet its objectives, and the effects of uncertainty are not the same for all organisations. Each organisation is, therefore, responsible for the extent to which it applies risk-based thinking and its actions to address risk, including whether to retain documented information as evidence of its determination of risks.

The quality manual may start with a statement that the organisation has established, documented, implemented and maintained a quality management system and tries to continually improve its effectiveness per the aerospace standard AS 9100 D requirements.

The management establishes the policy and objectives of the organisation. They determine the processes and responsibilities necessary to achieve the process objectives. They pick and provide the infrastructure and resources required to attain the goals. They establish and apply methods to assess each process's conformity, efficiency and effectiveness. They identify potential conflicts of interest at the level of the organisation and the individuals and establish means of dealing with the situations. They demonstrate methods of correcting nonconformities as and when they occur and take corrective actions to eliminate their causes so that such nonconformities do not recur. They take long-term measures, through risk management, to prevent or mitigate potential nonconformities from occurring. The organisation should have a plan and periodic reviews for establishing and applying a process for improvement (proactive or reactive) of all the above activities, which covers the total Quality Management System.

Armand V Feigenbaum defined total quality control as 'an effective system for integrating quality development, quality maintenance and quality improvement efforts of the various groups within an organisation, to enable production and service at the most economical levels that allow full customer satisfaction'.

'A systematic approach to implement the Quality Management System instead of a formal compliance with the requirements of the Standard' is what the QMS documentation (both documents and records) should reflect. Organisations should introspect whether the documented management system is applied effectively to maintain and continually improve the organisation's performance by evaluating its ability to deliver the products and services by meeting all customer expectations, needs and expectations of the interested parties and applicable statutory/legal and regulatory requirements.

The quality manual contains the scope, details of and justification for any requirements determined as being 'Not Applicable', identification/reference of the other documented information on processes and procedures and a Sun and Planets diagram or a business process map indicating the processes and their interactions within the organisation.

As discussed earlier, the 'quality manual' should contain all the other documented information established for the Quality Management System.

Since organisations like to present their quality manual to prospective customers and other stakeholders (for example, civil aviation authorities, defence industry and authorities, airlines, certification bodies, trade associations, suppliers, maintenance repair and overhaul service providers, distributors of OEM equipment and spares, IAQG member organisations and IAQG strategy teams), it is advisable to make it as a presentable, brief and adequately addressing the Quality Management System aspect of the organisation. Therefore, all the documented information, such as procedures, is prepared and kept separately for daily use within the organisation. While addressing the requirements of the Standard to meet the Quality Management System in the organisation, all the applicable processes and their interactions are to be described. The description must affirm that the organisation has already established, implemented and practised the system requirements in the international standard. It is desirable to identify and map the processes in a sketch, a graphical business process map or a model and show their interactions through arrow marks in either direction. This map is the business process map or Sun and Planets diagram. According to international standards, clauses 4 to 10 generally cover all business processes. A business process map/Sun and Planets diagram can be prepared by referring to the figures (Figure 0.2 Schematic representation of the elements of a single process and Figure 0.5 Typical process map/planet diagram for a product realisation process) in the Introduction, and customising it to the organisational processes.

The processes can be depicted in various ways such as process maps, turtle diagrams, octopus, SIPOC method (identification of supplier, inputs, process steps/tasks, outputs and customer), Sun and Planet diagram, etc. Process maps and process flow charts will reduce the amount of description. Other documents are, as determined by the organisation to demonstrate effective planning, operation and control of its processes/various functions. These additional documents could be referred to in the Apex Manual or included separately. Some examples of these documents are organisation charts, work and test instructions, specifications, process maps, process flow charts and descriptions, quality plans, test and inspection plans, production schedules, documents containing internal communications, approved supplier lists, etc. Large organisations may need several documents and operating manuals at the global, national or regional level and a more complex documentation hierarchy.

All documents must be controlled as per clause 7.5.3 Control of Documented Information. While creating and updating documentation, the document should indicate the changes and the current revision status. Relevant and applicable versions of documents must be available at the points of use, and suitable identification must be given to obsolete documents if they are retained for any purpose.

When the process elements are described or mapped, all the following requirements will be understood, i.e., 4.4.1 (a)–(e). The risks and opportunities can be identified and assessed per the risk register being periodically reviewed via clause 6.1 of the Standard (addressed via Chapter 6.1 in this book). Process evaluation for effectiveness is done by verifying the achievement of performance objectives against a set target for a period. If the process results are not achieved, the changes/improvements to achieve planned outcomes are to be identified and implemented. Monitor the improvements or modifications until the intended results are achieved. This is what the process owner regularly does to improve the process and the Quality Management System in the organisation.

4.3.2 Process flow and execution

It is necessary to keep documented information for complex processes in the organisation. All processes are not complicated. Many unique tools and software programs are required when too many steps are necessary. When the intervention of various skills is warranted, unique processes are included and the process can be complex. A guideline document, process flow diagram, description of step-by-step activities or work instructions for such processes is always beneficial. Maintain such documents as necessary, which the organisation decides. The document will support the operation of the process. Suppose a log is maintained against the step-by-step activities, with a positive indication that the processes are being carried out in the correct sequence and maintaining the parameters necessary. In that case, it conveys adequate confidence that the process is being carried out correctly and as planned. There is no room for the operator to change the process, sequence or parameters according to the operator's discretion. No such preference is to be used unless duly authorised by the designer of the process.

The documented information about the process must include a general description, relevant interested parties such as the previous internal process owner, internal customer of the process, external customer and product requirements, and how the process achieves the planned activities and results. Suppose there will be an impact of the processes, process results and product characteristics on any interested parties. In that case, that aspect must be documented with adequate description. The description must be within the Quality Management System's scope and applicability.

What is the role of the process for the business process in the organisation, sequence of the operations, interactions and interface of the process within the organisation?

The process owner's responsibility and authority can also be documented along with the information. All the processes in the organisation, their sequence and interactions, indicated in a business process map, are adequate to address all the above requirements. Such information was generally kept as documented; you may continue to call it a quality manual. The note indicated in the Standard below clause 4.4.2 confirms that. Nobody has ever prevented you from reaching the document quality manual, as nobody can object to you for maintaining a quality manual. Saying 'quality manual is not necessary' according to the revised standard is a matter of opinion. You have your freedom of not to have the freedom!

The general requirements of the Standard are summarised or paraphrased in this paragraph. The organisation shall establish, document, implement and maintain a Quality Management System and continually improve its effectiveness by the requirements of this standard. The outline or framework and the details can be found in the following clauses. The organisation's QMS documentation needs to identify the customer and applicable statutory and regulatory Quality Management System requirements appropriate to its products and services and integrate them into its resource management and product realisation activities. The organisation must demonstrate this and make it available to all interested personnel for compliance.

Auditors generally are aware of the general and specific statutory and regulatory requirements applicable to the products/services included within the scope of the QMS. They will verify the identification of the same within the QMS documentation, as while preparing for the audit, they obtain relevant information from internal or external sources concerning these statutory and regulatory requirements. This will allow them to judge the suitability of the QMS to address such provisions.

The organisation should determine the application of the Standard 'where appropriate' requirements, wherever they appear, since most of the requirements specified are generic and apply to any organisation. Since every organisation is unique, intending to use the system should consider what is appropriate for the process and what is not and whether it affects the organisation's ability to satisfy its customers' requirements. No excuses are permitted if reasonable and must be included to demonstrate its ability to consistently provide a product that meets customer and applicable regulatory requirements. Suppose the criteria are appropriate concerning the organisation's scope of operations. In that case, the requirements must be considered and established in all processes where such an option is given in the Standard. Therefore, organisations can assess whether this requirement adds value or benefits the process without considering the 'where appropriate' element. Any third party who decides what is appropriate should have the skills to understand how best the processes are established.

The organisation must first identify and determine all the processes needed for the Quality Management System and their application. Identify all processes related to the customer-oriented or the core processes with which the organisation tries to complete the job or produce the product required by the customers. These processes can be grouped into core, management and support. Once the main processes/groups of processes are identified, it becomes easy to determine the sequence and interaction of these processes. Each primary process may have several sub-processes with their sequence and interactions with other processes.

Drafting an organisation chart indicating hierarchy is not a requirement of the Standard. Hierarchy only means authority and responsibility at a broad level and the bureaucratic flow of information. Processes are not departments, sections or silos. Do not confuse the organisation chart for a collective list of all processes in an organisation. Even when an organisation chart is made, it may not directly reflect all the processes in the organisation. Otherwise, departmental designations should identify the process names. Who can be the departmental head for customer satisfaction? Who can be the department head for customer feedback? Who will head the department for handling customer complaints? The departments/sections in an organisation cannot be identified as processes because departments/sections in an organisation may be responsible for carrying out one or more main processes and many sub-processes.

Moreover, the departments/sections in an organisation may turn out to be closed compartments (organisational silos), where inputs from some departments and output to some other departments may not flow smoothly. In a process approach to management, it is mandatory to identify the process requirements. Process requirements are – what is the process, who is responsible, who has the competency, what are the inputs, what are the stages and their sequence, how the process stages/steps are carried out, what is the output, who is the customer – internal or external, how the planned activities and outputs are measured/monitored and how the objectives or process performance levels are measured and compared with the targets already designed. Plan-Do-Check-Act (PDCA) can be applied to all processes. Individual processes and their intended results are directed towards meeting the customer needs, expectations and requirements, thereby building customer satisfaction. The Quality Management System manages the processes in an organisation and integrates their output. When the processes are implemented as a system in the organisation, they become practical tools. Integrating QMS with business processes in the organisation is mandatory. Higher-level processes are usually described as a business process map. Appropriate descriptions of interactions between the processes will do. Lower-level sub-processes can be described or indicated as a process map (turtle diagram), planet diagram or operator/work instruction. A business process map/process map or planet diagram is not a requirement of the Standard. Still, it is an easy and effective method of describing processes and their interactions.

Every process, when defined, should determine the criteria and methods needed to ensure that both the operation and control of these processes are effective. The procedure is determined effectively when planned results are achieved and the process objectives are met.

The QMS documentation must also identify a customer and applicable statutory and regulatory Quality Management System requirements. The requirement is indicated at the QMS level and not only at the documentation level. The organisation must demonstrate that the statutory and regulatory requirements for its products/services have been properly identified, applicable and valid requirements are complied with and the relevant documents/laws/rules are available for reference. During third-party audits, the auditors will not ignore any nonconformity related to statutory/regulatory requirements of health, safety, environment and labour laws. Deliberate noncompliance may affect the image and credibility of the Quality Management System and the organisation.

It is necessary to state in the quality manual that all applicable Quality Management System processes/clauses of the Standard and their application throughout the organisation have been ensured. Sequence and interaction of the processes can be defined through a Sun and Planets diagram or a business process map, indicating the core business activity, management processes and supporting functions. It is necessary to identify criteria for and methods needed to ensure that the operation and control of these processes are effective. The processes can be considered adequate only when the objectives and performance measurements meet planned results. When intended results, measured at periodic intervals, are not achieved, the organisation should take appropriate actions to improve the outcomes and achieve the targets. The organisation should ensure the availability of adequate resources and required information to support the operation and monitoring of all the processes.

Understanding the difference between 'monitoring' and 'measuring' is essential. Monitoring implies observing, supervising and keeping under review by using monitoring devices. It may include measuring or testing at intervals, primarily to regulate or control. Measuring considers the determination of a physical quantity, magnitude or dimension using measuring equipment or device.

It may be necessary to monitor, measure and analyse the processes and implement actions to achieve planned results and continual improvement of these processes. Measurement is needed to ensure the effective operation and control of these processes.

It is to be understood that 'The International Standard specifies requirements for a Quality Management System where an organisation needs to demonstrate its ability to consistently provide a product that meets customer and applicable regulatory requirements and aims to enhance customer satisfaction through the effective application of the system, including processes for continual improvement of the system and the assurance of conformity to customer and applicable regulatory requirements'. A claim that an organisation conforms to the international standard should mean to its customers that it can consistently provide products that meet customer requirements.

We have seen that a process is a set of interrelated or interacting activities which transform inputs into outputs. An 'outsourced process' is a process that the organisation needs for its Quality Management System. The organisation chooses to have this process performed by an external party. An outsourced process can be performed by a supplier independent of the organisation. The source can be part of the same parent organisation, such as a separate department or division. Still, the department or division is not included in the Quality Management System established. The outsourced process 'source' can be within the organisation's physical premises or work environment, in an independent location or away from the premises. Suppose the process performed is not included in the Quality Management System. Still, the process is needed for the final realisation of the product or output being delivered to the customer or to meet the quality requirements of the product. In that case, it is identified as an outsourced process. Suppose an organisation chooses to outsource any such processes. In that case, the organisation first needs to identify the processes by name, list them as outsourced processes and establish control over such processes through process specifications, descriptions and purchase processes.

To meet the product/customer requirements, organisations outsource extraordinary processes (refer 8.5.1.2) for the following reasons:

a. The capability exists but is overloaded, or it is commercially beneficial to outsource the process

b. Capability does not exist within the organisation. Hence, it looks outside for a capable source (a supplier/an external provider), evaluates the capability and selects the source/supplier. (Refer 8.4.1).

The controls exercised for outsourced processes must be identified and documented within the Quality Management System. It is unnecessary to indicate the details of control of the outsourced processes in the quality manual. Emphasis is made on the need for the organisation to have adequate management for the outsourced processes and not merely depend on the controls exercised by the process owner to whom the processes are outsourced. It may also be mentioned that the outsourced processes need not be limited to the product realisation processes; they can be any other process, such as management or support.

It is also to be understood that ensuring adequate control over outsourced processes does not absolve the organisation of the responsibility of conformity to product requirements and other customer/statutory/regulatory requirements related to the product. It means the organisation cannot ignore any process affecting product conformity and requirements nor exclude it from the Quality Management System. Hence, adequate control over outsourced processes may include:

a. Entering into a contract with the source outside, depending on the nature and importance of the process and the risks involved
b. Insisting on the required competency of personnel in carrying out the process/process requirements
c. Giving them or approving the process requirements/specification/ process parameters or steps to be followed for satisfactory completion of the process
d. Indicating the routine and periodic checks or tests to be done, including process validation for unique processes (Refer to 8.5.1.2)
e. Requesting a certificate of conformity to the process
f. Issuing purchase order or related contract agreement for the outsourcing as a purchase process (this is described in clause 8.4.1, 8.4.3)
g. Any other information which adds to the type and extent of control to be applied
h. Determining and monitoring/reviewing the potential impact of the process on the product quality requirements
i. Determining and monitoring/reviewing the degree to which the control for the process is shared between the organisation and the outside source, the potential impact of the process on the product quality requirements
j. The degree to which the control for the process is shared between the organisation and the outside source
k. Considering a Quality Management System as a pre-requisite for the supplier to demonstrate confidence and the capability of the supplier to undertake the process.

Per AS 9110 C, it is a regulatory authority requirement that maintenance organisations obtain Quality Management System approvals and any other approvals, certificates, ratings, licenses and permits to carry out their business. The organisations must maintain them to undertake any tasks related to MRO activities.

The organisation's Quality Management System shall also address Quality Management System requirements imposed by the customer and applicable statutory and regulatory Authorities. These requirements are known as Production Organisation Exposition (POE) for production organisations and Maintenance Organisation Expositions (MOE) for MRO organisations.

Document that your organisation has established, documented, implemented and maintains a Quality Management System and tries to continually improve its effectiveness in accordance with the requirements of the Aerospace standard AS 9100 D. It is advisable to provide a Sun and Planets Diagram (business process map), which clearly shows the management involvement, core activity and support activities. Indicate how the customer requirements are consolidated, perhaps in a separate document, which may be referred to in the quality manual. Identify and document the statutory and regulatory requirements you comply with while undertaking the aerospace business of design/development/manufacture/servicing of components/assemblies or engineering services you provide to your customers. The list of statutory and regulatory requirements and their compliance status, kept separately, may be referred to in the quality manual. Identify the outsourced processes and the controls exercised to ensure the quality of the products/services outsourced. The details regarding outsourcing, including methods, specifications, sources, source controls, etc. can be kept as documented information separately and referenced in the quality manual. (Customise the statements as per the organisational practices. Write precisely what you do and confirm that what you do is in accordance with the Standard requirements.)

Maintaining (issue, revision, approval, updating and controlling its issue and use) a documented quality manual and procedures is not mandatory. However, the information must be available to see and verify in any media except the human brain.

5

Leadership

5.1 Leadership and Commitment

5.1.1 General

It is the most important clause of the Standard. Commitment to quality and efforts to satisfy customers begin at the top. The commitment must be at the peak whether the organisation is big or small and the product range is extensive or single. Top management should be the highest level of person or group at the organisation's location. Even if the organisation is part of a giant corporation, the head or the division establishing the Quality Management System at the site is considered the top management.

Leaders at all levels establish unity of purpose and direction and create conditions in which people are engaged in achieving the organisation's quality objectives. Leaders identify strategies, policies, processes and resources for an organisation.

With globalisation playing a big part and off-loading of jobs to small industries in India, the realisation has come to that implementation of a Quality Management System in an organisation dealing with engineering services, design or manufacture of a component or a sub-system for aviation, space or defence equipment is a customer-driven requirement. Supply chain management is an increasing challenge as globalisation continues rapidly. Prime organisations in the USA and Europe are offloading components and sub-systems of aviation, space and defence equipment to India for various benefits. Supplier-customer relationship is based on the foundation of mutual benefit. Therefore, during initial discussions with suppliers, every customer clarified that implementing AS 9100 QMS requirements is mandatory for business or even for giving a trial order. This is because if implemented at the suppliers premises, consensus-based best practices help improve the quality of a product, process or service. Moreover, the International Aerospace Quality Group (IAQG), which prepared the quality system standard, had set a goal that any supplier who follows AS 9100 quality systems will be acceptable to all aerospace prime organisations worldwide.

Supplier development undertaken earlier by major industries had cost limitations apart from the constraints of being a dedicated supplier for a limited quantity of products. The rapid growth of the aviation sector in the country has thrown up multiple opportunities for the industry.

Management commitment is derived from the fact that a long-term, mutually beneficial relationship can be developed with the world's prime aviation, space and defence organisations. Selecting and establishing a Quality Management System, which will satisfy the prospective customers and products while meeting the organisational objectives, are the most essential steps in planning the strategy for the business. Setting up a new facility or expanding the existing enterprise is thought of. A business model is selected, and a business process map/Sun and Planets diagram with details of the processes employed to realise the products are prepared. A broad plan of action is made for the infrastructure to acquire the land and buildings, set up the shops and other facilities, purchase/utilise machinery and find the appropriate human resources. Size and the organisational structure are planned. Management commitment is visible in all such activities, which means a responsibility of the senior executive management team. The management team needs to be committed and involved in the implementation and maintenance of the business, and the commitment must be visible to all personnel throughout the organisation.

The Standard requires that when planning its QMS, the top management must implement and promote the use of the process approach and a culture of risk-based thinking throughout the organisation to determine and address the risks and opportunities associated with it, assuring that the QMS can achieve its intended result(s), provide conforming products and services, enhance customer satisfaction, promote desirable effects and improvement and prevent, or mitigate, undesired effects/risks. Top management is a 'person or group of people who directs and controls an organisation at the highest level'. When there is a doubt about who among the top management is 'the Top Management', the answer points to the full management person who is empowered or authorised to decide on financial matters; in a small organisation, this may include the owner or partners and a few key people who report directly to them.

Management direction is vital if the QMS must deliver its intended results. Without this, the QMS will be overtaken by other priorities, and the benefits of continual improvement to focus on customer needs will be lost. It is imperative to ensure all employees are engaged optimally to achieve the common goals or intended outcome of the QMS. Therefore, the role of top management is to inspire and lead by example. Principal leadership is expected to be involved in everything to the extent management desires and ensure that the quality policy and objectives are consistent with the overall strategy and context.

The establishment of a Quality Management System should be announced. The employees at all levels should be invited to participate effectively in the implementation, even before documenting the system and all through the various phases of its implementation. Executive management should drive the system, participate in it and allow the practice of the QMS down to all the work areas and employees through their actions and not just their words. That is, the management should 'walk the talk'.

Management team expresses its commitment to the organisation, customers and other stakeholders through established methods and communication thereof, such as:

a. Quality policy,

b. They are setting up measurable quality objectives for all the functional levels in the organisation, for the products/services delivered to the customers and for meeting the customers' requirements. The objectives must be targeted towards improving the quality of products and services, individual and organisational performance and overall cost reduction. Executive management defines all the measurable objectives. These objectives bring out the organisation's strength when achieved against the set targets. When there are different divisions of the organisation, the result of all the divisions together measures the corporate objectives.

c. They provide adequate resources and ensure their availability for use. Resources include resources for implementing, maintaining and continually improving the effectiveness of the Quality Management System; human resources; providing necessary infrastructure, buildings, workspace and utilities; equipment, machinery and supporting services such as transport, communication or information systems.

d. Periodic health check-ups and effectiveness of the Quality Management System by monitoring and analysing results of internal audits

e. Monitoring and analysing the business processes at the functional levels through periodic reviews. The management identifies the vital operations and controls them to meet the customer requirements with the required quality and optimises them to make them cost-effective.

f. Directing/initiating improvement as a follow-up action of data analysis measured in continual improvement, corrective actions and preventive actions. This includes data analysis about internal and external rejections, customer complaints, cost of quality, etc. 8D methodology, a problem-solving tool, rewards the team for bringing in substantial improvements or corrective or preventive actions, which increases the organisation's reputation and effects significant savings in cost. The management reviews the findings and recommendations. When found suitable, the team is specially rewarded. That shows the management's commitment to continually improving the products or services.

Management may have short- and long-term strategies to further the business objectives. Periodic communication to the employees, such as encouragement, feedback, guidance or direction, is also seen as a managerial commitment.

In short, top management commitment is evident when the highest-level executives directly participate in an organisation's specific and critically important aspect or programme. Implementing the Quality Management System includes setting up and serving on a quality committee, establishing quality policies and objectives, providing resources and training, overseeing implementation at all levels of the organisation and evaluating and revising the quality policy and objectives in the light of results achieved.

Management commitment must also be visible in two aspects with regard to providing resources and training to personnel. Top management should have understood the importance of identifying the competencies of personnel required for the various processes in the organisation. Each of the functions has its specific competency requirements in the organisation. Training is provided to fill the gaps periodically reviewed for effectiveness.

An enlightened management establishes a mission and vision for the organisation and identifies the strategies needed to achieve them. The answer to why the organisation exists is a mission statement. The mission should describe the overall purpose of the organisation and the value it seeks to create for its stakeholders. The mission statement considers the organisation's unique products, services, markets, values, concern for public image and maybe priorities of activities for competitiveness.

The vision answers what the organisation wants to be in the future. The futuristic vision statement indicates the level at which the organisation wishes to grow, with the augmentation of resources and adequate process controls. Vision is an aspiration which should not be too idealistic to achieve.

The organisational objectives and targets are identified accordingly. The mission and vision are identified concerning the long-term goals, organisational characteristics, competencies and overall benefits to the society and stakeholders.

As per AS 9110 REV C, additional requirements are establishing the safety policy and ensuring the safety objectives are established. As in the case of quality policy and objectives, it is necessary to have a safety policy statement. The policy shall be a framework for achieving the safety objectives to be measured and monitored. International Civil Aviation Organisation provides 'Safety Management Manuals' with guidance material for service providers to use and implement.

As for the documentation, make a statement in the 'quality manual' that the top management is committed to implementing the Quality Management System and continually improving its effectiveness. The management demonstrates leadership and promotes the use of the process approach and a culture of risk-based thinking throughout the organisation. The importance of meeting customer statutory and regulatory requirements has been communicated throughout the organisation. The quality policy and the quality objectives have been established. Management conducts periodic reviews and ensures the availability of resources.

In a dynamic organisation, the senior management knows the business operations well. Most of the decisions are based on facts and figures. Regular reviews are held on business and their impacts. Management reviews are planned to address all business areas and organisational objectives. Customer satisfaction and performance (product quality, on-time delivery performance, adherence to technical specifications to meet customer requirements, post-delivery support to customers, operational costs, etc.) are the key performance indicators (KPIs). Project management, risk management and configuration management processes are in place where top management involvement is visible. Results of action plans are regularly reviewed and assessed to confirm that the desired results are achieved.

Top-level management commitment to performance monitoring is evident in resources management, including training of personnel, measurement of objectives at the functional level and the organisational level and management reviews for improvements – visual management, deployed through the display of periodic performance results and customer satisfaction measurements. Continuous performance monitoring drives action plans to improve the business.

Organisations may demonstrate management commitment through their knowledge of the quality systems and communication of the quality policy and objectives to the employees. Communication may be through the policy statement signed and displayed through posters, newsletters, periodic meetings and employee interaction. The employees should be able to mention the spirit of the policy and how they achieve the objectives. Positive and regular involvement of the top management in the review of the achievement of goals and their total commitment to meet the customer requirements of quality and on-time delivery performance of the products demonstrate management commitment. The management should be involved in continual improvement of the performance of the organisation as a whole and at a functional level for the processes.

Implementation is shown through the Quality policy available and displayed. Quality objectives are defined and measured against a target at various functional and organisation/corporate levels.

Resources are available (infrastructure, personnel, plant and machinery, supporting services, quality system documentation and implementation efforts, etc.). Internal audits are conducted. Management reviews are held. Continual improvement activities are planned and completed, including risk management.

Usually, auditors under training ask – How to evidence the management commitment and leadership?

One of the foremost key responsibilities of top management is to ensure integration of the QMS requirements into their business processes to ensure 'one organisation, one set of goals'. The top management is responsible for ensuring that a well-meaning quality policy and relevant and valuable quality objectives are established and meticulously monitored for implementation and achievement. Resources needed are planned and provided and periodically reviewed by management. Continual improvement is promoted and supported by management. Management should mention that there are processes to comply with customer, statutory and regulatory requirements. They are understood and met, and all the employees understand why this is important. This evidence can be obtained while talking to the committed leadership and management personnel. The management commitment should be visible, too.

There should be documentary evidence for the role of personnel in the organisational roles, responsibilities and authorities. This information should be known to all employees. Auditors usually observe visible proof of top management's commitment by the management's presence during the audits, opening and closing meetings, and their interest in knowing the outcomes of the audits. Top management usually implements recommendations from audits, corrective actions, management reviews, etc.

5.1.2 Customer focus

The whole message of the Standard is customer satisfaction. Like love is the essence of all religious teachings, customer satisfaction is the essence of the QMS standard. Customer requirements are determined per clause 8.1, and customer requirements related to the product are determined per clause 8.2 of the Standard. Customer and applicable statutory and regulatory requirements must be determined, understood and consistently met to keep customer focus. Customer satisfaction is determined as per clause 9.1.2. Customer requirements are to be identified in total with the aim of satisfying and enhancing customer satisfaction. Top management must provide for all requirements thoroughly and effectively. The organisation-wide risk management committee or ERM should have identified the risks and opportunities that can affect the conformity of products and services while planning to meet the customer requirements. The ability to enhance customer satisfaction is also determined and addressed. Improving customer satisfaction is not only by meeting customer requirements but also by adding value to the organisation's product or service. Adding value is in terms of unique organisational methodology, more efficient and fool proof processes, giving the brand value of the product, etc., which are unique to the organisation. The focus on enhancing customer satisfaction is to be maintained for products delivered every time.

The customer is satisfied when all the requirements are met on the product; that is when product conformity is realised. Customer is satisfied when the products are delivered on time, as specified by the customer; that is on-time delivery performance. Throughout the realisation processes and at the end of every production process, the organisation must measure and record all the product characteristics and ensure that the final product meets the requirements specified or agreed upon by the customer. Only the acceptable product, i.e., a product that meets the conditions, will be delivered to the customer. Delivery should happen on- time, on the specified date and time.

This can be achieved by establishing process capabilities from manufacturing and assembly to packaging and product delivery and installation. The computation of a simple indicator of process capability (Cp) or adjusting the process capability toward a specification (Cpk) would help managers quantify their process risk. The objective would be achieving each process's highest economically feasible capability, thus minimising the chance of producing unintended output.

You are a customer of print media, and if your favourite morning newspaper is delivered to you in the evening, will you be satisfied? Sometimes, things can go wrong, such as the product realisation process or the delivery performance. The process approach tells us to act on the difference when the measured results are not as planned. Find out the reasons for not achieving the outcomes as planned. It may be a technical or logistical reason for not completing the plan. It may be one or more of those 6M letter words (men, materials, machines, methods, measurement, (mother earth) or any of the risks).

Finding out the causes and taking corrective action to eliminate the reasons is the way for continual improvements. Taking appropriate actions and not finding excuses to postpone actions is necessary for progress.

The customer focus is to have customers and retain them forever for the business. Unsatisfied customers walk away. They have their reasons, which may be related to the quality of the product or the delivery performance. Both are controllable at the organisational level. Organisations that care for the customer focus on these directly measurable results of quality and delivery performance directly proportional to customer satisfaction. Such organisations retain and increase their customer base.

Communication with existing and prospective customers is one of the proven methods of keeping the customer in focus for the business. Communication includes obtaining customer feedback on the quality of product delivery performance, inviting suggestions for improvement, proactively inviting customers to project any requirement of servicing, warranty compliance, inviting complaints, if any, etc. If everything goes well, periodic appraisal of the business and re-assurance to the customers that everything possible is being done to satisfy them and further enhance their satisfaction level could be the communication matter. You could project your achievements to prospective customers concerning the product range and capability to meet specific or unique customer requirements as a marketing strategy.

Customer focus and customer feedback are processes. These processes are to be established in an organisation and appropriate actions will be taken based on the available performance data. The data may be related to on-time delivery, after-sale service, corrective action requests, results of customer satisfaction surveys, nonconformity, complaints regarding the product quality and its usefulness, responsiveness to internal and external customer requests, etc. It is necessary to introspect how these processes are managed. The ability to make meaningful decisions and implement actions to resolve issues effectively shows the overall effectiveness of the quality management system and organisational excellence in satisfying the customers.

Make a statement in the quality manual that the top management has ensured customer requirements are determined and met to enhance customer satisfaction. You can even spell out how you are doing it. The management has confirmed that product conformity and on-time delivery performance are measured. The organisation takes timely action if the planned results are not, or will not be, achieved. Planned actions concerning value addition to the customers, nonconformity management, response to corrective action requests and complaints and ensuring on-time delivery performance can be indicated or referred to in documentation. Show the performance by trend charts and measured results.

A department, generally the marketing process or the department which directly interacts with the customers, should have a system to identify and record the customer requirements in coordination with other functions/ departments. The department makes all enquiries, requests for quotations, initial order handling, etc. Details of all the requirements, including special requirements from the customer and customer requirements related to the product, are dealt with in clauses 8.1 and 8.2 of the Standard. Determinations of customer satisfaction handling of complaints/feedback are generally done by the same department/section. The requirements for determining customer satisfaction are in clause 9.1.2 of the Standard. Submit quotations to the customers and indicate order acceptance after the section does the required reviews. Communication with customers, obtaining customer feedback and initial handling/registering customer complaints are the other functions of the section.

The organisation identifies the list of essential customers explicitly. (Refer to context of the organisation.) All customers are important. However, the customer who delivers more revenue to the organisation must be retained to improve the business further. The organisation's method of engagement/ contract/dependency and communication with the customers, joint efforts to improve the product quality, periodic determination of their satisfaction levels, organisation's actions to reduce complaints or returns of the products, etc., demonstrate the customer focus.

'With dedicated customers, especially government customers, some organisations take undue freedom and postpone necessary tests and pending installations or works to the delivery stage of the product, manipulating virtual OTD performance. This is not customer focused. Waiting for work on the product to be completed later by the customer before the product is used neither enhances customer satisfaction. If analysed, it can be seen that the 'out of focus' is due to internal organisational issues.

Consider this situation. Engine B was 'ready' for the test bed run, except for the bleed valve assembly, which was not fitted, as it was unavailable. It was awaited from the feeder unit after overhaul. Top management, generally committed to complete customer focus, was informed that if the regulatory authority to sign off the acceptance of the engine test allowed the deficiency, it could be planned for a test bed run. It was a contingency plan and an ingenious one to cut short the delay in the delivery of the engine to the customer.

"To save time, we will fit the regular one later before dispatch. It is coming within two days. The engine B can be tested with a bleed valve assembly cannibalised from another accepted engine A, just for the test acceptance of the engine B."

A request was made to the regulatory authority inspector without revealing the consequences of testing the engine and replacing the bleed valve assembly before dispatch. The inspector insisted the engine be tested again to ensure the bleed valve operation. The production agency was aware of this requirement. Still, it was not willing to repeat the engine test after replacing the bleed valve, saying that the bleed valve is an LRU (line replaceable unit) on the engine. The request continued for a few minutes. Being pushed to the wall but committed to the production task, the manager expressed the anguish of the management – "It is not a big issue – you are putting spokes on the production's wheel!"

The regulatory inspector remembered the lines from the manual: 'While starting, the bleeding valve opens at 35–40% compressor rpm to divert the excess air and regulate the flow to the combustion chamber to ignite the air-fuel mixture. At 40–45% rpm, the valve closes, redirecting some airflow rearwards to pressurise the sealing rings and tapping air for aircraft air-conditioning and cabin pressurisation systems. Opening and closing of the bleed valve are set and locked on every engine during the test. After the test bed run, the bleed valve settings are "not to be disturbed". So, the bleed valve assembly becomes dedicated to the specific engine and remains so during the service use.' The engine must be tested with the replaced LRU when replacement becomes necessary.

Is it a difference of opinion? Is it about not helping with the production task? Is it not about meeting the requirements for the product? Does it not meet the needs of the quality system the organisation is committed to? Is it showing the managerial commitment and focus on enhancing customer satisfaction to leave the necessary test and adjustment of the bleed valve assembly to the user after delivery of the engine?

AQMS requires retesting the engine to be recorded and communicated to the customer or obtain the customer's approval. Clause 8.6 of the AQMS states that "the release of products and services to the customer shall not proceed until the planned arrangements have been satisfactorily completed, unless otherwise approved by a relevant authority and, as applicable, by the customer.”

The best course of action would have been to seek approval from the regulatory authority and customer, as the subsequent testing and adjustment of the bleed valve becomes the customer's responsibility after delivery of the engine. Instead, the manager retorted - “Keep seeing your boss occasionally. I will speak to him”.

The inspector withdrew silently and said while walking out - “Welcome. I think my boss will appreciate that I do my duty”.

The incident explains the need to look at the best practices per the AQMS, implement the system for the managerial commitment to the cause of quality, and focus on enhancing customer satisfaction. Ultimately, it is the effort to identify the root cause of the undesirable situation and implement corrective actions that would bring improvements. If a similar problem is likely, appropriate risk assessment and preventive measures would be necessary for improvements'.

Specific performance levels and trends of the quality of products and OTD performance should be measured regularly and recalled by the top management instantaneously and followed up by reference to the relevant measurements. The organisation should be able to demonstrate the level of performance relating to customer satisfaction, product conformity and on-time delivery. Customer response to the levels achieved and if there is a shortfall, actions taken for improvements to re-assure the customers may be presented to demonstrate compliance to customer focus requirements of the Standard.

Refer to functional objectives for key processes/key performance indicators (KPI) in **Annexure C**.

5.2 Policy

5.2.1 Establishing the quality policy

An organisation's top management declares its core values and activities through quality policy. It is an honourable task to draft the quality policy, which includes the mission, vision and values such as integrity, understanding, excellence, unity and responsibility of the organisation and convey the commitment to satisfy the requirements of the Quality Management System standard and to make efforts continually to improve its effectiveness. Quality policy needs to be established and communicated to all personnel within the organisation. The management should ensure that all the personnel understand it well.

The organisation shall align QMS policy and objectives with its strategy. Strategy is a plan of action designed to achieve the organisation's overall aim and is included in the organisation's context.

The organisation's overall performance is the true reflection of the quality policy and its effective deployment. The performance of the organisation and its Quality Management System can be generally assessed through the management review records of the commitment and involvement of top management in establishing, implementing, monitoring and updating the quality policy. The management must translate the quality policy into meaningful words and provide guidance/directions to all the personnel in the organisation. Quality policy needs to have corresponding objectives at all functional levels. The personnel in the organisation must be able to say in words what the true meaning of the words in the quality policy is, in terms of what they do to achieve the objectives in their functional areas and how they measure it. The employees' awareness of how their work affects the measurement results of objectives will motivate the employees who wish to see their organisation grow in stature.

Continual improvement of the Quality Management System and customer satisfaction should be inherent features of the quality policy. A quality policy statement provides a framework for establishing quality objectives. The objectives must be reviewed periodically for their relevance, suitability and results. The management does these reviews in consultation with other process owners.

5.2.2 Communicating the quality policy

Usually, the quality policy and the corresponding objectives are displayed at prominent places in the organisation to bring awareness and motivate personnel to achieve the objectives depicted by the policy statements. The policy statement must be signed by the head of the organisation when displayed.

Examples of the quality policy statement of some organisations are indicated in 'Documented Information' in Annexure C, with their scope of Quality Management System, for a clear understanding of how to word the quality policy.

Make an appropriate quality policy statement with a commitment to comply with and make efforts to improve the effectiveness of the Quality Management System. Define the relevant quality objectives under clause 6.2.1 of the Standard. It is desirable to make a clear statement that the policy is reviewed (give a periodicity) for continuing suitability, and this review can happen along with the management review as an agenda point (9.3).

Quality policy and objectives are documented in the quality manual and maintained as documented information.

Create opportunities to explain the practical ways and means of translating the policy into measurable objectives and share information about employee achievements.

The quality policy includes understandable words within the organisation, with corresponding objectives/key performance indicators (KPIs) at each process/functional level. The implemented status is that all personnel know the policy statements and their relevance and relation to their work. It is demonstrated by how the employees express their understanding of the policy and objectives in their own words. Methods of communication of the policy and objectives to the employees and the periodic reviews for suitability by the management demonstrate compliance with the requirements. The information is made available to the relevant interested parties when necessary.

AS 9110 C requires that the top management establish the safety policy and identify the safety objectives within the safety policy framework. The objectives shall include those needed to meet aircraft and personnel safety requirements. The objectives are to be established at relevant functions and levels within the organisation. As with quality objectives, the safety objectives shall be measurable and consistent with the safety policy.

5.3 Organisational Roles, Responsibilities and Authorities

'Duty' is obeying things given by God's commandments, which one is expected or required to do by moral or legal obligation. For instance, one must live, love, learn and leave a legacy behind when one stops to live. Violation of duty is a serious offence. Each one of us has a role to perform in this big world.

Responsibility is a commitment towards anything where one is free to obey or not to follow. Responsibility is the duty to accomplish the tasks and objectives related to the position of a person in an organisation. Managers work to meet the objectives, deal with the day-to-day work, supervise employees' work and report on the work progress to their superiors. Managers influence workers to carry out the job correctly and systematically, and they have the power to be informed, have the expertise and have the resources available to them as given by their top management.

All authorities flow from above and as given from above. Authority is the right to exercise influence or make decisions, carry out necessary actions and direct others. For example, managers can assign work or make a purchase request. A manager who shares power and authority will have the most significant ability to influence others to work towards the organisation's objectives.

Most organisations have a formal 'authority flow down' system that depicts the authority relationships between persons and their work. Some have functional authority in a specific area.

Top management defines the responsibilities and authorities of all personnel, at least up to the managerial/supervisory level and documents them. Responsibilities and authorities are to be defined concerning the position/designation/post occupied by the personnel and not by the names of personnel. It could be a description of duties, responsibilities and authorities for each post/position in the organisation. It could be a responsibility and authority matrix regarding the clauses of the Standard. It must be modified/customised for the scope of work carried out by different organisation members. The responsibility and authority and their interactions must be defined and communicated throughout the organisation for general information and specifically to the affected persons for compliance. This can be done in several ways, including job descriptions, procedures and instructions, organisation charts, etc. The communication can be a notification for compliance by the affected positions and general information. The organisation chart can be indicated in the quality manual in a separate section or as an appendix to the main document. Whatever the responsibilities and authorities are communicated, they must be practical and reviewed or amended periodically as and when changes occur in the organisation's structure or strength. Names of individuals are not necessary to be documented.

Aerospace Standard AS 9110 C for Maintenance Organisations requires that in addition to the above, responsibilities and authorities delegated to the 'Accountable Executive Manager' and 'Maintenance Manager' are to be defined, documented and communicated to all personnel within the organisation and customers and regulatory authorities.

The accountable executive manager appointed by the top management shall be with corporate authority to ensure that 'all ordered maintenance can be financed, the necessary resources obtained, and all ordered maintenance completed in accordance with all organisational, customer, and Authority requirements'. Similarly, the maintenance manager appointed by the top management shall be responsible for ensuring that all maintenance required is carried out in accordance with all organisation, customer and authority requirements. (Authority here means regulatory authority.)

This is to identify the personal and direct responsibility for the maintenance work carried out and certified by the organisation on civil aircraft.

Communication is effective when personnel can identify their roles and responsibilities and become dedicated to achieving the targets and objectives without doubts, inhibitions, hurdles or mental blocks.

Duties and responsibilities of process owners include reporting that the processes are delivering their process outputs. Any deficiency in meeting the performance targets and strategies to overcome the deficiencies must be communicated to the top management. This can be formal or informal. Just indicating the deficiencies and waiting for the management to do according to their wisdom is not desirable. Management wants the process owners to be responsible for bringing up issues and problems and being part of the solutions and the team's progress. Solutions should facilitate improvement in the direction of improving customer satisfaction, with a focus on customers.

When responsibilities and authorities of personnel are changed on a need basis, and sometimes the roles of essential process owners are changed by the top management, it is always done with a lot of thinking and planning. The issues will be discussed in management reviews, and appropriate decisions will be made and communicated to the stakeholders. They always ensure that the integrity of the Quality Management System is maintained and placed firmly on the road to improvement.

The requirement to appoint a management representative (MR) is not specified in the revised ISO 9001:2015. The reason is that the top management cannot entrust the role and responsibilities of management to a specific member of the management team and call him the management representative. The management team is responsible for directing and controlling the establishment, development and maintenance of the Quality Management System in an organisation. Appointing an MR and giving them the sole responsibility of implementing and improving the QMS does not absolve the responsibility of the top management. However, selecting a management representative (MR) is a requirement specified in the AS 9100D. MR needs to be a specific member of the organisation's management. He has the responsibility and authority to oversight the requirements and report to the top management of:

a. QMS conformity and effectiveness
b. Process conformity and effectiveness of all processes in the organisation
c. Ensuring the integrity of the QMS
d. Resolving quality management issues (with organisational freedom and unrestricted access).

MR is the interface between the top management and the rest of the organisation. All authorities and responsibilities of top leadership concerning the

QMS flow down to MR. MR interacts with different functions to orient them towards meeting customer requirements in the organisation. He monitors the health of the Quality Management System through continuous tracking of activities and achievements.

Management representatives must be employees and part of the organisation's management team, having separate responsibilities and authority in their area. The top management designates MR to take care of these additional responsibilities with authority to take the initiative and drive the Quality Management System throughout the organisation. Top management does this by issuing a letter in addition to his earlier appointment as an employee, addressed to him, with information to all in the organisation describing his duties and responsibilities and giving him the authority to resolve quality management issues. Liaison with external parties on matters relating to the Quality Management System, such as coordinating with the certification body, representing the management during certification and surveillance audits, conducting internal audits with the help of approved internal auditors, preparing the summary of audit findings/following up of nonconformity reports and report to the management review meetings etc., can also be included. MR duties include preparing the agenda points for the management review and record decisions, circulating the agenda points to all the members before the scheduled reviews or meetings (if held) or both, participating in the MRM and recording minutes of the management review meetings and circulating to all concerned.

Some organisations have previously outsourced the 'Management Representative' role to someone in a different organisation or a consultant. It is not a healthy practice. One example is the findings of an external auditor 15 days after an internal audit was conducted in the organisation. The consultant reported no nonconformities, did not maintain adequate objective evidence of the audits carried out (all ticks of yes), left out management processes and customer QMS/statutory/regulatory requirements were not audited. Sometimes, the annual internal audit was conducted precisely a year later on the same day and month! (Minimum changes in the record, to be appreciated!). The recent change in the Standard requirement is to eliminate/reduce the inefficient practice effectively and to bring in the positive intervention of the top management in all matters of conformity and effectiveness of the QMS/business processes in the organisation.

Persons appointed as MR may preferably be from the quality domain since they may have exposure and experience in Quality Management Systems, which is an added advantage. However, there is no restriction, nor is there any conflict of interest, for any other senior employee becoming the MR of an organisation, may it be from the personnel department or production department, provided they attend the work regularly at the site of the certified organisation and not an occasional visitor to the site. MR can also have his team/assistants detailed for regular work, connected with the Quality Management System in a large organisation.

6

Planning

6.1 Actions to Address Risks and Opportunities

Determination of risk and opportunities can involve

a. the identification of sources, events
b. their causes and their potential consequences
c. can apply historical data, theoretical analysis
d. informed and expert opinions.

Risks and opportunities can arise from:

a. organisation's internal and external issues,
b. interested parties' needs and expectations and
c. other means, such as departures from specified QMS requirements including in products, services and processes.

Risks and opportunities can also arise from applicable statutory and regulatory requirements related to the products and services.

a. failing to comply with legal requirements, including the changed ones and in a time-bound manner can damage the organisation's reputation or result in legal action
b. performing beyond its compliance obligations (which can enhance the organisation's reputation) – opportunity.

Risk management methods

Although risks and opportunities need to be determined and addressed, there is no requirement for formal risk management or a documented procedure for the risk management process. It is up to the organisation to select the method it will use to determine its risks and opportunities.

Depending on the organisation's context, the method may involve a simple qualitative process or a full-scale quantitative assessment. The risks and opportunities identified are inputs for planning actions and establishing environmental objectives.

Organisations may choose any methodology suitable for determining and assessing the risks. Some of the methods followed by various organisations, either individually or in combination, are as follows:

a. SWOT Analysis: Strength, Weaknesses, Opportunities and Threats related to issues, compliance obligations, needs and expectations of interested parties;
b. PESTLE Analysis: Political, Economic, Social, Technological, Legal and Environmental issues explained in para 4.2,
c. Brain storming: Information gathering technique,
d. Historical data on failures,
e. Professional expertise and
f. Defect investigations, etc.

6.1.1 Organisational risks

This clause addresses the risks and opportunities when planning for the organisation's Quality Management System. The top management must systematically plan to establish, document, implement and maintain the Quality Management System. Efforts are necessary to improve its effectiveness continually. Quality policy and relevant objectives will be identified, and a documented strategy will be placed in the organisation to measure the objectives. Management must implement the planned strategies to comply with the identified statutory and regulatory requirements while establishing the Quality Management System. Suppose the system has not yet been established in the organisation. In that case, the matter is discussed among the organisation's top executives, a core team is identified to carry out a detailed study about implementing the system and an action plan with targets is laid down. Top management appoints a management team member as the management representative (MR) to coordinate the activities and report the progress to the management. The planned QMS can only be implemented when the gaps and differences are identified. An awareness programme on the

Quality Management System requirements for aviation, space and defence industries/organisations must be conducted. The importance of complying with the requirements must be highlighted to the employees for business progress with the customers. The core team may carry out the tasks of bringing awareness and identifying the gaps. The primary job will be to identify, document and comply with the requirements of the Standard in the following areas, which have been described in the same paragraphs:

- 4.1 Understanding the organisation and its context
- 4.2 Understanding the needs and expectations of interested parties
- 4.3 Determining the scope of the Quality Management System
- 4.4 Quality Management System and its processes
- 0.3 Process approach (which includes plan-do-check-act cycle and risk-based thinking)

While monitoring and reviewing the internal and external issues identified vide para 4.1, it is necessary to identify the potential nonconformities related to those issues. Such risks may be related to not achieving the planned results, objectives/performance issues, internal or external matters, issues that impact business, customers, etc. Foreign exchange fluctuation concerning local currency value is an external issue that can impact business profitability with foreign customers. International Traffic in Arms Regulations (ITAR) controls the export and import of defence-related articles and services on the United States Munitions List (USML). The Export Administration Regulations (EAR) is a set of United States government regulations on the export and import of commercial items. These requirements affect the export of products to that country. Fumigation of packing cases is another external issue which impacts the business if not considered and managed before the finalisation of the contract. Identify any such requirements and evaluate whether such problems affect the QMS processes. This paragraph is particularly to integrate and implement the actions to reduce the risks and take advantage of the opportunities an organisation may encounter while planning processes into its Quality Management System.

While monitoring and reviewing the needs, requirements and expectations of interested parties, identified in para 4.2, it is necessary to identify the potential nonconformities related to those interested parties. Scope indicates the specific products and services offered by the organisation.

Each of the QMS/business processes is identified and planned as in the 'Schematic Representation of the Elements of a Single Process', Planet diagram, SIPOC diagram or flow chart. All the processes are arranged as in a business process map or a 'Sun and Planet Diagram', or like the 'Processes within an Organisation'. There is no prescriptive arrangement except that following the 'Process Approach' is advantageous as they are the current best practice. Risk-based thinking must be integrated into this planning process with appropriate documentation. Identification, assessment and management of risks related to each type of risks viz; QMS process, people, products/deliverables, etc., as described in 6.1.1/An item – entry in the risk register – 'Under the given circumstances, what can go wrong? What can be done to prevent errors/mistakes/potential nonconformities?' would be adequate. The risks during the operational phase (Operational Risk Management is dealt with in Chapter/para 8.1.1) can also be integrated into the risk register when the QMS/business processes are practised regularly. It will be the best thing to happen in the QMS planning actions that are practically thought about, systematically complied with and meticulously managed to prevent or reduce undesired effects and the risks identified in the planning stages.

Quality is always achieved through planning and thereby preventing problems through communication. It is cheaper and more effective to run preventive systems than inspection-based systems and corrective actions.

Once the quality system processes are identified, their sequence and interactions are identified and the operations are planned with risk evaluations, a fair amount of confidence is gained that the system can achieve the desired or intended results. Meeting the requirements of the Standard as per the respective clauses will ensure conformity, that is, the realisation of all the planned activities. Measuring each process's objectives/key performance indicators (KPIs) against a set target for a specified period will ensure the intended results' effectiveness and achievement. Enhancement of the desired effects can happen through continual improvements over some time. Incremental improvements happen in several areas, such as systems, processes, methods, technology, competence, motivation, etc. Better results and progress are achievable in an ever-changing world. Still, it takes time through recurring efforts in all those areas.

A committed management who earnestly desires to establish a Quality Management System in the organisation generally does not make frequent changes to the Quality Management System or changes in the personnel assigned the task before fully establishing the system. Some of the possible changes may be changes in the management, changes in the structure of the organisation, changes in the product line, introduction of ERP and other computer systems for product realisation, changes in the list of customers, organisational mergers, bifurcating, etc., which could affect the functions of the organisation. These are considered while identifying the needs and expectations of the interested parties, stakeholders and customers. No structural changes in the compliance of the Standard should be made. It must be ensured that no compromise is made on mandatory requirements specified in the clauses of the Standard. Individual requirements of the processes should not be overlooked or ignored. It also underlines that the Quality Management System is not personnel-dependent. Irrespective of personnel, the system should run smoothly and should be able to deliver benefits to the organisation and customer.

6.1.2 Risk identification and assessment

The organisation shall plan:

(a) Actions to address risks and opportunities

Using the PDCA cycle, the organisation must integrate the actions to address these risks and opportunities into its QMS processes. Not all processes of a Quality Management System represent the same level of risk in terms of the organisation's ability to meet its objectives, and the effects of uncertainty are not the same for all organisations. Each organisation is, therefore, responsible for the extent to which it applies risk-based thinking and its actions to address risk, including whether to retain documented information as evidence of its determination of risks. When planning its QMS, the organisation must consider the risks and opportunities presented by external and internal issues and the needs and expectations of interested parties relevant to its purpose and strategic direction. Means to address risks may include avoiding risk, taking the risk to avail an opportunity, removing the source of the risk, changing the likelihood or consequences, sharing the risk or making an informed decision to retain the risk. Opportunities can derive from favourable circumstances that can lead to the use of new practices, the launch of new products, entering new markets, addressing new clients, reducing waste or improving productivity, growing relationships, new technology and other desirable and viable opportunities to facilitate the organisation in achieving its strategic direction and enhancing customer satisfaction.

Actions addressing risks and opportunities start with identifying threats affecting the QMS processes. Determine risks and options, considering the issues raised and requirements identified. Plan appropriate actions to reduce undesired effects on the QMS and evaluate effectiveness. Manage risks at the organisation/processes level (such as new customers, new markets, company partnerships, business localisations, product configuration, critical items, product safety, internal audits, nonconformities and corrective actions).

Likelihood of Occurrence					
Often	5	10	15	20	25
Frequent	4	8	12	16	20
Probeble	3	6	9	12	15
Occesional	2	4	6	8	10
Remote	1	2	3	4	5
	Negligible	Morginal	Criticol	Hozerd ous	Cotastr ophic
	IMPACT (SEVERITY)				

RPN 1-5 = Low 6-9 = Medium 10 -25 = High
Acceptance Level of RPN = 9 with Severity Mnx 4 only.
Mitigation Action for RPN above 9 or Severity 5.

SCALE 1-5

RPN = IMPACT X OCCURRENCE, CHOOSE FROM TABLE ABOVE
SELECT ACCEPTABLE LEVEL OF RPN ACCORDINGLY: 9

Figure 6.1 Risk assessment (RPN).

It can be performed as a risk identification (What can go wrong?), assessment of impact (How severe will it be if the risk happens?) and probability of occurrence (What are the chances it will occur?). Using the scales of the assessment indicated below, arrive at the Risk Priority Number (RPN). Impact and probability of occurrence are evaluated/assessed on a scale (low-medium-high) or 1–5, 1 being lowest, five being highest, with no units, being a subjective matter) so that the multiplication of Impact and Probability gives a number. We call it the Risk Exposure Number or Risk Priority Number RPN). If the RPN is above the acceptable threshold level, it is mandatory to identify and complete mitigation plans. RPN is high if the risk level is high so that the unavoidable risks can be reduced or mitigated to a lower acceptable level. Detection is usually not used in the assessment of risk levels in ASD systems and products since even the remotest or the most challenging detection of a risk happening is of no use, as it may be too late to react by that time, the risk is likely to grow into a catastrophe.

The risk is always evaluated so that it does not result in a hazardous situation when faced with it. Adequate mitigation plans are to be identified and complied with as part of the preventive actions to reduce the impact of the potential nonconformity when and if it occurs for the first time. When mitigation actions are completed, the effectiveness of those actions is to be verified subsequently, as in the case of any corrective measures taken for a detected nonconformity. Periodic reviews of the risk register ensure that. When the mitigation actions are completed and found effective, the risk levels need to be re-assessed to arrive at the residual/remaining risk. The residual/remaining risk will be recorded in the risk register until the following review.

Preventive action is a design function. It is the designer's prerogative to assess the possibility of an error, deviation or nonconformity appearing at any time during the production or use of the product. This function, an essential part of the design of a product, is performed by the designer in the form of failure mode effect analysis or failure mode, effect and criticality analysis.

All health monitoring devices used on equipment are to detect the cause of any potential nonconformity; hence, it is a risk mitigation/preventive action. Once the causes are identified through health monitoring systems, necessary preventive action or action to prevent failures can be taken. One example on an aero-engine is the installation of magnetic chip detectors, which, when fitted on lubricating oil passages, would collect any iron filings/chips originating from the bearings and flow along with the lubricating oil. When the chip detectors are examined periodically, an accumulation of iron filings on the chip detectors indicates an impending failure of the bearings.

Redundancy offered by standby systems designed and established in aircraft systems is a preventive action/mitigation action to reduce risk. When one system fails, the other or alternate system automatically takes over the functions. Hence, there are no gaps or mistakes in the continued operation of the required performance. Another preventive action is an alternate source of power supplies, such as a stand-by direct current supply where a regular alternator power supply fails. These are examples of preventive actions established by the designer of the aircraft systems. Where corrective actions are not found effective or when the recurrence of snags or defects on any aircraft part or a system is noticed, certain modifications are designed and incorporated in the aircraft components/systems as a preventive measure/mitigation action to reduce risk. Improvement of the reliability of parts and assemblies is also seen as a preventive measure. Error prevention is by design for a proper sequence of operations so that without completing the previous operations, the subsequent operation cannot be performed. Interlocking the sequence of operations is another preventive action when certain operations cannot be achieved without following the correct sequence from the beginning. These are all examples of preventive actions incorporated into the design of aircraft and/or aero-engine operations. The designer built the standby arrangement or redundancy into aircraft systems with a study of failure mode and effect analysis, which is used for risk assessment and management. In a fly-by-wire aircraft, four parallel on-board computer systems are used to operate flight control systems to prevent any failure of the all-important flying control system. Such a system's reliability is almost one (1.0), and the failure rate is much less than one in a million flying hours. Service problems or information on product problems reported by customers and other external sources may prompt designers and manufacturers to evolve suitable preventive actions/operational risk management.

Simple Poka-yoke ideas and product and process design methods can eliminate human and mechanical errors. These preventive actions eliminate potential defects in products and processes and improve quality and reliability. It can also fine-tune improvements and process designs from six-sigma 'Define – Measure – Analyse – Improve – Control' (DMAIC) projects. Once the nonconformity has been identified or potential failures are analysed, the cause of the nonconformity is eliminated by the following:

a. Preventing the possibility of an error, limiting the probability of human error
b. Having an alarm, warning or indication to the operator that an error has occurred in the sequence of operations or the system
c. Introduce limit switches, gates, which open one after another or automatic shut-off devices (electrical, electronic, mechanical, procedural, visual, human or any other form that prevents incorrect execution of a process step) and
d. Using a checklist to ensure completion of all process steps in the correct sequence

The improvement must be tried, verified, reviewed, validated and measured for effectiveness. Actions to address risks and opportunities provide the chance to avoid risk, take the risk to pursue an option, eliminate the risk source, change the probability/likelihood or consequences/impact/severity, share the risk or retain risk/accept the residual risk by informed decision. Opportunities can lead to adopting new practices, launching new products, opening new markets, addressing new customers, building partnerships, using new technology and other desirable and viable possibilities to address the organisation's or its customers' needs.

Refer to Annexure H: Risk Register (Example). In that, PART I of the risk register may be maintained, dealing with risks and opportunities at the enterprise level. Use the FMEA/PFMEA template to address similar risks and options and find answers to the questions given below:

1. Based on the organisational strategy/plan, identify the current risks and opportunities. The organisation may do a SWOT analysis as part of its

business strategy to identify the external risks and opportunities and action plans to address them. The organisation may perform a formal business risk assessment, considering its context, associated risks and opportunities and mitigation plan.

2. What risks will impact our business and quality system, and our QMS may not achieve the intended results/KPIs? What are the opportunities the situation may provide? Examples of risks are:

    a. Product conformity is not likely to be completed, and there are some potential issues identified

    b. QMS processes and their effectiveness level may not be as desirable

    c. External and internal issues may affect the business, or new issues are coming up

    d. Achievement of QMS objectives/KPIs as planned may not be achieved (business level) in general

    e. On-time delivery performance (across the organisation at all project levels) may not be achieved in full

    f. Customer satisfaction may not be executed as desired

    An example of an opportunity arises when the customer wants to give us a big project involving a long-term engagement. While establishing the resources/support to mitigate the risks of such special requirements from the customer, the organisation can expand the business or diversify the operations. Is there any such opportunity which arises while addressing any of the risks above?

    A QMS's processes do not represent the same level of risk regarding the organisation's ability to meet its objectives. Due to this reason, the consequences of failures or nonconformities concerning processes, systems, products and services will not be the same for all organisations. When deciding how to plan and control its QMS, including its component processes and activities, the organisation needs to consider the type and level of risk associated with them.

3. Is our internal audit process not likely to result in an effective QMS?

4. Is our internal audit process not likely to bring out the actual measure of the effectiveness of the QMS?

5. Are performance results not likely to achieve the planned targets at the operations level?

6. Are our organisational performance results not likely to achieve the planned results at the engagement level?
7. Is the effectiveness of the corrective actions, improvements, etc., likely to have a shortfall during review by management?
8. Is process level effectiveness likely to have a shortfall?
9. There are pitfalls to our business strategy. What are the likely causes?
10. Does our marketing not get us more customers in more challenging work areas?
11. Out of the list of interested parties and their needs and expectations, are some needs and expectations of the interested parties likely to affect/impact our business/QMS?
12. Are some of the specific needs and expectations of the interested parties likely to affect:
13. Conformity of our products/services
14. Customer requirements
15. Legal requirements
16. Are any of our actions to enhance desirable effects likely to fail?
17. Are our actions to prevent or reduce undesired effects not likely to achieve planned results?
18. Can we achieve continual improvements in the QMS?

6.2 Quality Objectives and Planning to Achieve Them

6.2.1 Measurable objectives

You often see high-sounding mission and vision statements on the marketing leaflets, wall hangings, posters and business cards of organisations. If you ask a manager about the organisation's objectives, he often reiterates the same. Ask him how you measure them, and he fumbles, 'Measuring objectives, mission, vision, we want to be the leaders in our business, we want to be globally competitive'. Unless organisations can indicate their tangible or intangible achievements in measurable quantities, it does not get more profound than the superficial level. Nobody takes cognisance of or realises anything that data cannot express. The manager will not be able to assess or appreciate the hard work being done by his team in different functional areas unless he has set a target of measurable objectives, with which he will be able to monitor the progress periodically, perhaps, apart from the profit earned. A quality objective may be related to finances, customer feedback, safety, efficiency, speed, time, improvements in the market share of the product or improvements in the product itself. Organisations should select what matters most to their success and what impacts their customers. The objectives should not be mere high-sounding words. After realisation, they are the planned results of the process, otherwise known as the key performance indicators (KPIs).

While organisations are enthusiastic about writing down and publishing their mission (What they do?), vision (What they would like to be?), etc., in the form of a quality policy, they occasionally fail to reconcile the fact that the quality policy laid down is a framework of the objectives of the processes in the organisation. The objectives need to be identified in clearly measurable metrics, for example, in terms of time, on-time delivery, cost, customer satisfaction, quality, quantity and severity of defects, defect density, schedule adherence, effort variance (within the budgeted effort), etc. and monitored at periodic intervals to assess the level of achievements. It is easy to see that some of these parameters are more important in some processes than others. So, the objectives need to be identified depending on the importance of the methods and the products, which ultimately contribute towards customer satisfaction.

Management ensures the defined organisational objectives (KPIs) are specific, measurable, achievable, realistic and time-bound (SMART). It is necessary to document it and give it as targets for measurement by various process owners periodically. The organisation's performance reflects the aims of the quality policy, and the objectives demonstrate the achievement of the policy. The main objectives should be to achieve the customer-specified quality of the products and meet the scheduled delivery, apart from achieving the highest levels of customer satisfaction. The trends of the above objectives should be consistent at the higher levels or continually improving periodically to the top targets.

The measured results will be of interest to the organisation's stakeholders. Quality Management System mandates organisations to indicate the measurable objectives in metrics, with measurement methods, against set targets for a specified evaluation period. The objectives are all that is looked for in measuring the achievement by the organisation, may it be the quality, profit, share values, increase in customer base, customer satisfaction, safety, on-time delivery, productivity and other sublime objectives of social status, reputation, global recognition, innovation, new products development, etc.

Suppose any of them is not in a measurable form. In that case, quantitative measures in the form of data hold no importance/position to reach a level of perception that the stakeholders can appreciate or, based on, make appropriate decisions to continue or re-organise. Everything that the organisation does concerns achieving objectives and its success. Saying that we practice excellent quality systems and that the organisation is driven by quality and safety are merely subjective statements. Subjectivity or comparison is not here when evaluating individual or organisational achievements or excellence. What are the indicators of the organisation's excellence? They are the objectives. They matter most for the organisation. They drive the organisation and motivate the personnel. The objectives need to be measured against set targets periodically. Some consider that meeting the production task set for the year is the only objective of the organisation. But that is only one of the objectives. No doubt, it is the major one. Suppose the production task is the only objective. In that case, many other objectives mentioned in the quality policy may not be identified, measured or achieved, organisational excellence may not be visible, and efficiency and growth may not be achieved.

Some of these objectives are achieving the targets in terms of quality and quantity, on-time delivery performance, process time, key performance indicators, performance improvements achieved for the product, profit earned, savings effected, process capability achieved and process consistency achieved and increased level of satisfaction achieved by customers (internal and external). Other objectives are the satisfaction of stakeholders such as shareholders and statutory and regulatory authorities in the case of defence industries, market reputation, increased organisation capability, competency of its human resources and increased efficiency of business operations.

6.2.2 How to measure

Top management shall ensure that quality objectives are established at relevant functions and levels within the organisation. Management should define the appropriate roles and levels, department/section-wise or processes in the organisation, mainly the top value stream processes. The objectives related to meeting the product requirements, that is, product conformity, should be included, along with meeting the quality characteristics of the products and customer requirements, on-time delivery, customer satisfaction, etc. Individual process performances add up together for the performance results of the organisation. It is the relationship between the Quality Management System and organisational performance. It is the combination of the *realisation of planned activities*/completion of the process conforming to the requirements of the process/QMS and *achieving planned results in full*. Together, we measure the 'effectiveness of the process. On achievement of effectiveness for all the processes in the organisation, we say the QMS processes in the organisation are effective. Verifying the above, external auditors evaluate the 'Process Effectiveness Assessment Report' for each process published on the IAQG/OASIS website for the information of all the stakeholders and prospective customers.

Sustainability is the word used to express the ability of an organisation and its organisational processes to achieve, maintain and continually improve its overall performance in the long term. The performance of an organisation directly relates to the interest of all the stakeholders. The organisation should identify the key performance indicators of every process and, more importantly, the operational processes/product realisation processes. Examples of performance metrics include customer satisfaction levels, internal and external quality performance, failure rates, on-time delivery, lead time of processes, process costs, cost of poor quality, supplier performance, waste management, defect/incident/accident occurrences, market share of business, sales, profits and stakeholder/third-party ratings of the organisation.

Some examples of KPIs of some processes are indicated in Annexure C, 'Documented Information'.

To assess the performance and evaluate the associated risks or to utilise the opportunities, an organisation needs to identify a broad set of performance metrics, including measured results of external and internal processes. An appropriate strategy will be established for gathering external and internal information about the organisation's performance measures. Assess the risks of any identified differences or shortfalls and their impacts on the organisation. Superficial knowledge about the performance is gathered by comparing the organisational performance with the best practices or the best in business and considering the stakeholders' views. Process measurements give internal performance information. This way, organisations can evaluate and measure their leadership level in the industry to remain globally competitive. Strategic and operational maturity level assessments help assess the organisation's performance level. Refer to ISO 9004:2009: 'Managing for the Sustained Success of an Organisation – A Quality Management Approach'.

The organisation needs to define the overall quality objectives and performance targets. The defined objectives must be a true reflection of the quality policy statement. They must be meaningful, measurable and aligned with the words of the quality policy. The objectives must be coherent, harmonious and not in conflict with the organisation's values or other objectives. The objectives must be realistic and relevant to the organisation's operations or products. The objectives should be aligned and compatible with the overall business objectives, including meeting customer expectations. They may be known as business process quantifiers. Since the main aim of every business operation is profitability, assets/value added and the monetary/business gains during the period could be added as one of the objectives. The objectives, like the quality policy, must be documented information, and the policy must be reviewed periodically for suitability. Measurement of objectives could be in quantitative or qualitative units. However, there must be a clear link between the dynamic aspects of the quality policy (in terms of continual improvement) and achieving objectives through managerial initiatives.

The general strategic objectives of the organisation may be visible through the business plans, minutes of the management reviews, annual budgets, business scorecards being presented to the stakeholders, etc. However, there must be a link between the corporate/organisational objectives and those defined and measured at the functional/process/department levels. The department/function/process owner must periodically report the measured results of the KPIs/objectives to the head of the organisation through the management representative, who compiles them and relates them to the corporate/organisational goals. There must be no exceptions or exemptions to this, as the defined objectives can be achieved only through the total functional efficiency of the organisation.

It is a desirable practice that the organisational (top level) and departmental (functional level) objectives are defined well in advance by the 'Committee of Management'/'Management Review Committee' documented. Yearly targets are set at the beginning of the financial/calendar year for everyone to aim to meet the objectives by the end of the year. Monthly measurements, shown with the cumulative average for the period and the annual target and achievement, will reflect the organisation's performance at the review time. Where the planned results are not achieved, analysis of the relevant data for some time through trend charts will indicate the deficiency. The management/process owner is responsible for conducting the causal analysis and identifying/recording actions to be taken for improvement in a time-bound manner. The targets are revised for further improvements if the results are consistently satisfactory. The stakeholders measure organisational excellence by meeting the quality objectives/KPIs, assets/value added and the monitory/business gains.

Some common errors are seen in establishing measurable objectives. Writing objectives that are not specific enough concerning the measurement metrics is one, and confusing objectives with strategies is another. What the organisation wants to accomplish is the objective, and how the organisation intends to achieve the objective is strategy. The objective has two elements: the specific target the organisation hopes to achieve and the time frame. A poorly written objective such as 'to increase awareness of the product' from 20 to 30 per cent in the next year is vague and impossible to measure. Instead, it is a strategy for increasing market acceptance (sales).

Objectives measure the benefits or the feel-good factors of the organisational processes or how far the customer's expectations are met. External provider/supplier performance is evaluated in the purchase process to maintain a long-term relationship with the suppliers based on supplier performance. Quality of supplies and timely (on-time) delivery are the standard criteria for supplier performance evaluation. If their performance is good, say Grade A, the organisation will retain them, or they will be advised to perform better. Keeping supplier performance at a Grade A level cannot be the objective of the purchase process. It is not a pertinent objective for the organisation since it is the performance of the suppliers that makes the 'Grade A' for them and not any action from the organisation or meeting customer expectations.

Similarly, having an objective at the calibration process as 'No instrument cross calibration due date/found used after calibration due date' is a requirement of the Standard for conformity (shall be calibrated before use) and not a planned result of the process to be objective.

A 'Balanced Scorecard approach generally has four perspectives: financial, internal business processes, learning and growth (including human resources and infrastructure) and customer matters. Each of these perspectives is inter-process and inter-related. Metrics for measurement must be aligned with the organisation's strategic plan.

Earned Value Management (EVM) is a programme management technique that integrates technical performance requirements and resource planning with schedules while considering risk. The primary objective of applying earned value to a contract is encouraging suppliers to use an effective technical, cost and schedule management control system. This permits the customer to rely on timely data those systems produce for better management. EVM is a valuable tool in project management for monitoring project performance objectives. The metrics integrate technical requirements, cost, schedule and risk management. EVM provides useful, quantifiable performance metrics for forecasting cost and schedule for the project.

Process enhancement framework and repository of best practices and learning (PEARL) is another template for monitoring the performance objectives in an organisation.

The organisation determines what to do when planning to achieve its quality objectives. The organisation identifies the resources, the person responsible, when to complete it and how to evaluate the results.

Ensuring that all the processes conform to the Standard's requirements is necessary. Identify the process objectives and measure the process effectiveness. Act if the effectiveness is not fully achieved.

6.3 Planning of Changes

A committed management, which earnestly desires to establish a Quality Management System in the organisation, generally does not make frequent changes to the Quality Management System or changes in the personnel assigned the responsibilities before fully establishing the system. Some of the possible changes may be changes in the management, changes in the structure of the organisation, change in the resources/personnel and their responsibilities, changes in the product line, introduction of ERP and other computer systems for product realisation, changes in the list of customers, organisational mergers, bifurcation, etc., which could affect the functions of the organisation. However, when changes are inevitable and made, the management must ensure that the integrity of the quality system is not affected and that the integrity and cohesion existing in the international standard are maintained. Now that the high-level structure has been introduced in the Standard, no structural changes in compliance with the Standard should be made. It must be ensured that no compromise is made on mandatory requirements specified in the clauses of the Standard. Individual process requirements should not be overlooked or ignored. It also underlines that the Quality Management System is not personal or organisation-dependent. Irrespective of personnel, the system should run smoothly and should be able to deliver benefits to the organisation. The changes are discussed in management reviews and evaluated for their impact on processes and resources. If there is no negative impact, then the changes are accepted. QMS integrity will be maintained when changes are approved in management reviews and meetings. Records of the management reviews indicate the decisions made.

7

Support

7.1 Resources

7.1.1 General

Management's primary responsibility is to provide adequate resources to implement, maintain and continually improve the Quality Management System. Resources are supplied for meeting customer requirements related to the product and effectively managing the business to enhance customer satisfaction and achieve organisational objectives. Management also provides resources necessary for measuring, analysing and improving the quality system and the products/services to ensure the product delivered to the customer meets all quality requirements.

The internal resources provided generally are:

a. Human resources

b. Infrastructure facility (land, buildings, plant/equipment/machinery including information technology (IT) resources, utilities (electricity, water, compressed air, coolants, chemical products, waste management, communication and transport facilities and equipment maintenance, etc.)

c. Work environment (weather, temperature, relative humidity, lighting, vibration, noise, hygiene, cleanliness, housekeeping, etc.) and

d. Financial resources, including working capital.

The external resources generally are:

a. All types of products and services being provided by external providers/suppliers (through purchase/outsourcing)

b. Constraints and gaps exist in the capability/capacity of existing internal resources, which can be identified and obtained from external sources (through outsourcing/training of internal resources).

Resource management is the efficient and effective deployment of resources to achieve and continually improve the objectives of an organisation and make the resources available when needed.

In the case of project management, resource management may include estimating and making the human resources available with adequate skill, processes, techniques, tools, methodology, etc., as necessary. Resource management software tools are available to assist in resource allocation to projects. Enterprise Resource Planning (ERP) software automates and integrates core business processes such as taking customer orders, scheduling operations and keeping operational data, inventory records and financial data. ERP software can drive huge improvements in the effectiveness of the organisational processes. ERP is employed in manufacturing, distribution, engineering services and data analysis.

Ensuring the resources are adequately and effectively managed to achieve the planned results is necessary. This means that appropriate resources are to be identified, planned, made available, used, monitored and changed or improved as necessary by the organisation.

Management of resources is usually the responsibility of the top management. Adequate evidence in the form of records must be maintained for provision, effective utilisation, maintenance and upkeep of all the resources. Checks on the quality of utilities and work environment are necessary for ensuring that the processes/products are not adversely affected using such utilities and environment.

Every process needs resources as inputs. Reviewing the input, process performance and output will adequately reveal how the resources are managed. Apart from that, the management commitment and responsibilities, the management review process, product realisation processes including the measurement of product characteristics (inspection), control of nonconforming products, risk assessments and management, corrective actions and continual improvement activities, etc. all will confirm effective utilisation, maintenance and upkeep of all the internal and external resources.

If the organisation has not considered effective management of resources, it may result in not satisfying the product requirements. Due to inadequate resources or non-availability of necessary resources, and if the management commitment/attitude is to manage the processes somehow, the desired output may not be achieved. That will show deficiencies in the quality system effectiveness and product conformity measurements. Some deficiencies (which cannot be found by external measurements on the product) may be revealed only when the products are used, as in the case of unique processes. It is a (production-operation) risk, which should have been identified and appropriate mitigation planned and established before starting the production process.

AS9110 for maintenance organisations additionally requires that the organisation should have a system to continually assess the availability of tools, technical data and adequately qualified and competent personnel (approved by the regulatory authority) to ensure the safe completion of the maintenance, repair and overhaul activities on aviation, space and defence equipment.

Preparing an asset list and comparing it with the project plan can be done. Identify if there is any gap. Take action to augment the internal and external resources as necessary for the implementation, maintenance and improvement of the quality system and the organisational resource needs. Ensure that the process conforms to the requirements of the Standard, i.e., all the planned activities have been realised. Identify the process objectives/ key performance indicators (KPIs) and measure the planned results for the process against targets for a specific period. If the process conforms and the results follow the plan, the process is effective. If the planned outcomes are not achieved, actions to improve the process will be identified and implemented without delay. Optimisation of resources is usually possible with an increase in the level of learning associated with experience gained over time.

Management demonstrates the provision of resources to achieve planned business results. Allocation of resources is based on past performance and present needs, visible through the infrastructure, plant/machinery, equipment, people and environment.

7.1.2 People

Human resources and their competencies play a significant role in any business process. Therefore, human resources management and their continued development should be given maximum attention when an organisation establishes and implements a Quality Management System. The personnel management function includes recruitment and training of the employees, ensuring they perform to their potential and meet the organisation's highest expectations. Human resources development activities include developing personnel within the organisation for career advancements, man management and excelling in specific technical areas where higher knowledge and skill levels are required. It is necessary to identify and document the competency requirements of all personnel in the organisation for efficiently carrying out their task. Management must have a system in place to identify the gaps and take adequate measures to fill them up. The management must ensure that personnel carrying out work, which may affect the quality of the deliverables to the customer, have the necessary competence. The management assigns responsibilities to personnel to implement the Quality Management System effectively and for the operation and control of its processes. The training needs of personnel are to be identified so that customer and product requirements can be satisfied and continually improved. Not only providing the necessary training to persons but verifying the effectiveness of training as to whether the training has benefited the organisation in increased quantity and quality output. Periodic assessments of personnel are carried out to monitor their performance to ensure that people in the organisation are performing efficiently and in a highly fulfilling manner and giving their best to the organisation.

7.1.3 Infrastructure

We have seen that the responsibility of the management is to identify and provide adequate resources needed to achieve conformity to product requirements. After human resources/people, the following resource type needed is infrastructure facility. Land, buildings, workspace, plant/equipment/machinery, process equipment (both hardware and software) and associated utilities such as electricity, water, compressed air, coolants, chemical products, support services like waste disposal or management, communication, information technology (IT) resources, information systems and transport facilities are some of the infrastructure.

In modern manufacturing, SMEs commonly use high-speed machining, multitasking and multi-axis CNC machines. Such machines give up to one-micron accuracy and require minimum set-up changes. Overall, equipment efficiency is at its maximum, and the devices are reliable for a long time. Organisations provide these resources intending to use and maintain all the resources and utilities, primarily for realising conformity of products and services. If the resources can benefit the product and personnel, it is to improve productivity and efficiency, respectively. Determining adequate resources can be shown by having a list of all the identified resources, mainly the buildings and production equipment such as plant/equipment/machinery, process equipment and support equipment, all of which require maintenance for upkeep and serviceability.

Maintenance is an essential requirement. There are different types of maintenance for the infrastructure, such as daily/routine maintenance, periodic preventive maintenance, break-down maintenance, total productive maintenance, etc. In a factory, it may include maintenance of conventional machine tools,

CNC machines, test equipment, process equipment, special process equipment, services equipment such as power supply, communication systems, computers and servers (hardware and software programs) and utilities such as compressed air supply, air-cooling systems, air-conditioning systems, water chilling systems, water coolers and material handling equipment. Maintenance includes repairs, overhauling, reconditioning, retrofitting and up-gradation. In some large organisations, 'Plant Rehabilitation Committees' exist to plan and execute activities to erect, dispose of and replace equipment. Spares management, annual maintenance contracts and training of maintenance personnel are also maintenance responsibilities. The maintenance department undertakes energy conservation, environmental management and plant/equipment relocations to improve production systems, apart from installing and commissioning new machines/equipment. Fulfilment of safety requirements such as periodic testing of air receivers and overhead cranes, monitoring of temperature/pressure/flow rate/pH and conductivity/noise/vibration/voltage/current/frequency/motion and rotation, etc. and implementation of various safety measures including statutory requirements are also undertaken by the maintenance department.

'An Indian Auditor inquired at the site. "Is there a statutory requirement for safety certification of overhead cranes used in machine shops in your country? We have it in India."

The Managing Director of an SME in Komatsu-Shi, Narita, who has held the ISO 9001:2000 certificate for the past seven years, was surprised and answered the auditor - "Sure, we too have".

On request to show the current safety certificate for the overhead crane, while walking underneath the overhead cranes on the shop floor, the maintenance manager brought a six-year-old certificate and said, "Though it is an annual requirement, we have not got it certified after that date."

The aerospace auditor appreciated the frankness and politely and promptly issued a Major Nonconformity.

MD was slightly worried, "We have been using it daily. We need to use this every day. Now, what is to be done as containment for the nonconformity?"

Auditor: "Please stop using it until you get the safety certificate."

The following evening, they got the safety certificate for the crane in that shop and all the cranes and other lifting equipment in the SME.

They were also subjected to statutory periodic 'thorough examination' as per Provision and Use of Work Equipment Regulations (PUWER), and the maintenance manager recorded the corrections and corrective actions in the Nonconformity Report.'

The maintenance process takes adequate measures for the timely completion of all types of maintenance. The user of the plant/equipment/machinery carries out daily maintenance. The type of jobs carried out during such routine maintenance may be general cleaning and upkeep, replenishing of lubricants, checking for completeness and correctness of the equipment/machine status, checking for any damage or deterioration, checking the availability of all functional requirements, serviceability, ensuring the utilities and availability of inputs, etc. before using the machine. The maintenance process's ultimate objective is to achieve product and service conformity. Specific process objectives can be error-free maintenance, fewer breakdown/non-utilisation hours, timely completion of maintenance, etc.

Prepare a preventive maintenance schedule indicating the maintenance to be carried out on all machines/equipment spread throughout the year. For preventive maintenance of each machine, prepare a checklist covering all the checkpoints and points that can be used for maintenance. The list may vary from machine to machine depending on the type of machine and the recommendations given by the machine manufacturer in the user handbook.

Open a breakdown register to make an entry as and when the users request/call to attend to breakdowns. The register can also identify the reason for the breakdown, replacements/repairs carried out, the time taken to repair and the name of the person who did the rectification. The register can be used to calculate the breakdown hours of each machine if the period of breakdown until the rectification is noted. Restoration of serviceability of the machine should be accepted by the user who reported the breakdown, a closed-loop control system, and closing the ticket by the user who raised it so that there is accountability, traceability and responsibility for the actions available in the register. And the register is an Excel spreadsheet on the PC!

The maintenance department should maintain a list of essential spares stocked to use during breakdowns without waiting for the procurement. A crucial spare list can be worked out from the machine manufacturer's recommendations and the history of spare parts consumption for any machine. Spares and additions, if any, can be replenished well in advance to reduce downtime.

VED analysis of spares is usually applied for spare parts to be stocked for maintenance of machines and equipment based on the criticality of the spare parts. 'V' stands for vital, 'E' for essential and 'D' for desirable. The stocking policy is based on the criticality of the items. The necessary spare parts are known as capital or insurance spares. The inventory policy is to keep at least one number of vital spare irrespective of the lead time required for procurement. Essential spare parts are those whose non-availability may not adversely affect production. Such spare parts may be available from many sources within the country, and the procurement lead time may not be extended. Hence, a low inventory of essential spare parts is held. The desirable spare parts are those that, if unavailable, can be manufactured by the maintenance department or may be procured from local suppliers; hence, usually, no stock is held.

Reducing downtime or calculating machines' availability throughout the year could be a measurable maintenance process/department objective. Records of routine serviceability checks and preventive and breakdown maintenance must be maintained.

Aerospace Standard AS9110 specifies that maintenance organisations shall have suitable facilities for undertaking MRO activities of specific aircraft types. The organisation establishes the facilities, including a suitable runway, hangar space, ground handling equipment and movement facilities, suitable ladders, electrical and mechanical ground servicing equipment, avionics laboratory, battery charging bay, etc., essential for maintenance activities. Appropriate inspection and test facilities should also be available to ascertain the serviceability of structures, aero-engines, equipment to facilitate washing, cleaning, fuel and oil replenishment, fire extinguishers and other safety equipment. Similar facilities are required where aircraft servicing is undertaken away from their usual fixed location. The regulatory authority verifies the adequacy and serviceability of all the facilities before accepting and approving the scope of work.

7.1.4 Environment for the operation of processes

The third type of resource is the environment for the operation of processes. The term environment relates to the conditions for achieving conformity to product requirements or the environment under which the service is delivered. The environment includes physical factors such as weather, temperature, rate of change of temperature, heat, relative humidity, lighting vibration, dust level, noise, hygiene, housekeeping, cleanliness, electromagnetic interference, etc. Production processes and measurement of product characteristics are also to be carried out in the proper environment. Some of these conditions may affect the quality of the product and, therefore, need to be controlled as necessary for the conformity of products.

A clean room may be suitable for assembling certain equipment/ products. The epoxy floor makes it easier to maintain a clean environment. Certain items must be stored in the appropriate temperature and relative humidity storage conditions so that the physical or chemical characteristics of the objects can be adequately maintained until use. Raw materials for composites, adhesives, sealants used for metallic fuel tanks of aircraft and rubber or elastomeric seals are some of them. Conformity to final product characteristics can be achieved only when these materials are stored in their recommended environmental conditions. Some materials have a specific storage life, beyond which the physical or chemical characteristics are likely to deteriorate and, if used after the prescribed storage life, cannot meet the final product requirements. Therefore, in production, the shelf life of these materials is to be monitored and their use controlled so that what was produced/stored first must be issued for use first – 'First in, first out'.

Some of the controls established for the environment necessary to achieve conformity to product requirements during production are listed below:

A clean room is a controlled environment where some aerospace products are manufactured and assembled. It is a room in which the concentration of airborne particles is controlled to specified limits. People, processes, infrastructure facilities and equipment generate the contaminants/particles. ISO 14644-1 is a standard for establishing air cleanliness classes for airborne particulate levels in clean rooms. Class 10,000 (equivalent ISO 14644-1 Class 7) clean rooms are designed to limit the number of particles to 10,000 (0.5 microns or larger).

Specific precision assemblies must be carried out in a dust-free environment, where dust particles per million (ppm) are controlled and maintained to an acceptable level. Precise, accurate measurements will be carried out in controlled conditions (of temperature and relative humidity) while using coordinate measuring machines.

As per 'NIST HANDBOOK 150-2G, National Voluntary Laboratory Accreditation Program, Calibration Laboratories Technical Guide for Mechanical Measurements', recommended environmental conditions during calibrations are:

a. 23 +3, –5 degrees C temperature,
b. 101.325 +1, –3 kPa barometric pressure and
c. 50 +25, –35 per cent relative humidity.

The organisation must determine what environment is required to meet the product requirements and accordingly provide a controlled environment. Typical working conditions will suffice when the climate is not likely to affect the product requirements. A secondary need is providing comfortable conditions and environment for the personnel to work efficiently.

Electrostatic discharge (ESD) is a prevalent and dangerous problem affecting modern electronic/avionics parts and systems. ESD is the sudden transfer of electrical charge between two objects at different potentials. Removing your polyester or nylon shirt in dry weather clings to your body due to static electricity, or you can see and hear even mild electric sparks in the darkness. The discharge of electricity or the mild shock you feel and hear is a form of ESD. The human body or other conductive objects can become electro-statically charged if not correctly grounded. Suppose this charge meets or passes near an ESD-sensitive (ESDS) device. In that case, ESD damage can occur. Even low voltages are capable of damaging ESDS devices, and the damage leads to early failure of electronic/avionics equipment. Standards like ANSI/ESD S20.20 Electrostatic Discharge Control Program Standard, ANSI-ESD Association standards, D-1348 rev. F, Standard: Electrostatic Discharge Control, etc., prescribe the minimum requirements for industry dealing with electronic, avionics and electro-mechanical components and sub-assemblies for protection against electrostatic overstress (EOS) or electrostatic discharge damage to ESD sensitive devices.

Compliance with ISO 14001: 2004, Environmental Management System Standard helps in creating an appropriate environment in an organisation, as it:

1. Demonstrates management commitment to caring for and improving the environment.
2. Provides a framework for an organisation to manage its statutory and regulatory compliance
3. Improves environmental performance, target setting and measurement.
4. Demonstrates the organisation's responsibility towards environmentally friendly business practices with minimal or no environmental impact.
5. Facilitates reductions in pollution, waste, injuries and incidents and conservation of energy and
6. Establishes environmentally friendly working conditions, emergency preparedness and response.

We have been talking about the physical environment. Organisations should focus on managing the environment to use human resources better. Human factors also play a prominent role in having a controlled environment. People want to work for an organisation that provides an appreciation for the work done, ample opportunities to grow, a friendly, trustworthy and cooperative environment, and a feeling that the organisation is a second home to the employee. The organisation environment includes culture, values, reputation, quality of people, employee development and career growth and leading technologies. The work environment should be such that the employees feel connected to the organisation in every respect when they are among efficient managers, supportive co-workers, challenging work, involvement in decision-making, clarity of work and responsibilities, and recognition.

The requirements for a work environment include employees' health and safety, though the Standard may not explicitly indicate it. Appropriate workplace ventilation is also a requirement for employees' health. OHSAS 18001, Occupational Health and Safety Management Systems – describe workplace health and safety requirements.

Social factors within the organisation influence human behaviour concerning quality service delivery. Working in a team of competent personnel, one gets motivated if there is no discrimination among the team members. Different opinions and methodologies are discussed amicably, politely and non-aggressively. The organisation (and the co-employees) must provide a conducive environment for the employees where the product/service realisation process does not cause stress, fatigue or emotional conflict.

> *'We were reluctant to spend for the additional infrastructure necessary to provide barriers, earmarking controlled space to prevent dust, moisture and foreign object debris from entering the work environment. It was for building aircraft integral fuel tanks, where sealants are applied on metallic joints. We learned great lessons when, after some period of aircraft usage, fuel leaks were noticed from locations of integral fuel tanks on the aircraft. The added lifecycle cost of the aircraft due to unscheduled repairs does not justify the reluctance to invest in mitigating the risk of fuel leakage while manufacturing integral fuel tanks in the hangar'.*

Therefore, the environment for the operation of processes combines physical environment, human factors and social and psychological factors.

Specific work and storage facilities (ex, composite, rubber and sealants), use storage areas for solvents and petroleum products, the ventilation system in case of welding, pollution-controlled painting booth, etc., demonstrate organisational arrangements for a proper operational environment. Clean rooms, electro-static protection for electronic assemblies, temperature, dust and humidity-controlled areas and the related product or process requirements and specific environmentally controlled storage areas for constituent parts/materials of the final product demonstrate organisational arrangements for the proper environment conducive to conformity to product requirements.

7.1.5 Monitoring and measuring resources

7.1.5.1 General

There are several monitoring and measurement requirements during the product realisation stages. There may be different types of measuring devices, in adequate numbers, needed for monitoring and taking measurements at various stages of production. Since measurements can be taken with multiple devices/instruments/gauges/equipment, etc., the word 'resources' describes different monitoring and measurement equipment, devices and instruments.

The product requirements/quality characteristics are known from the customer requirements/specifications. The organisation determines how the product requirements/characteristics are to be measured and recorded so that there is evidence of conformity to needs available with the organisation before delivering the product to the customer. The suitability and capability of the monitoring and measuring device or equipment must be verified before its actual use.

Verifying a measuring equipment/device for its 'fitness to use' is known as 'calibration/verification'. Measuring or monitoring equipment is verified for metrological confirmation. This verification includes any necessary adjustment, repair and subsequent re-calibration, and it compares the measurements with metrological requirements. This comparison is between two measurements: one made with a standard known as the master equipment and another with the actual device (unit under test). When the exact measurement is compared with the measurement made with the master, the measurement error is known. The instrument is suitable or fit for use if the error is within the permissible limit. Subsequently, the equipment is sealed and labelled. The requirements for intended use include such considerations as range, resolution and maximum allowable errors. Metrological confirmation is documented in calibration reports, and the reports are to be verified for the continued suitability of the measuring equipment.

Calibration is required for a new instrument and when a specified time has elapsed. When an instrument has fallen or had a shock or vibration, which might have caused it to be out of calibration or not fit for use, or whenever measurements appear to be erratic, the instrument needs to be re-calibrated.

Calibration is sometimes regarded as adjusting the measurements or indication on an instrument to agree with the measurement on a standard/master to be within a specified accuracy. For example, after a pressure gauge is calibrated and the error of indication or the correction required is determined, the indicator could be adjusted to show the actual pressure in kg/cm^2 at specific points on the scale. The adjustment should be safeguarded from further adjustments that invalidate the measurement result. No other disturbance in the adjustment is permitted until the next calibration.

For the metrological confirmation process for measuring equipment, refer to ISO 10012:2003 Measurement Management Systems – Requirements for measuring processes and equipment. American National Standards Institute (ANSI)/National Conference of Standards Laboratories (NCSL)'s Z 540-1 'General Requirements for Calibration Laboratories and Measuring and Test Equipment' may also be used.

7.1.5.2 Measurement traceability

The organisation must maintain a register of all the monitoring and measuring resources (instruments/equipment/devices). Monitoring and measuring equipment includes, but is not limited to, test hardware, software, automated test equipment (ATE) and plotters used to produce inspection/verification data. It also includes personally owned and customer-supplied equipment to provide evidence of product conformity.

The following details of the measuring instruments/devices/equipment are required to be indicated in the register:

a. The equipment type (digital, analogue, electro-mechanical, hydraulic, laser, etc.)

b. Unique identification (number or code allotted to the instrument by the organisation or the serial number of instrument/equipment given by the manufacturer, not the type/part number.),

c. Location (where is it customarily located/used in the production shop or the inspection booth, tool crib, etc.),

d. Frequency of checks (interval between two calibrations, say, once every six months or not exceeding six months),

e. Check method (refers to the master equipment used as a standard for verifying/calibrating the unit under test; the master equipment should be calibrated and the calibration traceable to national or International standards) and

f. Acceptance criteria (what is the accuracy required for the measurements while using the instrument, or the smallest measure to be taken on products compared with the instrument's precision).

The process employed for their calibration/verification is to be defined. The procedure of calibration may be calendar-based or use-based. The date of calibration, the date due for the next calibration and the method of recalling instruments for sending them to the calibration agency well in advance need to be indicated.

Since the product requirements are known, and the organisation must determine and establish processes for monitoring and measuring all possible calibrations or verifications of the requirements of the product, the organisation has to arrange for all such measurements. The organisation should decide all actions necessary, verification calibration or both.

Metrological Traceability can be defined as an unbroken chain of measurements and associated uncertainties.

The organisation's responsibility is not only to arrange for the measurements but also to ensure that the measurements are carried out in such a manner that it meets the measurement requirements of the product. A state of having limited knowledge, where it is impossible to describe existing status exactly, comes when specific measurements are not recorded in the result sheet. Suppose the final inspection report does not record the monitoring of requirements like cleaning, surface finish, burr removal from a metallic component and rounding off sharp edges. In that case, the organisation's claim that these requirements/ measurements meet the drawing/product characteristics requirements cannot be justified by mere assumptions or excuses like not having proper measuring devices/equipment to measure the requirements accurately. The organisation should decide on verification methods/actions and certify compliance.

Some organisations have a facility or standards room to calibrate monitoring and measuring equipment. Some others get the calibration services done from external providers through outsourcing. Regardless of how the calibrations and verifications are done, the organisation must ensure that the environmental conditions suit the calibrations, inspections, verifications, measurements and testing. Suppose measuring devices/equipment is recommended to be in a controlled environment like a coordinate measuring machine (CMM). In that case, a controlled environment is to be provided for using the equipment. Equipment manufacturers usually provide specifications giving ranges, maximum loads and environmental conditions limitations for correctly using the equipment. The recommended environment for engineering measurements, including laboratory calibrations/verifications, is in the Standard ISO/IEC 17025:2005 – general requirements for the competence of testing and calibration laboratories.

Measuring equipment is deemed to have traceability to internationally or nationally recognised measurement standards when it is calibrated by one of the calibration laboratories accredited for the measurement in question by a national laboratory accreditation body. In that case, the measurements are reported by the laboratory on a report or certificate that contains the logo of the accreditation body, including the Standard ISO 17025.

Depending on the type, measuring equipment/device must be calibrated and verified at specified intervals. Before the use of the equipment/device, calibration is mandatory. Calibration should always be against measurement standards traceable to international or national measurement standards. That means the master equipment used for calibrating the equipment/device should have a valid calibration status traceable to international or national measurement standards. Traceability is usually achieved through reliable calibration laboratories with their traceability to national measurement standards. For example, a laboratory complying with the requirements of ISO/IEC 17025 (NABL accredited Lab in India/SAC-SINGLAS in Singapore) could be considered reliable and suitable for selection as a source/supplier/external provider.

Measurements are to be traceable and not the measuring equipment. Metrological traceability requires an established calibration hierarchy. Organisations should ensure that all measurements carried out to evaluate the characteristics of their products (including measuring devices) are traceable to System International (SI) units or an agreed testing method. Traceability may also be achieved using validated test methods, calibrated master equipment and certified reference standards. The requirements for measurement traceability are detailed in ISO 17025: 'General requirements for the competence of testing and calibration laboratories'.

When it is mentioned that the calibration should be traceable to national standards, the equipment should be able to produce measuring results traceable to national standards. Where no traceability requirement exists, as for a local/tertiary level facility, the basis for verifying measuring devices or calibration needs to be identified and recorded. One way of ensuring that the measurements are accurate is by tracing them back to national standards with known higher accuracy. Traceability is a method of guaranteeing a measurement's accuracy through an unbroken chain of reference or comparisons. National standards are accurate to 0.001%, Masters in calibration labs are authentic to 0.01%, and the instruments in a production shop are accurate to 0.1%.

Once the calibration is completed, primarily by outsourced calibration agencies/external service providers, the organisation must verify the calibration results to ensure that the errors indicated by the calibration are within acceptable limits. Error is the difference between the measured and valid values of the measured object.

High precision means good repeatability, and low precision means poor repeatability. Accuracy is a qualitative term relating the average of the measurements to the actual value. Increased accuracy means the measured result is the actual or very near the actual value. Low accuracy means the measured result is away from the actual value.

No measurement gives perfect results. Uncertainty of measurement is the doubt about the result of any measurement. We can say how confident we are about the outcome by quantifying the possible measurements spread. 20 mm ± 0.1 mm dia of a shaft at a confidence level of 95% measures 19.9 to 20.1mm dia (spread of measurements = precision). Accuracy is the ratio between the average of the measurement results and the actual value.

Even when the measured result is precise and accurate, the measurements will still be uncertain. Uncertainty is the quantification of the doubt about the measurement result. When the uncertainty of a measurement is evaluated and stated, the fitness of the purpose for a particular application can be adequately understood.

The user must maintain a record of this verification for suitability for continued use. When an instrument is accepted as suitable for usage, there is a system to identify its calibration status. The status is usually determined by a label indicating three pieces of information – instrument identification number, the date calibrated and the date next due. All tools and devices are protected from damage and deterioration during handling, maintenance and storage.

The organisation must have a process for recalling, monitoring and measuring equipment requiring calibration or verification. The recall is for getting the equipment from the point of use and calibrated before the previous calibration status expires.

When instruments are used for measurements regularly, some instruments may likely become unsuitable for use due to damage, wear and tear or misuse, and error creep could be noticed only when the instrument is calibrated subsequently. The organisation promptly rejects the unserviceable instrument or gets the instrument repaired or reconditioned for use again after re-calibration. Usually, the matter ends there. However, nothing should be left to chance when the products are critical to safety. It is a probability that if the unserviceable instrument was used for some time, inadvertently or unknowingly, before detecting the measurement error, there is always a possibility that some parts have been cleared. Since nothing should be left to chance, there must be deliberate action to verify all such measurements made, if at all made, until the error was detected on the instrument. Verification must be made on all parts, equipment/assemblies if assembled and cleared from when the instrument might have become 'unfit for use' until it was detected unfit. If any accepted parts are not assembled, verification is simple using another similar calibrated instrument. Where more than one instrument of the same type is used in an area, how can the 'unserviceable' instrument be distinguished from others? A detailed analysis is required to decide on the recall of parts cleared if the measurements were made on the part by using similar instruments, including the 'unfit' instrument. How do you check whether the 'unfit' instrument was used to measure the parts? The instrument was there but may not have been used! A possible permanent escape from the dilemma is to have the instrument identification noted in the inspection records whenever any measurement is recorded. That can specify whether the instrument was used or not. Therefore, it is necessary to establish and implement a system of recording the instrument identification whenever any measurement is made using that instrument. The second dilemma is whether the error noted on the 'unserviceable' instrument could have affected the measurements on the already cleared part. What could be the consequences if the previous measurements were not verified on the part since the part has already been satisfactorily assembled on the final equipment? This should be analysed and appropriate decisions should be made. The Standard requires that the organisation take appropriate action on the equipment and any product affected. Records of the verification results of the previous measurements must be maintained.

In software organisations, since the product/service delivered to the customer is in electronic media, no physical measurements use measuring devices/equipment. Therefore, the software industry generally has too few or no actions concerning measuring device calibration. If no measuring device/equipment is used, the Standard clause 7.1.5 could be declared as not applicable, with justification.

Some of the customers insist that the calibration systems shall meet the applicable requirements of ISO 10012, ISO 17025 or American National Standards Institute (ANSI) and the National Conference of Standards Laboratories (NCSL) International announce that ANSI/NCSL Z540.3-2006 Requirements for the Calibration of Measuring and Test Equipment.

Training personnel in Measurement System Analysis (MSA) and Gauge Repeatability and Reproducibility studies helps improve the effectiveness of the measurement process. Ensure that the process conforms to the requirements of the Standard, i.e., all the planned activities have been realised. Identify the process objectives/key performance indicators (KPIs) and measure the planned results for the process against targets for a specific period. If the process conforms and the results follow the plan, the process is effective. If the planned outcomes are not achieved, actions to improve the process will be identified and implemented without delay.

7.1.6 Organisational knowledge

The next level of the Quality Management System, which is likely to dominate and control the future business, it is said, is 'knowledge management '. 'Monopoly' industries exist in every sector and are the only suppliers of a particular commodity. Monopoly industries depended on exclusive knowledge and management to produce products and services. Over time, improvements in processes and technologies have emerged. Based on improved techniques and technologies, multiple industries have emerged and made various products for the same purpose. Until and unless the monopoly industries gain more knowledge about improved methods and technologies and improve their products to benefit the customers, they will vanish from the scene. Knowledge management is an enabler to achieve the organisation's objectives better and faster through an integrated approach. Knowledge management is to provide improved processes, technology and quality systems combined with the organisational behaviour and culture to smooth the transition to improved products/services. Knowledge management establishes knowledge sharing that is relevant to the organisation among its employees. It eliminates total restructuring to improve its products and services.

Organisational knowledge of the business/QMS processes is available at different levels. Some of them are:

- *At the individual level, explicit knowledge is found in documents and tacit knowledge is* gained by experience.
- *At the group level*: Knowledge that is explicit, implicit and embedded, found in groups but not shared with the rest of the organisation. The group members share common values, language, procedures, know-how, etc. They are a source of learning and a repository.

Organisational knowledge is knowledge specific to the organisation, generally gained by experience. Information is used and shared to achieve the organisation's objectives. Knowledge management (KM) is the process of creating, sharing, using and managing the knowledge and information of an organisation. It refers to a multi-disciplinary approach to achieving organisational objectives using the best knowledge. Knowledge management efforts typically focus on corporate objectives such as improved performance, competitive advantage, innovation, the sharing of lessons learned, integration and continual improvement of the processes in the organisation. Knowledge management is a strategic asset that focuses on acquiring and sharing knowledge.

Organisational knowledge comprises all the knowledge resources within an organisation. The knowledge can be realistically used for all organisational processes. This knowledge can, therefore, be available to individuals and groups or exist at the corporate level.

Organisations must help develop expertise at all levels to enhance organisational knowledge. It must facilitate and promote its diffusion to individuals, groups and across the entire organisation in accordance with the organisation's requirements. Knowledge management portals help storage and retrieval capabilities, creating an environment conducive to learning and sharing (employees seeking and employers providing knowledge). Knowledge sharing depends on the willingness of the employees and their assimilation capabilities. Organisational culture and traditions, human resource development, career promotions and incentive schemes, etc., help in organisational knowledge management.

Every organisation doing business has determined the knowledge necessary to operate its processes. They have already established the means of achieving conformity to their products and services. However, to continue the business, it is necessary to maintain and share the knowledge with the relevant stakeholders to the extent required. This is best done by maintaining a knowledge bank, repository or documented information on critical processes and their controls.

In reputed software/engineering services organisations, a knowledge management portal, a knowledge bank, a repository and a central place in the local area network, where all the positive contributions, innovative ideas and improvements made on the products/services are stored. It may include multiple databases or files for distribution over a network or a location directly accessible to the user without travelling across a network. The employees even patent some of the improved knowledge or processes. Such a knowledge bank facilitates the flow of knowledge from where it is located to where it is required or can be applied to achieve organisational objectives.

Suppose someone in the organisation has successfully created an innovative solution. In that case, the information must be shared by all interested parties for profitability and organisational excellence. The knowledge repository is how the 'pockets of excellence' are spread throughout the processes for the organisation's benefit. Another person in the same organisation trying to figure out the same kind of solution through a trial and error method is not profitable and a waste of time and resources.

Organisational knowledge can be based on the internal resources such as intellectual property available to the organisation. Knowledge gained from experience in project management or product realisation processes is organisational knowledge. Lessons learned from failures and successful projects in the past (risks and opportunities) impart profound organisational knowledge. Capturing and sharing undocumented knowledge and experience during the execution of the tasks is a suitable method of acquiring organisational knowledge. Organisational knowledge includes the results of improvements in the processes, products and services realised because of improvement measures taken in the past.

Organisational knowledge is also gained by personnel by participating in seminars/conferences and workshops on the relevant themes/subjects. External standards, guideline standards, books, research papers, etc., are sources of gaining the appropriate knowledge and gathering knowledge from the customers or suppliers, like the quality function deployment (QFD), which enhances organisational understanding.

Knowledge management actively promotes innovation. It helps the organisation to institutionalise best practices identified through the knowledge management programme by making them part of standardised business processes. Once the critical business processes are identified and documented, it is easy to focus on the improvement efforts of these processes. Knowledge management is an enabler of organisational learning.

Some of the employee suggestions/ideas have been quality-improvement projects due to the knowledge management programme initiative. Successful quality-improvement projects were often initiated as pilot projects in one part of the organisation.

Changes in processes or technology do happen over some time. When such changes are considered to improve the processes or products, the organisation makes appropriate decisions on acquiring or accessing additional knowledge and associated process improvements, product improvements and managing the changes with an impact study.

Management may issue an 'Instruction' (indicated below as an example, with suitable modifications) for implementing the requirements of 'Organisational Knowledge':

'*Knowledge Repository/Bank files*

Organisational knowledge is the knowledge specific to our organisation. The organisation generally gains from the experience of all personnel. It is information that is used and shared to achieve the objectives set out by the organisation. It is the wealth of knowledge continually built up by the employees while doing business, improving the company, prospering in the industry and later passing it on as a legacy to the next set of personnel, who will inherit the wealth of knowledge and flourish.

Our organisation holds certain (specify): -

- *Ethics (moral principles that govern an organisation's behaviour)*
- *Ethos (the set of beliefs, ideas, etc., about the social behaviour and relationships of a group of people and*
- *Values (Important and lasting thoughts or ideals shared by the members of an organisation, culture about what is good or bad and desirable or undesirable).*

All business activities include the reflections of these ethics, ethos and values. Values influence a person's behaviour and attitude and are broad guidelines. Organisational values are the collective behaviour of all personnel in the organisation. Some common business values are fairness, innovation and community involvement. Knowledge management combines organisational learning, memory, knowledge sharing, knowledge creation and corporate culture.

Organisational knowledge is based on internal resources such as intellectual property, knowledge gained from experience, lessons learned from failures and successful projects, capturing and sharing both documented and undocumented knowledge and experience, the results of improvements in processes, products, etc. Knowledge gained from external sources such as standards, academia and conferences; gathering knowledge from customers or external providers is also to be maintained.

The knowledge repository of our organisation includes the following:

1. *QMS Documentation*
2. *Business Processes and Procedures (indicated in the Quality Manual)*
3. *Products and their features, the core competency of our business*
4. *Projects and their domain have been managed so far. Lessons Learned in the past with respect to the business processes completed, products delivered and support/services provided to the customers after delivery of the products/services – Project-wise information maintained.*
5. *Accident Investigation reports, Defect Investigation reports, Special study reports, etc., are made available to all the employees*

of the ASD organisation for the lessons learned improvements made on products like aircraft, aero-engines, space and defence equipment.

6. *Research studies, technical papers presented/published and suggestions implemented for the benefits it brings to the products, organisation customers, etc., are made available to all the employees for reference.*
7. *Innovative ideas, automation and specially developed and patented software tools, which are designed for one project, are made available at the knowledge portal of the organisation for use in similar projects within the organisation.*
8. *Enterprise-level and operational-level risks were identified and mitigated (FMEA/PFMEA/Risk register) when the implementation of AQMS 9100 was established and reviewed periodically.*
9. *New Technology was acquired while doing the scope of work established through the QMS documentation and specific business with customers.*
10. *Diversification plan/New Business Plan and the feasibility study/reports generated there on and progress made*
11. *Our organisation has a (specific) Strategy, Vision, Mission, Policy and Objectives appropriately documented in the QMS documentation.*
12. *Group Knowledge of interested groups and Knowledge Communities, where the top management participates.*
13. *Suggestions were given, suggestions made by the shareholders, methods of solving the business's external and internal issues, needs and expectations of relevant interested parties that were addressed, and concerns affecting the industry.*
14. *Mentor communications, guidance and directions*
15. *Any other relevant information that the management desires to be maintained for future reference*

The Top Management/MR keeps the master files/copies. The knowledge repository/'Knowledge Bank' files will be updated as and when necessary at periodic intervals, decided by the management. Sharing information with relevant parties is at the top management's discretion.

All personnel are requested to contribute towards building up the legacy of the organisational knowledge for posterity. '

7.2 Competence

Human resources and their competencies play a significant role in any business process. Therefore, human resources management and their continued development are given maximum attention when an organisation establishes and implements a Quality Management System. The personnel management function includes recruitment and training of the employees, ensuring they perform to their potential and meet the organisation's highest expectations. Human resources development activities include developing organisational personnel to carry out the processes conforming to the customer requirements effectively and efficiently. The management ensures that personnel carrying out work, which may affect the quality of the deliverables to the customer, has the necessary competence. The competency of personnel should be such that the processes carried out must give the desired process outputs and performance results. The competency of personnel should be such that the performance and effectiveness of the Quality Management System should not be affected. This may further help the personnel advance their career and excel in specific technical areas requiring specialised knowledge and skill levels. It is necessary to identify and document the competency requirements of all personnel in the organisation to efficiently carry out their processes. The management establishes a system to identify gaps in the competency of personnel. They take adequate measures to fill up the gaps. Any process within the Quality Management System may affect the product/service requirements, not only the direct core production operation or service provision. All competency gaps are filled up by training. Training is provided to personnel so that the processes follow the requirements, ultimately meeting the customer's requirements and the product/service delivered. The management has the responsibility of not only providing the necessary training to persons but also of verifying the effectiveness of training. Effectiveness verification ensures that the training has benefited the organisation in increased quantity and quality output. Periodic assessments of personnel are carried out to monitor the performance of its employees to ensure that people in the organisation are performing efficiently and in a highly fulfilling manner and giving their best to the organisation.

Competence is determined based on appropriate education, training, skills and experience. An efficient way of assessing the competency of the personnel is to map the competency of all personnel in a matrix and evaluate their performance towards the organisational goals/objectives. Once the evaluation is done, appropriate changes could be made in the work area of personnel depending on their strengths. That will ensure competent personnel are placed to perform work. That work/performance will not adversely affect the product quality.

Further personnel recruitment and placement for any specific task could be entirely based on the basic mapping for the competency needs and other attributes needed for satisfactory process compliance. Documented information needs to be maintained as evidence of the available competency of personnel, including the training given to them. Periodic review of the necessary competence and training given must also be evidenced in the records.

Competency management is mainly used to develop operator versatility, mitigate risk for shortages of adequate skill levels and reduce waiting time in realising products and services. It will be helpful to list the process steps requiring specific technical skills for operators. Plan the training needed wherever a gap is noticed. The skill matrix could indicate the skill levels on a scorecard, with the score gradually increasing from untrained learner, practitioner and developer to coach.

The phrase 'soft skills' describes the skills characterising relationships with others or how you approach life and work. Some of the essential soft skills necessary for competency are communication skills, the skill to make appropriate decisions, self-motivation, leadership skills, skills to work in a team, creativity and problem-solving skills, time management and the ability to work under pressure.

Hard skills are specific, teachable abilities that may be required for a particular job and are physically executed. Hard skills include proficiency with software applications, operating machinery and technical trades such as NDT, inspection, welding, riveting, etc.

Adequate competency is required for specialised skills such as soldering, different types of welding, various NDT methods, visual checks, etc. while manufacturing and inspecting components/equipment for aviation, space and defence use. Apart from the organisation employing persons with such specialised skills, periodic evaluations and certifications of the personnel are carried out by the organisation under the supervision of regulatory authorities to ensure consistency of quality of the skilled jobs carried out by the personnel.

An organisation can do the following things area-wise to achieve the competence/effectiveness requirements:

a. Identify the competencies required by personnel performing work, which affects the product's quality and effectively perform the assigned processes. This is by maintaining a register, database, human resources plan, competencies development plan, project plan, etc.

b. Identify which personnel are already performing the work and have the required competencies. This will be shown in their job. The technicians should know the actions to take when any deficiency in the process is found. Checking on that aspect periodically will ensure the competency of the personnel. Process validation records, nonconformity records, internal and external audit reports, customer complaints, etc., will also indicate the competency of the personnel working on those processes.

c. Decide what additional competencies are required.

d. Decide how these additional competencies will be obtained – possibilities are training personnel (external or internal), theoretical or practical training, recruiting new competent personnel, assigning existing qualified personnel to the required work area, etc.

e. Review the actions taken to satisfy competence needs and periodically review the same. HRD functions of the continued development process for the technicians and regular periodic appraisals of the personnel and their performance; Quality control functions like routine inspection and testing of products for which the persons are responsible, etc., are all instruments through which the continued competence of personnel can be verified. Changes made periodically in the competency matrix of personnel may indicate that the organisation is proactive in maintaining and updating personnel performance levels.

Suppose it is necessary to train technicians in any process or unique process. In that case, the management must identify the training needs periodically. The management must provide training or take any other action to achieve the necessary competence. The training could be in-house or external, where such competencies can be developed in the person doing the job. Organisations should have an established procedure for training need identification with respect to the competency matrix of personnel employed in respective areas of work. Usually, this is done as part of all personnel's bi-annual or annual appraisals. HRD managers should have a consolidated plan for training personnel in the identified areas. That is the training plan for the organisation. Then, a calendar is prepared to conduct the training in-house or in external institutions where the training courses are available. The calendar is ready so that adequate personnel will always be available for the organisation's routine jobs without postponement of everyday work. That means there should always be a second line of competent personnel developing in the organisation for every process.

After providing internal or external training, feedback is usually obtained. It is required, from the personnel who underwent training, with inputs and suggestions for improvement of the type of course/training. The HRD manager or the training manager compiles this feedback, and appropriate actions are taken to improve the training/course for next time.

Training records may indicate that a training course has been completed. That does not mean the trainee has achieved the necessary competence. The person who underwent training needs to be kept under observation by the supervisor or manager for his continued display of competence during everyday work. This is to evaluate the effectiveness of the training.

Obtaining feedback from the personnel who underwent training is not the evaluation of the effectiveness of training. Training was provided to an employee to utilise him for a specified task or process. How good the employee is because of the training supplied to him, or how much he has benefitted or how much his training has been beneficial to the organisation in carrying out or improving a process for which he was trained – that assessment is the evaluation of the effectiveness of training. The review is generally done by the superior/higher authority in the hierarchy and reported as an appraisal record from the person's place of work to the HRD/personnel/training manager. The right way of evaluation is at four stages:

a. Reaction: How do the trainees feel about the training?

b. Learning: Did the trainees achieve the training objective?

c. Transfer: Can the trainees apply their knowledge and skills to the job? and

d. Results: What is the impact?/Was the training beneficial to the organisation?

If the training provided to the employee was found inadequate to meet the requirements of the product and the organisational objectives, re-training or any other appropriate actions are to be taken by the HRD manager.

'Just examine, will the organisation benefit if they keep a 'documented information/record of training and effectiveness verification of the training' in the following manner?

- *The internal Training plan indicated "To develop the skill of turning/milling". Three newly recruited employees having no experience attended from 0900 hrs to 1100 hrs on a date (xxx). Attendance sheet signed by all trainees and trainers,*

- *Review effectiveness – Evaluation method indicated as (discussions/issues raised during/after the session), Evaluated on the same training day.*

The external auditor raised a doubt: "Can the evaluation of the effectiveness of skill training (turning/milling) give a favourable result "found effective" within four hours of training on the same day?"

The organisation may not have realised the purposeless "documented information" unfolding its ineffective QMS!'

Persons doing work must be aware of the importance of their activity and how they contribute to achieving the organisation's objectives. Sincerity, discipline and dedication to the job will come to a person when the management considers him to be an essential link in the organisation. Personal attributes such as dedication to work and discipline also come from the awareness that the person is spending a day of his life for the organisational goals for an agreed remuneration. These matters grow as a culture in the organisation, which gradually encompasses all the personnel. Creating this environment is the responsibility of the management and function of the HRD department.

'As mentioned earlier, competency mapping of all personnel in the organisation would be a significant reference document for all training-related activities.

Some managers may say, "Why should I send them for training? They get trained and leave the organisation. The training does not add value to their processes; they enjoy and return. I don't believe in anything that does not add profitability to my business". Training is sometimes considered for 'production personnel only since they bring profitability' to the organisation or quality control personnel since they are the 'custodians of product quality'. These are all misconceptions. All personnel must be included in the training calendar based on the objective of human resources development and improvement in every organisation function.'

Large organisations employ competency mapping to understand how to deploy the competencies and strengths of workers. They may also use competency mapping to analyse the combination of strengths in different employees to produce the most effective teams and the highest quality of products and services.

National Aerospace Standard, NAS 410 establishes minimum requirements for the qualification and certification of personnel performing non-destructive testing (NDT), non-destructive inspection (NDI) or non-destructive evaluation (NDE) in the aerospace manufacturing, service, maintenance and overhaul industries. For example, to carry out different welding processes for aircraft parts made of metals, the persons doing the job should have approval from the regulatory authority for the type of welding and metals used. Regulatory authority issues a Certificate of Competency (C of C) to a welder employed to perform specialised processes, such as welding, non-destructive testing and electroplating. The competency matrix kept by the organisation for such personnel doing work affecting conformity of products should indicate the qualification of the person as eligible to carry out the process, periodic examination by the regulatory authority for the maintenance of the required skill level, records of following the correct process consistently and test results on the representative processed specimen/samples to confirm that the process is satisfactory. The product quality meets the desired requirements.

If the HR process is combined with administration in addition to Training, the following sub-processes are to be included:

a. Recruitment and selection of personnel based on the requirements or competency needs of the position in which recruitment is being carried out.

b. Post-recruitment training to personnel for orientation towards the functions and responsibility

c. There must be career planning for personnel inducted into the organisation. Everyone expects an improvement in their career and job satisfaction. If there is hope, they will stay. If encouraged and motivated, they will be more dedicated to their work. If they find greener pastures, they may leave for better prospects.

d. As a dynamic organisation, there must be continuity of excellence in everything the people in the organisation do. Therefore, succession is necessary to be established. Replacements for those who leave on superannuation or for those who go for better prospects are to be found and deployed without a gap in the activities of the processes. Performance evaluation and related records become the reference for selecting successors or promoting the positions for career advancements of personnel.

e. Rewards and recognition are how the people in the organisation are encouraged, satisfied and motivated to make the processes effective.

The sub-processes, like administration, recruitment, career planning, etc., can have their objectives/key performance indicators identified for which appropriate targets may be fixed. Measurements may be carried out periodically to ensure these sub-processes are effective.

AS9110 C also requires that the personnel performing maintenance, repair and overhaul services and aircraft release are qualified and certified per regulatory authority and customer contract requirements. Top management should ensure this always. Suitable training programmes are to be established to ensure that personnel performing specific tasks remain current regarding procedures, human factors, technical knowledge and applicable regulatory authority requirements. Initial training and periodic refresher training are to be planned and established. While refresher training is done periodically, the changes in the regulatory authority requirements and the organisational procedures (reviewed or improved) during the intervening period are to be included in the training programmes. Revisions or changes in the maintenance standards and systems/tasks of aircraft being maintained, repaired and overhauled are also to be included in the training.

Some of the specific customer requirements are as follows: ‘Unless otherwise specified, procedures shall be implemented to ensure that eye examinations, including visual acuity and colour vision, as applicable, are administered by a medically qualified/trained person to all individuals performing visual inspection. Intervals shall not exceed one year. Individuals shall be tested in at least one eye, corrected or uncorrected. Colour perception testing is required one time only. Individuals shall be capable of adequately distinguishing and differentiating colours used in the method for which certification is required. At the same time, the process is being performed during the inspection.

Specific additional requirements are also mentioned in the Standard AS 9110 Rev C. They are as follows:

a. Regulatory authorities have a system of certifying the personnel in an MRO organisation after assessment for their capability to carry out our MRO activities. They are certificated persons, and their certificate is valid for specific work on aircraft and a valid period. In addition, if there are persons in the organisation who carry out MRO activities without obtaining a certificate from the regulatory authorities (like persons under training), the organisation should have a system of assessment of these non-certified personnel for their capability to carry out work satisfactorily. This assessment should be carried out before assigning any work to them since they are non-certified personnel.

b. Personnel who finally release the aircraft or aircraft components after MRO activities in the organisation should possess the prescribed qualifications and a certificate from the regulatory authority for the aircraft type's release scope.
c. Organisations need to establish a continuation training programme to ensure that personnel performing specific tasks remain current in terms of aircraft maintenance procedures, human factors and technical knowledge and
d. When continuation training is provided to personnel in the organisation, it should be ensured that the training covers changes in relevant regulatory requirements, the organisation's procedures and the maintenance standards of the type of aircraft or aero-engine being maintained.

Regulatory authorities have a system of issuing periodic regulations or advisory notes for every specific aircraft type and aero-engine. It emphasises that the personnel working on the aircraft/aero-engine must be updatedin their technical knowledge about the class of aircraft/aero-engine and regulations or regulatory advice. The regulatory authorities either conduct initial training or refresher training. Institutions approved by the regulatory authorities sometimes conduct such training based on specific criteria for approval.

7.3 Awareness

All personnel performing various processes in the organisation must be aware of the organisation's quality policy. Every function has one or more identified objectives/KPIs, which must be known to those working there. All personnel must be aware of their contribution to achieving conformity to the process, i.e., realisation of planned activities and effectiveness of the processor, in other words, achieving planned results/objectives. The employees become aware of the benefits of improved performance through the organisation's progress. When all the processes in the organisation are conforming and practical, the Quality Management System is effective. In such a scenario, there would be no complaints from the customer, delivery of products/services would happen as planned, no internal rejections or re-work, the management would be happy they got repeat orders, and the employees would get better bonuses! All the happiness and peace fade when the Quality Management System is not sustained as effectively and continual improvement is not forthcoming.

Access to Quality Management System documentation shall be provided to all personnel by the organisation. It must be ensured that personnel are aware of the documented information, i.e., all documents, procedures and records relevant to them and their processes. Any changes introduced to the documentation must also be brought to the notice of all personnel.

It will be a good practice to highlight all the steps, activities and controls to be exercised by everyone in the organisation and their processes in particular. The specific information may be a work instruction, a flow chart or anything else that helps the person/persons execute the process carefully and meticulously. It is to ensure that the process is carried out in a planned manner and the activities are fully realised in conformity with the documented information. The specific details highlight the performance results of the process, and the process owner will be able to measure the results and monitor their achievement, i.e., the process's effectiveness.

By definition, safety is freedom from those conditions that can cause injury, illness or death; damage to or loss of equipment or property. Or harm to the environment. Employees must be aware of the safety features designed and manufactured on the product. There may be some safety issues that the employees must consider during the execution of their work. The most important contribution of employees to product safety is to comply with all safety rules during the process. The safety equipment/practices appropriately used during work are:

a. Wearing personnel protective equipment (PPEs) like helmet, gloves, goggles, protective apron/clothing, appropriate shoes, respirators, harness

b. Ensuring the safety of handling equipment like cranes, forklifts, trolleys, trays,

c. Careful handling of monitoring and measuring equipment, protecting it from damage

d. Machine platforms, ladder, overcurrent protection for equipment and conductors

e. Compliance with the relevant statutory and regulatory requirements intended to promote safety by eliminating or mitigating conditions that can cause injury, damage or death, damage to or loss of equipment or property or damage to the environment.

The organisation ensures that its employees know the importance of ethical behaviour. Ethical behaviour is how the people in the organisation, wherever they are, work and interact with customers, suppliers and colleagues across the business in such a way that helps it to shape and strengthen its organisational culture, principles and values. The management is aware that ethical behaviour is good for business.

The organisation issues a code of ethics and policies for the employees to read, understand and adhere to. Every organisation wants employees to follow the highest ethics, honesty and integrity standards. It recognises that every employee is essential to achieving the organisation's objectives. Human rights are fundamental principles which allow an individual to lead a dignified and independent life, free from abuse and violations. The organisation will not tolerate human rights abuse within any part of its business or supply chains. Disciplinary procedures will be rigorously applied for gross misconduct. Employees may be asked to sign an agreement for non-disclosure of information.

It is the organisational policy to promote an environment free from discrimination, harassment and victimisation where everyone will receive equal treatment. The organisation has a responsibility under legislation to protect the health, safety and welfare of all its employees, customers and contractors on their premises. All employees are responsible for complying with safety rules and procedures and taking reasonable care of their health and safety and that of others who may be affected by their actions or omissions, e.g., customers, colleagues, visitors and suppliers.

Global principles of business ethics for the 'Aerospace and Defence Industry' based on best practices include implementing the policies through appropriate communication and training, zero tolerance for corruption, managing conflict of interest, Respect for proprietary information and applying dissuasive sanctions for evidenced cases of non-compliance.

7.4 Communication

> *'The most important body part of a person in power, be it a king or a king's mistress or a chief executive officer or his strongest assistant, is the ear. Win access to the ear and whisper the right message, and your influence will find its true measure'.*
>
> *- James O'Shea and Charles Madigan, 'Dangerous Company'.*

Good communication is about getting the right message to the right person in a suitable medium at the right time. One who communicates (wants to exchange or obtain or pass on information) decides what to communicate, when to communicate, whom to communicate (internal or external parties) and the medium for communication. Suppose the data is related to the Quality Management System in the organisation. In that case, the sender may have the authority and responsibility to do so from the management. The communication is effective if the recipient understands the information and responds appropriately.

The top management is responsible for communicating with people in the organisation concerning the quality policy and objectives, the importance of meeting customer and statutory and regulatory requirements and promoting awareness of customer requirements throughout the organisation. Personnel in the organisation have defined responsibilities and are aware of the relevance and importance of their activities and how they contribute to the achievement of quality objectives.

The organisation's management should facilitate effective communication among all the employees regarding the quality policy, objectives and achievements, and customer requirements. That should motivate personnel in the accomplishment of the quality objectives. Management should encourage communication among the personnel and invite feedback from personnel to involve them directly. The process must be effective throughout the organisation. It is necessary to ensure that the subject of communication is related to the effectiveness of the Quality Management System. Examples of internal communication media could be e-mails, circulars, notices, workgroup meetings, bulletin boards, meetings led by management, quality alerts, trends, bulletins and advice issued by appropriate authorities. There can be not one but several ways for effective communication to take place.

An effective communication process contributes to the success of the organisation's Quality Management System. Many problems that occur can often be traced back to poor communication. Defining and implementing an effective and efficient process for communicating the quality policy, requirements, objectives and accomplishments is necessary. This helps the organisation improve its performance and allows people to get involved in achieving quality objectives.

Appropriate communication processes that may be established within an organisation include identification of the process owners between whom the communication should occur; information to be communicated; means of communication, the method selected to monitor its effectiveness; and documentation and records necessary to verify that the communication has occurred and is continually improving. Some of the following means can demonstrate an effective communication process:

a. The top management, employees at all levels in the organisation and the suppliers all generate, receive and respond to communications;

b. The information to be communicated is clearly defined, appropriate and accurate to the purpose of the communication;

c. The means used for communication is appropriate to the literacy and other skills of those expected to receive and act upon the information provided;

d. Monitoring takes place to ensure that the information communicated is acted upon and the desired outcome is achieved;

e. The corresponding procedures, if any and the records maintained demonstrate that the communication has occurred; it is effective and is continually improving.

There are various means of communication in a dynamic organisation. Adequate communication facilities will be provided depending on the need to communicate with customers, suppliers and other stakeholders. The infrastructure facilities for that may be telephone, electronic communication such as e-mail, intranet and websites. In contrast, internal communications are through in-house magazines/newsletters, individual notices or letters, management and employee interaction meetings, etc. There may be formal or informal management-led communication in work areas, team briefings and other meetings, such as those for recognition of achievement meetings, production review meetings, management review meetings, etc. Records are maintained for formal meetings like MRM and production review meetings. Shop floor meetings conducted by the top management to resolve day-to-day issues are generally not recorded, being an internal communication process. Proactive management does not find faults with the process owners but tries to understand the gravity of the situation and facilitate solutions, or work for solutions, in the light of the Quality Management System, where a solution is indicated for every problem. It is not finding faults or firing that helps the organisation to grow but finding corrections and corrective actions that lead to improvement. It depends on how effectively the management can eliminate the root cause of day-to-day problems. Whether the meetings are formal or informal, brief records of such meetings, giving a gist of the points discussed, could be beneficial for future reference. That could convert the notepads carried by the process owners into a valuable record of effective communication within the organisation. When tasks are allotted to personnel, though there may be personnel preferences, management needs to ensure that personnel within the organisation work as a team and not in isolated compartments.

The effectiveness of the organisation's internal communication processes cannot be determined solely from one source. It is the combination of all means of communications recorded that can be visible to a third party, for example, correspondence, minutes of all meetings, evaluation of training and induction programmes within the organisation, the feedback mechanisms within the organisation, the current status of communications displayed on notice boards, awareness of Quality Management System status to the employees, actions taken on identified nonconformities during internal and external audits, etc.

Communication sent to the customers with accurate, precise, appropriate language, replies received for the desired actions from customers and suppliers, etc., are all visible testimonies of effective external communications. Telephonic conversations with customers/suppliers, especially regarding product-related or changed requirements, are to be recorded on file and confirmed to the customers/suppliers through a letter or e-mail. Here, there is no question of doubting the integrity of the customer/supplier, who found it easy and quick to communicate on the telephone. There is no doubt about the integrity of the person who recorded the customer/supplier input. Moreover, there is objective evidence that such communication occurs when the organisation confirms it through a letter or e-mail. The letter-mail is a record for future reference.

7.5 Documented Information

7.5.1 General

Generally, all the documented information of the Quality Management System could be placed at four levels depending on all the required documentation necessary by the organisation to ensure the effective planning, operation and control of all its processes. The pyramid structure indicated vide Figure 7.1 (Document levels I to IV.) indicates the documented information at four levels.

Here again, some may doubt about how much documentation is needed. The standard answers are as follows. The extent of the Quality Management System documentation can differ from one organisation to another due to the following factors:

a. the size of the organisation and type of activities

b. the nature of the organisation's processes, products and services

c. the complexity of operations and their interactions and

d. the competence of personnel.

The best practices suggest that the documentation should be crisp and minimal. Documented Information clearly defines the procedures and avoids

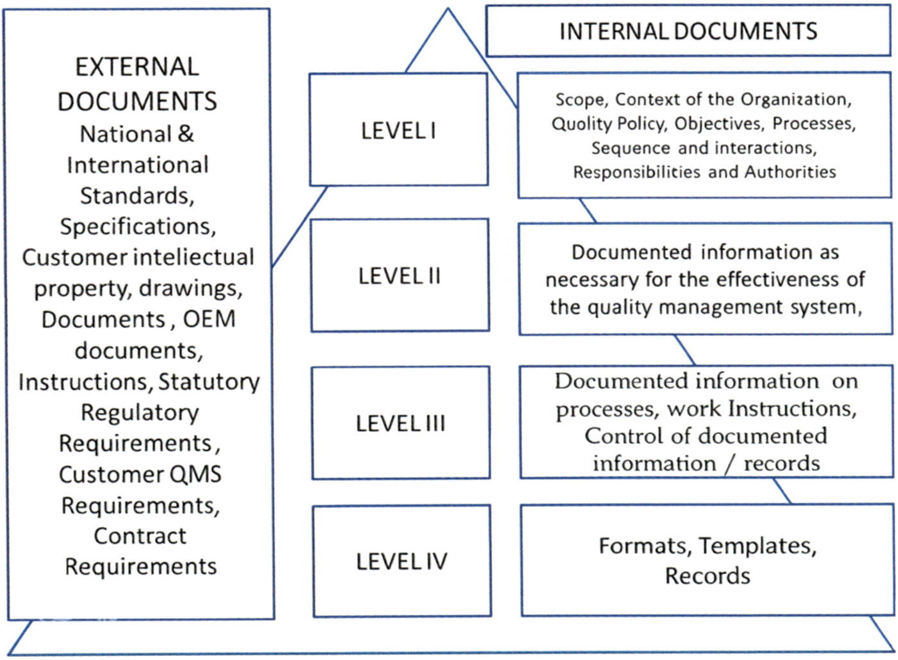

Figure 7.1 Document levels I to IV.

any doubt about acceptable practices when faced with unfamiliar situations or risks. Documentation is a tool for the communication of information. Documentation also means documents and records; the word documentation is replaced by documented knowledge per the revised standard ISO 9001:2015. Records are needed to show evidence of conformity through actual practice concerning what was planned.

When ISO and similar QMS requirements and practices were initiated three decades ago, there was a saying: 'Document what you practice, and practice what you document'. Quality Management Systems for aviation, space and defence sectors are very stringent and vital to ensure safety, quality, reliability and maintainability; apart from meeting statutory and regulatory requirements, there must be a modification to the saying – 'document what you practice, and practice what you document, but practice you shall, as required by the Aerospace Standard'.

Meticulous practice is more essential, and practice must benefit us in achieving the results and effectiveness of processes. It is another best practice to maintain all documented information in soft copies, with read-only access to the personnel in the organisation. Distribution through local area networks ensures it is available to all. It avoids printing copies on paper and issuing several copies. Moreover, it reduces the use of paper, leading to fewer trees being cut, thus saving the environment in the long run. While establishing the documentation, all personnel in the organisation must be taken into confidence, as they are the ones to understand the information and put it into practice in their respective processes. Not only that, but they are also the ones to show evidence of the practices through a special kind of documentation, which we may call records.

The term 'documented information' that appears within the international standard at several places means the information that needs to be established, documented, implemented and maintained or retained; i.e., some documents are maintained, and others are retained. The documented information retained is called records.

A single document may address the requirements for one or more information. The requirement for documented data can also be covered by more than one document. In short, it means that standalone information for a need may be documented, or one document may contain more than one set of written data.

The word 'shall' means mandatory, and therefore, establishing the following documented information, either defining the methods of implementing the requirements (documents) or retaining the documented information (records), is mandatory as per the Standard.

7.5.1.1 Mandatory QMS documents

The following are the mandatory documented information (maintained as documents) as per the Standard:

a. Context of the organisation that includes a general description of relevant interested parties (Clause 4.2 a)
b. Scope (Clause 4.3)
c. Operation of its processes, including processes, description of interested parties, scope, definition of processes, sequence and interactions, responsibilities and authorities (quality manual) (Clause 4.4.2)
d. Quality policy (Clause 5.2.2), Objectives (Clause 6.2.1)
e. Identify documented information for effective QMS (as required)
f. Control of documented information, including records (Clause 7.5.3.1)
g. Quality plan maintained (Clause 8.1)
h. Configuration management maintained (8.1.2)

i. Design and development requirements are met – Considerations (8.3.2)
j. Production records, product details, drawings and specifications (8.5.1)
k. NC control process (8.7.1)
l. NC and corrective action management process (10.2.1)

7.5.1.2 Mandatory QMS records

The following are the mandatory documented information (retained as records) as per the Standard:

a. Operation of its processes completed as planned	(Clause 4.4.2)
b. Fitness of purpose MandM Resources, Calibration reports	(Clause 7.1.5.1, 7.1.5.2)
c. Competence,	(Clause 7.2)
d. Production operation, planning and control	(Clause 8.1)
e. Review of requirements	(Clause 8.2.3.2)
f. Design inputs	(Clause 8.3.3)
g. Design review, verification and validation	(Clause 8.3.4)
h. Design outputs	(Clause 8.3.5)
i. Design changes	(Clause 8.3.6)
j. Supplier evaluations	(Clause 8.4.1)
k. Supplier records, retention periods	(Clause 8.4.3)
l. Inspection records	(Clause 8.5.1.c 1)
m. Validation of special processes	(Clause 8.5.1.2)
n. Production process verification (FAI)	(Clause 8.5.1.3)
o. Traceability	(Clause 8.5.2)
p. Customer property	(Clause 8.5.3)
q. Control of changes to production	(Clause 8.5.6)
r. Release of products	(Clause 8.6)
s. Control NC and results	(Clause 8.7, 9.1.1)
t. Audit and audit results	(Clause 9.2.2)
u. Results of management reviews	(Clause 9.3.3)
v. NC and results of corrective action	(Clause 10.2.2)

7.5.1.3 Other necessary documents and records

The organisation is at liberty to maintain any other documents or records as deemed necessary and continue to call them documents and records. They don't need to call it 'documented information' as given in the Standard.

Establishing and maintaining a process or determining the methods of implementing the processes and controlling them to get the desired effects are mandatory for most of the requirements as per the Standard, though specifically not called for documentation. In addition to those documents and records indicated above, the best practice is to document and establish the QMS practices to achieve uniformity throughout the organisation for the following essential processes, especially when the organisation is large. The implementation of the functions is complex.

a. Actions to address risks and opportunities (risk management – planning and operational)
b. Configuration management
c. Product safety
d. Recall of measuring devices for calibration, verification and status identification (monitoring and measuring resources and measurement traceability)
e. Operational planning and control of the production processes
f. Performance evaluation: customer satisfaction, effectiveness of the quality management system
g. Contract review (determining the requirements for products and review of requirements)
h. Design and development of products
i. Control of externally provided processes
j. Production process verification

The organisation ensures the availability of relevant Quality Management System documentation at the location of the process, and the process owner shows them to appropriate interested parties when required.

7.5.2 Creating and updating

Several sources of documented information (internal documents and records) are generated within the organisation, such as quality manuals, procedures, departmental manuals, work instructions, various forms and formats for creating records, quality plans, inspection instructions, etc. Some of the product-related documents may also be generated internally.

S1000D (Issue No. 4.1) is an International specification for technical publications using a common source database developed by the Aerospace and Defence Industries Association of Europe (ASD) for use with military aircraft and later modified for use with land, sea and commercial equipment. The United States' Aerospace Industries Association (AIA), the Air Transport Association (ATA) and national industry and defence representatives from several countries maintain the standard.

Technical publications of indigenously designed and developed ASD equipment can follow the above specifications (free download from http://public.s1000d.org/Downloads/Pages/S1000DDownloads.aspx)

Whenever a new document is generated within the organisation, the following points are to be ensured on the paper:

a. Identify the document by a name or title and a reference number which belongs to the organisation,
b. Identify the organisation or authority issuing the document,
c. Every document issued by the organisation must have a format which identifies the size, shape, medium on which the document is issued (print or electronic), language, revision status or version of the document when issued for the first time,
d. Identify the authorised person who reviewed the document for suitability and adequacy and indicate the method of review,
e. Identify the authorised person or authority that approved the document for issue and the method of approval indicated by date and signature on hard copy (master) or authorised signatory or by electronic signature,
f. Whenever any update is issued to the original document, all the points from (a) to (e) above must be repeated,
g. A centrally managed periodic Publication Amendment Management List (PAML) is issued to verify that all the amendments (Internal and External documents) are incorporated.

7.5.3 Control of documented information

7.5.3.1 Control of documentation

The standard requires that all documents created or generated within the organisation and updated subsequently be controlled. Control of documented information establishes a planned activity and desired results.

> *'You have a look at it. You only looked at it twice, looking for evidence of its control.*
>
> *The document holder says, holding a printed copy, "What control? This is the right document. I got it Xeroxed from the manager. We all use it; I shall try to get it when there is a revision."'*

Whenever you get hold of a drawing, a process specification or a product-related communication from the customer, your common sense warns you to ensure that it is the correct, complete and current document for production. What is the risk if it is not the 'right document'? What can go wrong? Does it doubt the integrity, intelligence or sincerity of the person using the document? What objective evidence is available to ensure it is the correct and relevant document? You can trust the beholder, but that does not give evidence. One of the most prominent concerns found in most organisations (using hard copies of documentation) is failing to understand the type of documents they use and the importance of controlling the same, which are available at different locations within the organisation, especially in planning and production areas. Product-related documents like drawings, specifications, customer-furnished data or information, physical characteristics or performance characteristics of products near test rigs, sampling plans, test procedures, etc., are found pasted on walls, machines, under glass on tables, personal files and cupboards in production/inspection areas. The documents may have no identification of control. Therefore, it cannot be ascertained whether the papers' applicable/latest/valid versions are being used for product realisation. Even obsolete and old versions of the documents are kept and referred by personnel for their routine use without regard to the consequences, which include nonconformity on the products as per the revised and applicable requirements of the product or purchase orders. Personnel keep obsolete or un-controlled/unidentified copies of essential documents for their study and reference without marking them as outdated and not caring to update the information as per the latest internal or external document versions. As and when amendments are issued to internal documents generated within the organisation, it is necessary to ensure that the pre-revised documents or pages are withdrawn and identified as obsolete. When critical information is distributed, actions may be taken. Still, amendments to the relevant documents are quickly forgotten, which can cause problems later. Mistakes and nonconformities on the products inadvertently happen when documents are not controlled. The customer will not accept any excuse for the lapse. There have been customer complaints, such as the part was not manufactured per the latest rev status of the drawing (not as per the part number and rev status of the drawing indicated in the customer PO).

> *'In another case, the process carried out by the external provider was not to the latest rev status of the specification (For example, say Black anodise as per specification MIL-A-8625, (plate to latest rev F was not mentioned in the PO, and not certified by the external provider). While "The organisation shall be responsible for the conformity of all externally provided processes vide (8.4.1)", the organisation did not have a copy of the specification to show or share it with the external provider! Referring to the latest rev status of the specification comes next!'*

It is necessary to make sure that all documents are controlled. Verifying that the document control process is systematic and practical is essential. All the documents (external or internal) are identified as master copy or controlled copy, current, valid for the relevant customer order, authorised and issued for reference or use, evidence of, which can be shown, verified and available.

There are documents from outside the organisation, such as national or international standards, other publications, reference books and customer-given documents. Let us call them external documents. Some product-related documents may be external, if generated outside the organisation, before being brought inside. Suppose the product-related documents were from OEM, external sources, licensors or customers. In that case, these documents will be treated as external product-related documents. Product-related documents include specifications, drawings, bills of materials, process specifications, test schedules, quality plans related to product measurements, assembly instructions, installation details and test schedules. The organisation is not empowered to amend, change, update, withdraw or re-issue any external or product-related documents without the customer's or external sources' permission. When there are different divisions of the same organisation, each will be treated as an independent organisation where the scope of the Quality Management System or certification is relevant. Therefore, if a document is issued from one division of the organisation and used by another, the document shall be treated as an external document. Therefore, no document amendment, change, update or re-issue is permitted by the user division unless authorised by the source division or all divisions' corporate offices.

So, it places some extra responsibility on the organisation to define those external documents that are important for their business operations.

This change may affect their process/operation. The control levels depend on the document's importance to the process and how the change impacts the customer or product.

7.5.3.2 A control of documents

Defining the controls necessary for the documented information/documentation is by documenting as:

a. Who has the responsibility to (by office/designation, not by name) approve the documents for adequacy before issue?
b. Who reviews updates and re-approves the documents as and when necessary?
c. Who ensures that changes and the current revision status of documents are identified on the documents (replacing revision status or version of the document in the original format)?
d. Who has the responsibility to ensure that relevant versions of applicable documents are available at points of use?
e. Who ensures that documents remain legible and readily identifiable, even after revisions?
f. Who ensures that documents of external origin, determined by the organisation to be necessary for the planning and operation of the Quality Management System, are identified and their distribution controlled? and
g. How to prevent the unintended use of obsolete documents and apply suitable identification if retained?

It is desirable to list the internal QMS-related documents, external documents and documents related to the products/services (internal and external) separately. Apply the control procedures as described above. When a distribution is in soft form, all documents distributed must be in 'read-only' format so that no editing is possible by the reader/user. The process owner or the system administration uploads the internal documents to the server after the head of the organisation has approved the same and duly authorised by MR for uploading to the server. If anyone requests a change, the change is reviewed and approved by the designated authority. When the head of the organisation or MR gets approval for the revision of the documents, the same will be uploaded to the server after withdrawing the earlier version. Suppose the original documents are printed on paper before distribution. In that case, MR will retain the signed/approved first copy as the 'Master Copy', and all other copies will be identified as 'Controlled Copy when in Red' with the date of issue and signature, Refer to Figure 7.2 Document control stamps. The usual practice is to mark them with a stamp identifying the organisation and using a red ink pad. Distribution of controlled copies will be noted in a register with acknowledgement so that when amendments or changes are made in these documents, the MR will take action to circulate the amendments to all controlled copy-holders. MR will do so after completing the approved changes in the Master copy. All amendments will be chronologically recorded in the amendment record sheet of all the documents. Soft copies, when printed or hard copies, when xeroxed, are to be treated as uncontrolled documents and shall not be used or retained for work related to the product/maintenance of the product. Documents in the system are usually given a watermark stating that they are 'uncontrolled when printed'. However, for day-to-day work/reference, if a print is necessary, ensure that the printed document/extract is disposed of at the end of the day. Some organisations specify 10–15 days for the expiry of control for the printed copies. If uncontrolled copies of documents are used, inadvertent errors or acts of omission cannot be avoided, which may consequentially affect the quality of the products/service. All documents that are superseded or obsolete later must be marked as such and may be preserved for reference, separately if required, away from the workplace. Distribution of soft copy documents is preferable since it avoids a lot of hassle.

Whether the organisation is big or small, someone should be responsible for controlling internal and external documents (usually the planning department), whether they are product-related documents or otherwise, and that comes to the organisation from outside. The responsibility for control of internal documents related to the products may be that of the design process, while the responsibility for external documents related to the products may be that of the planning process. The planning process may control customer-given drawings and other documents. In small organisations, the responsibility may be rearranged for convenience and ease of control. The persons responsible for taking action should periodically check all the areas for the availability of documents, withdraw uncontrolled documents or documents with unidentified statuses and take appropriate action to control and make relevant documents available to the users.

When hard copies of internal documents such as drawings, specifications, test schedules, quality plans, etc., are issued to external parties like suppliers/external providers/outsourced process/service providers, customers and other stakeholders, the documents are to be identified, controlled and cross-referred in the communications including purchase orders being issued.

ABC Engg Pvt Ltd
MASTER COPY
ORIGINAL
NOT FOR RE-ISSUE
Date Received..................

ABC Engg Pvt Ltd
CONTROLLED COPY OF MASTER WHEN IN RED
Date Issued.................
Xerox copy is controlled only when this stamp is available

ABC Engg Pvt Ltd
OBSOLETE COPY FOR REFERENCE ONLY
DO NOT USE FOR WORK
Date Withdrawn......................
PLEASE ASK FOR CURRENT VALID DOCUMENT

Figure 7.2 Document control stamps.

Being the intellectual property of the organisation, there should be an established procedure for the return of these documents to the organisation after their use by the suppliers if not permitted to be retained by them. Control of suppliers, who deal with the organisation, includes control, care and undertaking non-disclosure of information in the documents issued to them.

All electronic documents are also subject to the control procedure. Organisations need to ensure adequate review, approval, publication and distribution of their management system documentation and other product-related internal documents in electronic formats. There should be consistency in the methods for developing and modifying electronic records, which also needs to be specified in the documented information. When documented information is managed electronically, the system administration must take care of data protection processes such as backups for preventing losses, preventing unauthorised access or access to only designated locations, anti-virus protection for the files or programs and any physical damage.

Document modifications by users are to be prevented, and obsolete documents are to be withdrawn or deleted or identified as outdated and located separately. Available features of the software should not interfere with the control procedures established. Therefore, read-only access is generally given to the documents in electronic form. Care is to be taken to identify the revision level of the documents while uploading the documents to the server. Documents in electronic media may be controlled through individual passwords for uploading/editing and read-only access. Even print/edit/copy permission can be denied through software settings. When print is allowed, it is a practice that the hard copies are treated as uncontrolled beyond a period of check/servicing of the aircraft, say for 15 days from the date of receipt/ print of the documents from the customer.

Electronic communication with external parties like suppliers, customers and other stakeholders may involve the exchange of documents. A non-disclosure agreement with the parties is desirable when the organisation's internal records are shared in soft or hard media.

7.5.3.2 B control of records

A record is a document that states results achieved or provides evidence of activities performed or events. While implementing the Quality Management System in the organisation, it is necessary to provide evidence of meeting the quality requirements of the product you provide to the customers and the proof of effective implementation of the Quality Management System you practice within the organisation. Records you create give that evidence to the interested parties. Records must be legible and straightforward. Therefore, no corrections or overwriting are permitted while they are made, approved and accepted. The organisation needs to define the controls established for creating the relevant records, identifying them and storing them for future reference. Store the records in a medium so the media does not get corrupted or lost, protecting them from damage or deterioration. Retrieval of the records from storage must be possible as and when necessary. Organisations should decide on the period of retention and, after the period, how the records are going to be disposed of permanently. These activities are the controls that the organisation establishes on product-related records and Quality Management System-related records. For effective and uniform practice within the various sections or departments of the organisation, these activities must be defined in documented information.

The records can be created and stored in electronic media so that, in the long run, the records do not get piled up and deteriorate as in the case of paper records. The retention time can be as long as the product's lifecycle. Irretrievability is better in the case of electronic storage of records. Records in electronic media may be controlled through individual passwords by the creators of the record. It must be remembered that when changes are made in computer systems software, retrieving records created by earlier software versions must be possible.

When the organisation receives products or bought items or supplies of raw materials, components, equipment or services from various suppliers, it is the practice that the organisation demands and receives several records such as test or conformity certificates, warranty certificates, release notes, invoices, waybills, etc. The suppliers create these records, and some are shared with the organisation as evidence of conformity to the product or service requirements they provide as part of fulfilling the contract. The organisation needs to define the controls exercised concerning these records received from suppliers. The controls are generally like those exercised by the organisation for their records. The organisation's contract with the suppliers should define the controls necessary for the records retained by the suppliers. It is essential to monitor the quality of the products or services given by the suppliers throughout the product's life and call for evidence of supplied quality when required by the customers during the use of the product. Aerospace suppliers are, therefore, unique because they retain the relationship with the organisation throughout the life of the product or services they provide. When necessary, the long-term connection is contractually bound.

Records assure the user that the machine/equipment they use will be reliable and safe. Therefore, the customer may be interested in reviewing them, and the organisation must provide for the review as and when requested. The non-availability of past production or maintenance records for reference has sometimes caused inconveniences for the investigation teams in establishing objective and conclusive evidence for fatal aircraft accidents. Crucial proof of installation was lost because of not retaining the records of the production of an aircraft for five years since the system of retaining the records of production was not implemented by the organisation. The customer or the regulatory authority has not specified the requirement for record retention. That brings up the importance of retaining records for adequate periods.

Organisations are always doubtful about determining how long the records retention period should be. The standard does not prescribe any period. Organisations may keep the retention period as the minimum feasible. However, if specified by the customers, the retention period requirement prevails. It may generally be seven years for records of structural parts and the life of the aircraft for other equipment, ratables and the aircraft. Based on how long the organisation must show evidence of having completed the process as planned or as required to the customer or other authorities, the period of retention of records must be decided. The decision should be to convey the confidence and records of the organisation that manufactured the product according to the requirements, if the customer wants it or expects it to be available with the manufacturers.

Responsibility for control of records may be assigned to the respective process owners who create the documents as evidence of completing their activities concerning the product or Quality Management System. MR generally makes records in connection with the management review and internal audits. Records of measurement, analysis and improvement activities in connection with the Quality Management System are gathered from the respective process owners for submission to the management review meeting for further discussion and deciding on the course of action.

Adequate resources will be provided for storing and retrieving the records after a specific retention period. Organisations may need dedicated IT resources to ensure the Enterprise Resource Planning (ERP) system operates efficiently and effectively.

Examples of records to be retained are listed in para 7.5.1.2 above.

Some of the specific customer requirements are indicated as follows (Please confirm the retention periods from your customer/regulatory authority):

Electronic imaging/microfilming of records instead of storing actual inspection records is permissible. All electronic documents must be controlled, retained and retrievable per the requirements identified for hard copy records. For electronic records that are transferred from computer files, the storage media must be capable of maintaining the data integrity for the entire retention period.

The following identified quality records shall be retained for the minimum retention periods specified below: -

a. 40 years for flight safety, critical/major rotor parts (i.e., turbine and compressor disks, hubs, shafts, free turbine couplings and turbine disk side plates), space shuttle fuel cells, as well as serialised major engine (cast/fabricated) cases, (i.e., inlet fan, compressor, intermediate, diffuser, combustion, turbine and exhaust cases), and main shaft bearing supports, which are not integral to a significant case and engine components traceable by Engineering Drawing/Quality Assurance Data required serial numbers

b. Ten years for Records of castings or parts shall be retained only if no other inspection record, such as completion and final acceptance of radiographic inspection, is retained.

c. Five years for Military hardware – Turbine airfoil (blades) casting radiographs for initial casting quality.

8

Operations

The standard requirements/clauses are arranged in accordance with the bigger picture 'PLAN-DO-CHECK-ACT' Cycle (Deming's Wheel). Chapter 8 is the 'DO' section or how the 'PLAN' is executed/implemented. The requirement details are indicated in the respective sub-paras of the chapter.

8.1 Operational Planning and Control

Operational planning, implementation and control for the processes in the organisation are dealt with in this Operations clause.

Processes in the organisation are identified as per para 4.4.1.

For example, let us say the identified processes are as follows:

1. Management processes
2. Contract review process (project management)
3. Design and development
4. Purchase
5. Planning
6. Production
7. Quality control
8. HR, training and facility management

For each process, a process map can be prepared. You call the process map a 'Turtle Diagram', 'Planet Diagram' or a 'Schematic Representation of a Single Process' as indicated in Figure 8.1.

All the inputs, activities, output and relevant information can be incorporated in the 'Turtle Diagram', 'Planet Diagram' or 'Schematic Representation

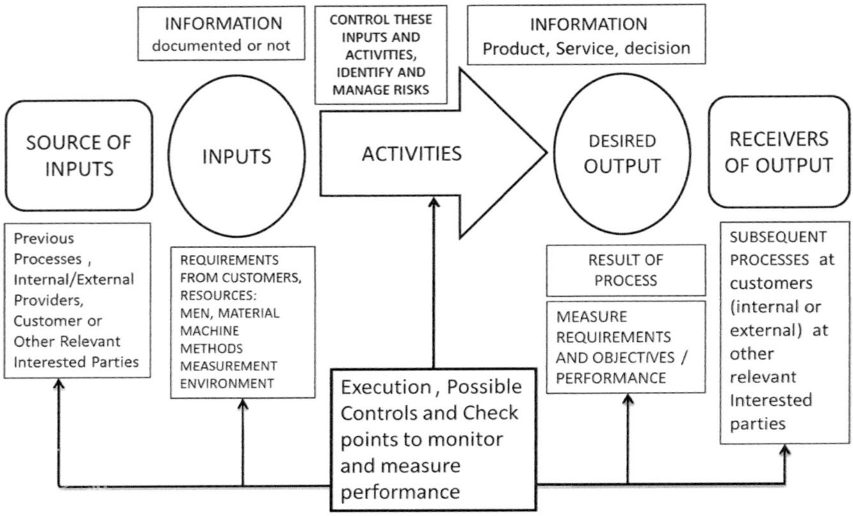

Figure 8.1 Schematic representation of a single process.

of a Single Process' as indicated in the example below for one representative process, say 'Purchase'.

The process information is determined as per clause 4.4.1 (a)–(e), and in the representative process 'Purchase', it is populated with the process information as follows (Refer to Figure 8.2):

a. Determine the inputs required:-
   - Evaluation and selection of suppliers/external providers,
   - Request quotations from them as per the request for purchase (size, quantity) from planning,
   - Material specification from customer drawing,
   - Requirements from customers,
   - Resources: Manager, computer, internet, information on material specification, source, quality, grade, quantity, size, quality plan, purchase process, --PO to external providers, evaluation of supplier performance;
   - Information to quality for verification of purchased product, -measurement process and controlled environment, and
   - The outputs expected from these processes:
   - Purchased/ received correct material meeting requirements.

b. Determine the sequence and interaction of these processes:
   - Inputs, activities such as checking material specification, source, quality, grade, identify and manage risks

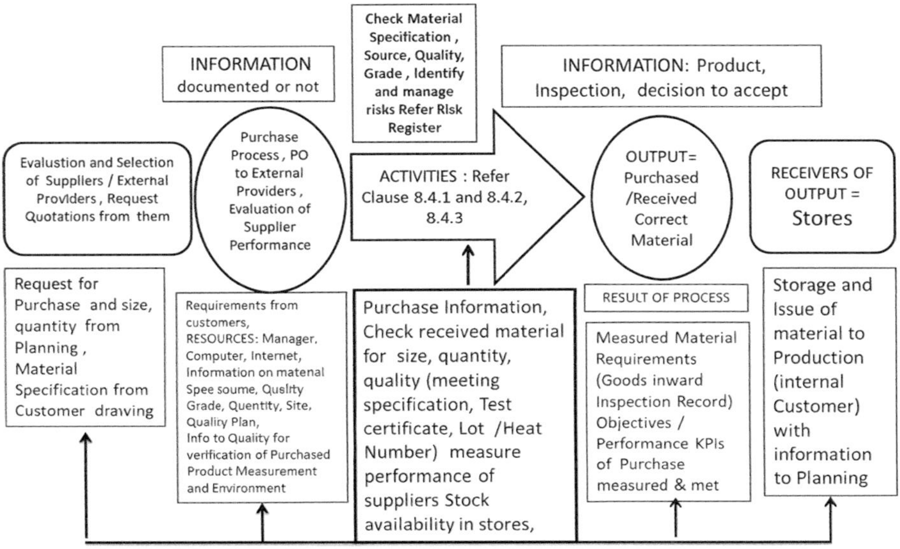

Figure 8.2 Example process – purchasing.

- Refer to risk register; on receipt of materials as per purchase information,
- Check received material for size, quantity and quality meeting specification, test certificate, lot/heat number (for traceability) and measure the performance of suppliers (for activities, refer to clauses 8.4.1, 8.4.2 and 8.4.3 of the Standard).
- The receiver of output is stored, and subsequent processes are storage and issue of material to production (internal customer) with information to planning.

c. Determine and apply the criteria and methods (including monitoring, measurements and related performance indicators) needed to ensure the effective operation and control of these processes: Measured material requirements (Goods Inward Inspection Record), objectives/performance KPIs of purchase: measured and met.

d. Determine the resources needed for these processes and ensure their availability: Manager, computer, internet, information (request for purchase) on material spec, source, quality grade, quantity, size, quality plan, stock availability in stores, follow the purchase process.

e. Assign the responsibilities and authorities for these processes: Process owner (the Manager).

Activities as per clause 4.4.1 (f)–(h) must be completed for all the processes after completing actions vide clause 4.4.1 (a)–(e).

Clause 4.4.1 (f) requires organisations to address the risks and opportunities and implement them as determined in accordance with the requirements of clause 6.1 (Actions to Address Risks and Opportunities). The same has been discussed in para 6.1. The resulting documented information – 'the risk register' – indicates all the risks identified and assessed and shows actions completed to reduce the risk to an acceptable level. Opportunities identified have been used for continual improvement of the processes.

We have already discussed the other actions planned to identify the quality objectives and plan to achieve them vide para 6.2. We also discussed the considerations when planning changes in the Quality Management System vide para 6.3 of the Standard.

Clauses 4.4.1 (g) and (h) require the organisation to evaluate and improve the process. Evaluation of the above operations individually (functional level) gives the evaluation result, whether the procedure has achieved its intended outcome. The result is the effectiveness of the process, and the performance results measure it.

Ensure that the process conforms to the Standard's requirements, i.e., all the planned activities have been realised. Identify the process objectives/key performance indicators (KPIs) and measure the planned results of the process against set targets for a specific period. If the process conforms and the results follow the plan, the process is effective. What remains to be done is to put our efforts into continually improving the process and the Quality Management System as a whole. Suppose the planned results are not achieved for any particular process. In that case, actions to improve the process will be immediately identified and implemented.

All the processes in the organisation can be schematically represented similarly to the 'Purchase' process. A description can also be written for the process, including all the information. The description is commonly known as a procedure – a specified way to carry out a process. If the procedure is written or documented, it may be called a documented procedure or documented information.

When all the processes are arranged sequentially, 'Plan-Do-Check-Act', we get the 'Business Process Map' or 'Sun and Planet Diagram' depicting all the processes in the organisation.

The stage is set, the entire Quality Management System framework is implemented and the business processes are ready to be integrated into the Quality Management System. The customers and interested parties will identify the specific product/service requirements. The operations are to be carried out by the organisation. The product/service has to be delivered to the customer, meeting all requirements. This is for deriving customer satisfaction and business excellence to achieve effectiveness for the processes and the Quality Management System.

Planning is the most essential function in a production organisation. If planning is meticulous and foolproof, no hassles will occur in producing and delivering products or services. In different organisations, the planning activity is generally known (name of the process) as production planning and control (PPC), production engineering, process engineering, planning and methods department/innovation engineering, etc. PPC envisages all the related functions from the identification of product, customer and other requirements to the delivery of the product to the customer and, if required, post-delivery support of the delivered product during their usage.

For this entire activity, let us assume the organisation has already documented the Quality Management System, identified and established the processes, and provided the resources as necessary. Product realisation/ operational processes are established. With adequate competency, the human resources required to carry out the procedures are identified. If it is a machining process, it is necessary to ensure the availability of appropriate machines on which the components are made. In the case of CNC machined components, the required software programs from the design/drawing/CAD model for use on CNC machines are developed. In assembling parts into a sub-assembly or leading assembly/equipment, the necessary details of components, hardware and consumables that go into the assembly are identified in a bill of materials. Suitable facilities and measuring devices for measuring product characteristics, whether a component or sub-assembly/assembly, throughout the production processes are arranged.

The necessary test facilities are identified and provided if the individual component or assembly must undergo specific tests to confirm product functional characteristics. Inspections are arranged at various stages of production – in-process and final inspections, including tests – to ensure that the product finally meets all the customer/drawing/specification requirements. The final product is dispatched to the customer, and the necessary documentation is planned. If after-sales servicing, product operation with customers, warranty servicing or maintenance of the product after a certain period of use (time between overhauls) is undertaken by the organisation, planning needs to consider the resources necessary to support operation and required servicing while planning for product realisation.

There are different types of production processes in industries. Job works are when one or more products are produced per the customer's requirement and specification. Production is to meet the delivery schedule, and the relevant costs are planned before the contract. Batch production is when limited quantities of each of the different types of products are manufactured on the same set of machines. Various products are produced separately, one after the other. Mass production is when the production run is conducted on a collection of machines arranged according to the sequence of operations. A considerable quantity of the same product is manufactured at a time and is stocked for sale. Different products will require different sets of machinery. Since one production line can produce only one product type, this process is also known as a continuous production line. Suppose the production line is only for a processed material. In that case, the production line is run indefinitely until the required quantities are produced.

Aircraft production or maintenance is done in a batch. Various equipment or components may be under batch production. Still, the hardware used in aircraft, such as rivets, bolts, screws, etc., is always through mass production. Volume and variety of the product, capacity of the machines available, lead time required for completing the production task, and flexibility and efficiency of the personnel and machinery are all factors considered before choosing the manufacturing process.

Scheduling is with respect to the capacity of facilities installed and the human resources available, keeping in view the agreed delivery schedule and estimated man/machine hours necessary for the entire process, from identifying customer requirements to delivering the product to the customer. It is a significant planning task that PPC needs to work out and document. PPC has a significant role in working out the cost of the product, considering the total expenditure and the desired profit. While planning the product realisation, the PPC considers two important aspects: meeting the organisational objectives and the product requirements defined by the customer. Meeting the product requirements in a better way enhances customer satisfaction. The entire planning needs to be consistent with the requirements of the processes of the Quality Management System. Measurement of the quality objectives defined for the process is envisaged at this stage. Process output measurement is planned for final deliverable quality, control limits, process capability, cycle time and effectiveness. Controls and methodology are established for achieving them. The rate at which a product needs to be produced to meet the customer delivery schedule is known as takt time, which is the ratio between the time available to do the job and the time allowed by the customer. Lean management is based on takt time.

Producibility is the ease with which a product, say an aircraft, can be manufactured or produced in batch quantities. Producibility considers interchangeability, ease of assembly and disassembly and non-interference with other systems and components/structures. Inspectability is that every component, equipment, sub-assembly and assembly can be reached for access, visual and further checks can be carried out *in situ*, the result of checks/inspections can be measured and the condition/status can be assessed. Assessment of individual components or assemblies is completed earlier in case the same cannot be reached for access and inspection.

Design and production of aircraft components, sub-assemblies and systems are such that suitable parts and materials ultimately contribute to achieving the desired quality characteristics, features and requirements. It is ensured that the materials have no adverse effect on humans or the environment. The selection and development of embedded software (such as those currently used in flight control, navigation, armament, communication and other aircraft sub-systems) should also meet the process objectives and aircraft performance requirements. Aircraft must be reusable after completing required maintenance, repair and overhaul activities. When finally needed to dispose of it after the end of its useful life, it should be possible to do so in accordance with the relevant statutory and regulatory requirements.

The design and development of aircraft and aircraft systems, sub-systems and components are verified to meet their design/specification requirements. Verification is inspection, checks and tests against the criteria defined in design documents. All monitoring and measurements are specific to each component, assembly, system and aircraft as a whole for acceptance against the given standards for product acceptance.

The equipment and systems are validated before actual use. On completion of the required flight tests and performance evaluations, the certificates of airworthiness and quality are obtained from the regulatory authorities. The certification enables the aircraft to be produced and used for the intended purpose.

Pilots report subjective feelings during de-briefing after test flights in a fighter aircraft, such as break is sluggish, load for movement of control column is high, break-out forces for the control surface movements are high, canopy opening/closing loads are high, canopy glass is showing distortion or hazy and blurred, etc. All these snags must be rechecked against the acceptance criteria, and the loads, visual checks, etc., must be re-measured against design criteria while taking corrective actions.

Operational planning, implementation and control for the processes include:

a. Determining the requirements for the product: Product requirements are specified by the customer in the purchase order. Purchase order normally mentions the part number and rev status in case of a component, and assembly part number and associated drawings/specifications in case of an assembly/equipment. In the case of a complete equipment or product, its configuration is generally indicated in the purchase order.

PPC has to determine the configuration management process appropriate to the product. Configuration means the physical and functional features of any product. If a component is to be produced, PPC identifies the component's part number and drawing number in the planning documents, i.e., the controller process sheet/work order. Configuration of the element is determined and controlled by the issue/revision status of the part number and its drawing number. Therefore, the configuration of a part or assembly is always defined by the part or assembly number and their respective issue/revision status as indicated in the part or assembly drawings. When raw material is used for manufacturing, the raw material traceability is also identified and recorded to control the configuration.

PPC releases the complete production and control processes in the form of a document called a master process sheet (MPS), quality plan, route card, job card with bill of materials, router, work order or a combination of these, with all the relevant information and indicating the quantity to be produced. The quality plan may be a document specifying the process/method of doing the job, meaning the resources, instructions to be followed, tools and measuring devices to be used. This document is the output of the planning process. The output of quality planning includes documented quality plans, processes, procedures and design outputs. A quality plan can be a single document or a combination of documents, such as a process flow chart, work instructions, control plan, project plan, etc. The output document is prepared in a format suitable for the organisation's operations method. It may be an internal document whose master copy is held by PPC, and an authorised/controlled copy is routed to the production department for use. Where 'enterprise resource planning' (ERP) – a software architecture facilitating the flow of information among different organisational functions – is in use, all the relevant information is indicated for read-only access by the production shops through the local area network. Once the production task is completed, production updates the information field for further processing. In the case of MRO organisations, the necessary maintenance schedules and associated detailed instructions, as per the OEM requirements, are issued for the shop to carry out these activities.

Records are necessary to provide evidence that the production processes and the resulting product, be it a component, assembly or software, meet the specified requirements. The records are made as a result of activities performed, checks/verifications completed and measurements taken throughout the production operation stages, wherever specified (quality plan) in the planning document.

Ensure each process conforms to the requirements of the Standard and established documented information. Identify the process objectives and measure the process effectiveness.

In a software design or engineering services organisation, the customer requirements are determined from the customer PO issued along with an agreed 'statement of works' (SOW), which generally defines the work to be performed by the organisation. Other customer requirements, special requirements, etc., are also identified while determining the needs for the product/service. The information may be conspicuously identified in any of the planning output documents for the production operations to be taken note of and executed. Notes added to the Standard emphasise organisations to do a better job by looking at the various considerations while understanding the requirements. There are several considerations to be given while determining the requirements of the product. Product safety is defined in the Standard in clause 3.4 as 'The state in which a product can perform to its designed or intended purpose without causing unacceptable risk of harm to persons or property damage'. Planning, implementing and controlling the processes are necessary to ensure product safety and personal safety during the product's design, development, manufacturing and use. Organisations must assess hazards and manage associated risks while designing or manufacturing the product. These and similar requirements are discussed here as sample question–answers, just like the auditors verify compliance with the process requirements:

i. What are the personal and product safety features you consider while determining the requirements of the product?
   Tooling and component manufacturers should answer (and show the evidence of compliance, which the auditor may ask): 'If we have the choice of materials, we select the materials meeting the regulatory requirements of RoHS (The Restriction of the Use of Certain Hazardous Substances (RoHS) containing more than the agreed levels of lead (Pb), cadmium (Cd), mercury (Hg) etc. by manufacturers,

importers, exporters. If the customers identify or give us the raw materials, we machine them as per the drawing requirements. Component safety features include blending knife and sharp edges, ensuring corner radii and other human factors and ergonomics, and occupational health and safety. As far as personal safety is concerned, PPEs have been provided to the employees'. (Note: For more on PPEs and safety, see Chapter 7, Section 7.3, 'Awareness', personal safety (a)–(e).)

ii. How do you ensure producibility and inspectability while determining the requirements of the product?
The designer or process planning engineer should answer (and show the evidence of compliance, which the auditor may ask): Producibility measures how easy or difficult it can be manufactured/machined. The designer confirms this aspect (verified and validated), and the drawing or process sheet indicates the material and process we can employ for manufacturing. Similarly, inspectability is how easy or difficult it can be to check or verify for quality characteristics. The quality plan indicates the steps/activities and the measuring resources/instruments that can be used for inspection.

iii. How did you ensure reliability, availability and maintainability while determining the requirements of the product?
The designer or process planning engineer should answer: 'Reliability of the product, to my understanding, is - when you want, it must work. That is what the customer wants. During the design of this equipment, reliability was identified as one of the critical parameters. Reliability of the equipment, systems or aircraft as a whole, availability for carrying out the mission when required, and maintainability of the plane and systems are all factors considered during the design and development of aircraft.

They have carried out reliability studies of the leading equipment, apportioned specific reliability figures like one failure in a million hours of use, predicted the reliability and designed the product. They carried out FMEA, FMECA and PFMEA. They managed risks with redundancy and other features like mistake proofing (poke-yoke) to improve design and reliability. Carrying out the endurance test and user trials were other reliability testing methods. All the reports, if needed, are available with the design. We are gathering data to calculate the field reliability data to compare with the prediction and establish post-delivery support for the equipment. This component or sub-assembly (if airborne) we manufacture is part of that equipment cleared by the regulatory authorities and meets airworthiness requirements from the points of safety and quality. Design and development activities are completed with the prime objective of production, manufacturability, and inspectability.

Reliability also ensures the equipment's availability for use or the component's assembly. Maintainability is a feature of the equipment that can be repaired or serviced with minimum time and effort. It can be restored to operational status immediately whenever a failure occurs. Here, my job is to ensure consistent quality of the part/assembly by meeting all the product requirements- and delivering it on time to the customer. Reliability, availability, and maintainability have all been ensured by the designers. Accordingly, the process is defined, and production is happening.'

iv. How did you ensure the suitability of parts and materials used in the product while determining the requirements of the product?
The designer or process planning engineer should answer: 'Designer /Customer has issued (approved by authorised persons before release) the design output documents, including the drawings, parts lists, Bill of Materials (BOM), standard parts, fasteners and consumables for use. All the components and materials have been identified, manufactured, fabricated, bought out from external providers, verified, and issued. Data, instructions, and specifications are also provided for production assembly instructions, handling, packaging, and preservation data needed for the finally acceptable product. All the parts and materials are issued as a kit from the stores; we verify the kit for completeness and satisfactory conditions before we take it to the shop for work.

The suitability of the parts and materials are decided by the designer/customer, and the parts and materials as per the list have been provided to us. The designer knows the criteria and considerations used for the suitability of the materials. The equipment is performing satisfactorily with the customers'.

v. What are the considerations for selecting and developing embedded software while determining the requirements of the product/service?
The designer, process planning engineer or project lead (who provides software engineering services) should answer: 'During the design phase, there were two teams, one going into the hardware

and the other trying to embed the software for the correct equipment functioning. They have designed, integrated the software and hardware, verified, tested and accepted the software and hardware. Some modifications and changes have been introduced subsequently, and the current 'Operational Planning, Implementation and Control' as per the customer's 'Statement of Work' (SOW) for the testing project. SOW clearly defines the customer requirements in detail. See this SOW, and it indicates the milestones and deliverables, Inputs including the supply of software from the customer, roles and responsibilities, skill set required, development environment, assumptions and external dependencies, our project management efforts including risk management, configuration management, testing project milestones etc.'.

vi. You have been producing this electronic product for the last four years. What are the actions you plan to take to see that your product does not fall into obsolescence while determining the requirements of the product for current production?
The designer, process planning engineer or project lead should answer: 'We are producing this electronic assembly. The design is for our customer, who always supports us. Our customer has an R and D Laboratory. They have also developed a methodology for generating algorithms that can be used to predict the obsolescence dates for electronic parts. They have a 'Strategic Obsolescence Management', which may be why we often get approved design changes. Our job is to accept the design changes after an impact study on the assembly process and test performance. So long as the current technology lasts, we will have a business, and when it changes, we will cope'.

vii. How do you plan to prevent, detect and remove foreign objects while determining the requirements of the product?
The process planning engineer or production process owner/quality inspector should answer (and show the evidence of compliance, which the auditor may ask) (hardware products/components/assembly): 'On components: Moisture, dust, sand, fibres of waste cotton, cloth and stockinet, metal shavings, machining chips, burr, scores, nicks, dents, surface cracks, material de-lamination, corrosion, pitting, even food items are foreign objects.

A significant portion of component rejections are due to the presence of foreign object debris. When the customer noticed a metallic burr sitting on the internal threads of a component we dispatched, he made a poster out of it. He returned the part with the poster. A foreign object was detected! We have made a lot of improvements since then.

The quality plan indicates that we use a checklist for visual inspection before final inspection and dispatch (what to look for and remove). For visual checks of minimal parts (for auto sector/medical/aerospace equipment), we use a magnification lens of up to 40X under proper lighting. We have trays for parts so they do not touch or rub each other when handled during processing or when packed for dispatch. Bubble wrapping is done for parts, which are ready for dispatch' (more on FOD in Chapter 8, para 8.5.4).

'On software programs/design/engineering services: Foreign objects in our services could be virus infections and computer systems and programmes data corruption. Appropriate anti-virus software is installed, and we get periodic updates for the protection'.

'On assembly of equipment/aircraft: FOD comes in all shapes and sizes ... a hand tool, dust, oil, loose nuts, bolts, spanners, cotter pins, lock wire remnants, drill bits, pencil, pen, packing material, metal or wire clippings, wire scraps, connectors, components, rags, loose coins, parts, gloves, caps, cups, eye-glasses, tape, staples, paperclips etc. We check every system/assembly for loose or missing parts, riveting or peening correctly performed or not, whether split pins are wide open after inserting, whether safety wire or locking has a positive locking or not, ensuring adequate clearance between moving parts and structure (like clearance between moving control rods on the spine of aircraft), unusual rattles during movements, etc. A checklist (part of the FOD Programme) will be followed during the final build and before the closing of panels on the aircraft. Signing the checks in the build book/ daily checks is a compliance record' (for more on FOD, see Chapter 8, para 8.5.4).

viii. How do you ensure proper handling, packaging and product preservation while determining its requirements?

The process planning engineer or production process owner/quality inspector should answer the following: 'On components – The raw material is drawn from the stores. For small parts, the raw material is cut to the required size and quantity per the route card, and the batch/lot/heat number is recorded in the route card before the first operation. The parts are handled in plastic picking bins/small parts storage bins/trays throughout the machining process. Larger parts, castings and forgings are handled in individual trays or more robust lift trucks or other material handling equipment in manufacturing plants, warehouses, and retail stores. Packaging is per customer advice or standard packing cases designed and fabricated to prevent damage during shipping, transport or movement. Fumigation (shipping and cargo fumigation treatments) for wooden packing cases is a regulatory requirement we follow, ensuring shipments are transported safely and without risk of contamination' (Registration, Evaluation, Authorisation and Restriction of Chemicals (REACH) is a European regulation that applies to substances manufactured or imported into the UK or Australia in quantities of 1 tonne or more per year).

'Preservation applies to all constituent parts of the product. The storage life of two-part Polysulphide PR-1422 Class B sealant is about nine months when stored at temperatures between 5°C and 27°C in original unopened containers. There is a prescribed pot life for the sealant after mixing. We control the storage conditions as recommended by the manufacturer. There are several items, mainly adhesives, sealants, chemicals or rubber products and other perishables, for which shelf-life expiry is marked on the products. Stock rotation uses the 'first in, first out' (FIFO) rule. The material safety data sheet (MSDS) provides information on health, physical and environmental hazards, handling precautions and first aid recommendations' (read more on preservation in Section 8.5.4, Chapter 8).

ix. How do you plan to ensure recycling or final disposal of the product at the end of its life while determining the requirements of the product?
Process planning engineer or production process owner/quality inspector should answer: 'Oh, end of life? That is a long way ahead! We have not planned anything regarding these manufactured parts and assemblies or software engineering services. We are not taking back anything that we deliver. Our contracts with customers do not bind us to do anything except that we are asked to maintain our production records for 10-40 years. Well, aircraft is built and operated for a lifetime of 40 years. Let the OEMs decide. We will continue to support with spares and technical support'.

b. Operational planning, implementation and control for the processes include establishing criteria for the process parameters and acceptance criteria for the result of the process (product/services). Criteria for the process parameters are the specific steps, activities carried out during the process and the actions that lead the process to its final result. Conformity to the requirements is the ultimate result. Suppose the planned activities are completed exactly as planned, moving towards meeting the product requirements. In that case, the result of the process is the product. Acceptance criteria for the product (the result of the process) are the characteristics we monitor in the process and the characteristics we measure on the product. Monitor the process for controls and measure and record the product characteristics at the end of the process. The measured results must show that the quality of the product is met in full. Measure the time taken right from the beginning to produce the product, meeting all the quality parameters in full, and check whether the actual time taken is as per plan and the 'On Time Delivery' performance is measured. Quality and 'On Time Delivery' are the functional-level organisational objectives (beneficial to customer). Deliver the product to the customer and determine the customer's perception of whether the organisation (process) has met the customer (and product) requirements. Customer satisfaction is the organisational objective (beneficial to the organisation).

While planning the process to implement and control the procedure and achieving the intended results, the criteria for the process parameters (what we monitor) and the criteria for acceptance of the product (what we measure) can be laid down. It is laid down in the process description or documented information (procedure).

I understand you have specified requirements to be met during product realisation. Could you elaborate on the statistical techniques you plan to use for product acceptance?

The designer, process planning engineer or project lead (who provides software engineering services) should answer appropriately, as practised:

1. Design verification (e.g., reliability, maintainability, product safety): While verifying the design, we generally look at the life cycle management of the product, which, of course, includes a detailed study of the safety, reliability and maintainability aspects. MIL-HDBK-217F has helped us predict Reliability and the DOD Guide for Achieving Reliability, Availability and Maintainability. Engineering changes were subsequently made to improve the safety and reliability of aircraft systems. In the past, acceptance of products was done after design verification using the above methodology. We are now reviewing for adequacy and suitability

to use similar plans for the ongoing project. Aerospace design verification involves reliability studies of aircraft systems and the aircraft as a whole; maintainability aspects including time between overhauls (TBO), time to repair, life-critical replacements and the convenience of replacements, etc.; and safety aspects with respect to design loads, failure modes, vulnerability, etc. Statistical techniques may support process controls during manufacture, such as selection and inspection of crucial characteristics, process capability measurements, statistical process control, etc. Statistical techniques may also help evolve experiment design (Taguchi method of identification of customer preferred requirements), inspection techniques and studies concerning failure mode, effect and criticality analysis.

2. Process control: Following the process steps and activities as established or documented as per the quality system ensures process control. In 'Purchase', definite steps are indicated in the correct sequence. In 'Planning', the inputs are processed through definite steps and activities to bring out the process outputs. Similarly, process control follows the planned stages and actions for any process. The process is monitored and controlled as per the specification, as in the case of heat treatment or plating process. Proper heat treating of metallic parts requires precise control over temperature, i.e., heating to a specified temperature, holding the parts at a specific temperature for a limited time, cooling to ambient temperature at a specified rate and using a specified medium. The process control follows the step-by-step activities precisely as specified in the specification. We monitor the process by using the gauges, timer and other devices. If the process is controlled, we will get the desired result, like the required hardness on the part.
3. Selection and verification of key characteristics: (Read more on key factors, special requirements and critical items in Chapter 3.) Design output documents like drawings and specifications identify the key features. Suppose it is not indicated like in older drawings. In that case, planning can identify them conspicuously and communicate to production to control their variation, etc., while producing/manufacturing. While carrying out the process, the executive verifies that there is no variation in those essential characteristics, critical items or special requirements. We provide for a repeat verification or inspection of the artefacts by a reviewer so that the result of no variation in the essential characteristics can be confirmed.

4. Process capability measurements: Process capability measurements give the confidence that the variability is within the specification limits. It is easy to plot a normal distribution curve and populate it with measured results of any in-control process. We get Six Sigma performances when 99.73% of the plots fall within the curve. Over time, we have completed capability studies for processes like heat treatment and machining. Our software processes are CMMI level 5 compliant (1 = Initial, 2 = Repeatable, 3 = Defined, 4 = Managed, 5 = Optimising – process management includes deliberate process optimisation/improvement). We can have fewer controls, reducing the cost of inspection if the processes are capable, and we have a better confidence level if we get a Cpk index of more than 1.33. A Cpk of less than 1.0 means the process cannot meet its requirements. The higher the Cpk, the better the process's capability to meet its requirements.

5. Statistical process control: A quality control method that uses statistical techniques. Statistical process control may be applied to industrial processes where continuous production happens. (Not normally applied to particular processes. The process monitoring tools like control charts are used to understand the variation in a process.)

6. Design of experiments: Design of experiments (DOE) is a systematic method to determine the relationship between various factors affecting a process and the resulting output of that process. It is a method to find cause-and-effect relationships to manage process inputs in such a way as to obtain the best result. With the knowledge of statistical tools and experimentation concepts, we can use a 'design of experiments' to analyse software programs, solve a product development problem or accept software testing.

7. Verification: (Verification is confirmation with objective evidence that the requirements have been met. Results of an inspection, performing alternative calculations, reviewing the documents, qualification or acceptance tests, etc., are all verification forms.) Every process is verified using the criteria for acceptance of the product. Verification is the most common activity carried out during or at the end of the process to confirm the product requirements are met and that the product is ready for customer dispatch. In the case of manufactured components, we have the drawings as acceptance criteria. How to measure the results is indicated in a quality plan. It

is similar in the case of an assembly. Test schedules are approved and issued to verify the performance results if equipment is assembled or overhauled. Complete equipment like aero-engine or aircraft undergoes a series of tests as per qualification test procedures or acceptance test procedures depending on whether it is for the qualification of the equipment or normal routine acceptance after production. Customer/product requirements are verified during contract review. Product configuration is verified during planning, production and final inspection before delivery. Every product characteristic is verified during/after production, and records are maintained that the result of verification or test meets the acceptance criteria.

In software design or engineering services, independent verification and validation (IV&V) is carried out as per the test procedure by a third-party organisation not involved in developing the product. The primary verification is to check whether the user requirements are incorporated while ensuring the software product is structurally sound and built to the required specifications. They may perform reviews of the source code, examine the associated product documentation, analyse associated algorithms for static verification and analyse dynamic verification by different testing methods such as integration testing to ensure that all software units (modules) are meshed as one, tested and properly work together. IV&V helps to find deviations and deficiencies in the product quality and specifications, helping the developers improve the product to meet all the user requirements and enhance customer satisfaction. IV&V also ensures that the design adheres to regulatory requirements. Functional testing ensures functionality meets user requirements if a facility is available with the organisation. Otherwise, the delivered artefacts go to the customer facility, where further verifications and validations are carried out.

8. Failure mode, effects and criticality analysis: (Refer to para 6.1, 'Actions to Address Risks and Opportunities' and 8.1.1 'Operation Risk Management'.)

c. Operational planning, implementation and control for the processes include determining the resources needed to achieve conformity to the product and service requirements and to meet on-time delivery of products and services. Meeting conformity to the product requirements and meeting 'on-time delivery' of the product are primary objectives of the QMS processes and for which adequate resources are identified, planned and established. Refer to Chapter 7.1, paras 7.1.1–7.1.6 People are the primary resources. The infrastructure necessary to carry out the processes, the environment for operating operations, and monitoring and measuring resources are also identified.

Process-wise, the requirements of the resources are planned while looking at the specific process and, similarly, for every process in the organisation. The process owner and the team members in the process execution are given the necessary responsibility and authority to proceed with the process. The process identifies conformity and OTD as the main objectives. It lays down any other goals, sets targets for a specific period, determines the measurement method and measures the achievement throughout the period. Details of the procedure must be known to all members of the team. The process owner may decide whether documented information is necessary for the process. Accordingly, the system is reviewed periodically based on the process results. If the process is conforming, product requirements are conforming and OTD is achieved as per plan, the review process is only for continual improvements.

d. Planning considers and establishes the correct method for implementing the process monitoring and controls according to the criteria for getting the desired process results. This is specifically true for the processes where continuous monitoring is necessary for maintaining the process parameters in accordance with the specifications, like in heat treatment or melting or cooking process, where the realisation is through a series of steps or passing through a series of equipment for a longer duration. It is possible to have alerts triggered if any particular events occur or if the time/duration/deadlines are exceeded. Some of the safety devices work on this automatically. In the software industry, project management identifies the deliverables and milestones in a business process. The project manager monitors it from time to time.

e. Planning must identify what documented information is necessary to establish confidence and consistency that the process is being carried out correctly and as planned. If the process is not complex and the skill level required from the resources is not very high, if the machine operations are not complex, need not many adjustments, and if the monitoring is possible, simple and does not need complex calculations, the whole process needs no continuous or recurrent supervision and maintaining a documented information of the process documentation may not be necessary. If you think having a document is better to ensure that planned activities are realised, have documented information. Follow the control procedures for the document. However, it is mandatory to demonstrate the conformity of products/services by ensuring that the process/customer requirements are met. Is there a better way to maintain and retain documented information? Those records will demonstrate the quality/conformity of the products and services.

f. The planning function includes determining the processes and controls needed to manage critical items, including production process controls when key characteristics have been identified. Identifying and managing key characteristics, critical items and special requirements is an essential step towards meeting important product characteristics and ensuring that there is no deviation happening to those characteristics, which control crucial features of the product. (Function, life, interchangeability, maintainability, safety and strength (FLIMSS). For every product - hardware, software, processed material or service -, there is always one or more essential characteristics as important features of that product. Hence, the designer, project team or planning process owner leaving the onerous duty to the team members declaring that there are no key characteristics or indicating that identification of key attributes as not applicable for the process, etc., does not fit the bill, nor augur well for the reputation of the organisation! Before starting the work, PFMEA studies by project teams or planning should deliberate and establish appropriate mitigation actions.

'In a software organisation doing engineering services for ASD customers, "Project Management Plan" is an excellent guideline to help identify operational planning and control for the projects. However, there were some undesirable statements in the plan's compliance record. For example:

Determination of requirements: Identify statutory/ regulatory requirements/ special requirements/ key characteristics /critical items/ value additions (needs not stated by the customer) that you provide to customers.

The answer given in the review record showed that there were no special requirements/ No National Aeronautic Association (NAA) /FAA/ EASA /DGCA requirements/ No key characteristics/ Nil / None /Not Available /Not Applicable /Customer will provide with SOW, etc.

Such organisations are at risk and cannot absolve their responsibility by the statements made in their records. The organisation can ask their customer to give the required information since critical characteristics are identified to facilitate/add value to/enhance customer satisfaction and do a better job for the customer'.

g. Engaging representatives of affected organisation functions for operational planning and control is a matter to be decided in consultation with the customer. Unless there is an agreement between the two organisations (the affected organisation and the organisation that plans and controls operations before delivery of the products /services), it is a potential risk that the customer-provided operational planning and controls may affect 'the conformity of services provided by representatives of the affected organisation. The organisation shall identify and manage the risks associated with the external provision of processes and services and the selection and use of sources identified by the customer). The organisation shall require that external providers apply appropriate controls to their direct and sub-tier external providers to meet requirements' (8.4.1).

h. Determining the process and resources to support the use and maintenance of the products and services is necessary only when the organisation has the ability, infrastructure and adequate money for post-delivery activities. The organisation must consider collecting and analysing in-service data and providing technical documentation and controls required for off-site work, warranties, maintenance, replacement parts, etc. Leading equipment manufacturers and MRO organisations only undertake such activities. However, every manufacturer or service provider is responsible for taking action, including investigation and reporting for any problems reported by the customer.

i. Planning is responsible for determining the products and services from external providers, i.e., when outsourcing to suppliers. Any process for which capability or adequate capacity does not exist within the organisation can be outsourced. While doing so, the associated requirements, such as evaluating and selecting suitable sources, including the risk assessment, implementing the process and controlling the process through the purchase process, are to be looked into. It may also be noted that ensuring adequate control over outsourced processes does not absolve the organisation of the responsibility for conformity to product requirements and other customer/statutory/regulatory requirements related to the product. The organisation cannot outsource without establishing the verification or other activities necessary to ensure that the externally provided processes, products and services meet requirements (8.4.2 d). The organisation cannot declare the outsourcing as 'not applicable' to their Quality Management System. Planning must consider the potential impact of the externally provided processes, products and services on their business.

j. The organisation as a whole and the planning process in particular has to plan, implement and establish controls to prevent the delivery of nonconforming products/services to the customer. Actions are envisaged in the Quality Management System to address nonconformity (refer to para 10.2, Nonconformity and Corrective Action).

Project or program management is one method to achieve operational planning and control, which design and engineering services in the software industry use.

There are well-established project management methodologies; some are in-house, specific to the organisation, and others are like the agile methodology. Project management may use integrated phased processes.

Projects are delivered in stages, and specific standard project management processes run across these stages. The gate at the end of the stage is the approval of your high-level project proposal and the business case that validates the approach you want to use. You must also show that you can achieve the project's goal within the required timelines and budget. Thinking in terms of stages can ensure that the deliverables produced at the end of each stage meet their intended purpose and that the team members are adequately prepared for the next stage. At the end of each stage, the project lead verifies the deliverables from that stage. Once the stage is completed, the project team can pass through the 'gate' to the next stage.

The project management process kicks in when there is an indication that the organisation is winning a project. The procedure is sequentially described in most parts to make it easily understandable. However, project management has many dependencies and rarely resembles the ideal process in real life.

The requirements of the project are first understood. A contract review is carried out, and clarifications are sought before accepting the order. Terms of engagement/statement of work are established and signed off with the customer. Roles, responsibilities and methods of communication, including data transfer, are established and agreed upon. Customer guidelines and practices, software tools to be used, standards to be followed in the project and applicable regulations are identified and made available as required. Training for the project team members is arranged. Even as these things are put in place, resources – people, hardware, software and infrastructure – are planned and obtained for the project.

Risks associated with the project are identified. A risk mitigation plan is put in place. The project charter provides a general overview of the project. The verification and validation methodology, including testing and integration where required, is determined and confirmed with relevant stakeholders. The part naming convention, revision control and data management methodology are also established within the team and with the customer.

The effort required to execute the project is calculated. A detailed project plan is prepared and communicated to the relevant stakeholders. Tasks are assigned, completion is tracked and progress is monitored. Reviews are conducted at pre-determined milestones, and course corrections are carried out as required. Risks and other project parameters are reviewed periodically. Changes and issues are managed and tracked to analyse the impact on schedule, effort and quality. The project plan is modified if required. Defects and problems are captured and investigated internally, corrections are verified and defect-free delivery of artefacts to the customer is ensured. Corrective actions are deployed horizontally after carrying out a detailed root cause analysis.

Input requirements checklist (domain-specific) duly completed with objective evidence, planning sheet with milestones to be achieved, similar to Gantt chart, list of project management tools, project tracking, milestones achieved, measurement results of effort variation and schedule adherence against planned targets, etc., are some of the records maintained in computer systems/local area network.

Project deliverables align with the requirements and delivery method agreed to. The customer accepts the deliverables after confirming that the deliverables satisfy the acceptance criteria, which were initially agreed upon. Records and logs of inputs and deliverables are maintained after due sign-off by the customer. A periodic determination of customer satisfaction is carried out to determine the customer's perception of the services and artefacts delivered to the customer. Steps are initiated to address any concerns indicated by the customer in his responses and feedback. Towards the end of the project, there is a session with all the stakeholders to deliberate on the lessons learned, risks and opportunities for improvements. The performance of the project participants is documented. The achievements, risks, mitigation and learning are kept in a repository of knowledge portals to spread the benefits across the organisation.

The planning output documents in a hardware industry suitable for the organisation's operations are: the quality plan, master process sheet (MPS), manufacturing process flow chart, work instructions, route card, job card with bill of materials, special instructions for manufacture, installation, storage and handling, router with sequence of operations and list of tools and required quantity of products, traveller sheets, detailed manufacturing instruction sheets, drawings and specifications, record of traceability during manufacturing and testing, manufacturing and test tools including software, acceptance test and delivery procedures, drawings, or manufacturing work order, etc., or a combination of these types of documented information.

It is interesting to note that certain unexpected situations/risks, which the organisations may experience during their business, are addressed here under the process of planning and controlling the work transfer.

In an established organisation, work transfer usually does not occur. As a result of risk evaluation, say for the risk of 'approved external provider shuts down operations temporarily due to labour unrest' (it can happen anywhere), the identified mitigation plan is to temporarily get the work done by an unapproved source if the situation is a temporary transfer of work, that is from an approved supplier to an unapproved supplier.

Suppose there is more than one division of an organisation within a complex, say. In that case, Division A (certified for QMS as per AS 9100) gets special processes (like hard chrome plating, aluminising, shot peening, argon arc welding process, etc.) carried out at Division B (not certified for AS 9100), and through inter-divisional transfer order (IDTO) of works, the situation is a permanent transfer of work. Division B has the facility to carry out special processes, and they have excess capacity above their needs. There is no intention to get the functions done through outsourcing since the organisation has a policy of not outsourcing Special operations and utilising the facility available at the sister divisions.

The above cases can be considered as transfers of work. However, only a thin line separates the transfer of work and outsourcing. The criteria to distinguish between a transfer of work and outsourcing is not whether it is a temporary or permanent transfer of work but whether the process follows the Quality Management System addressed elsewhere in the Standard. In all cases of work transfers, it should satisfy conformity to product requirements. Therefore, it is necessary to establish, implement and maintain a process to plan and control the approach and verify work conformity to product requirements. It means that even if the organisation does not usually transfer work, there must be a plan in place; it is a contingency plan for action, like the mitigation plans in case of an identified risk. Implementing the mitigation plans is unnecessary until the identified threat is assessed with high severity/impact. Otherwise, it should be considered as a preventive measure and implemented. Work transfer is the movement of the work (products and associated activities) from one manufacturing location to another. The plan is to control the processes when temporarily transferred to an unapproved supplier or permanently transferred to a sister division. The control plans are to be established. The plans may be revised according to the situation. The situation could be that the order received is beyond the installed capacity of the organisation (contract review should bring out the need for a temporary risk mitigation plan for meeting the requirements of the customer while taking action to augment the capacity), or there is a temporary setback for own production facilities. (This also could be identified and addressed in the risk mitigation plans for all contract reviews.)

However, the most critical aspect is that the plans are laid down to verify or ensure conformity of work carried out or conformity to the quality requirements of the product.

No relaxations in Quality Management System requirements can be made because the unapproved supplier has undertaken the work from the organisation for the first time, or the responsibility for the quality of the product is shared between the two divisions of the same organisation as in the example given above. In the case of unique processes carried out by the sister division, all the requirements of the Standard under clause 8.5.1.2, Validation and Control of Special Processes, shall be included in the control plan. Evidence of compliance, including the process's effectiveness, must be obtained from the division. All the controls exercised on an approved (outsourced) supplier must be implemented. However, the source is a sister division of the organisation. Control of nonconforming parts and resolution of customer complaints, if any, and improvement actions are to be effectively followed by Division B (in the given example) and verified periodically by Division A. The corporation of the organisation may say that the divisions share the responsibility for quality. Still, 100% of responsibilities rest with the division, which transfers work (Division A in the example). The corresponding IDTO should contain all purchasing information per clause 8.4.2 of the Standard.

While deciding to temporarily transfer work to an unapproved supplier, all the related evaluation, selection and approval based on their ability to meet requirements, including 'Production Process Verification' of the product (8.5.1.3, Production Process Verification) (because of source change) and periodic review of the performance of the new source (8.4.1.1), have to be initiated. Product requirements, including special needs and critical characteristics, must be indicated to the new source. If the work transfer takes place from an approved supplier to a sub-tier supplier, the flow down of all the requirements has to be insisted (8.4.3).

The transfer of work management process commences with the proposal supported by sufficient information, including risk assessment, the reason for transfer such as cost reductions, cycle-time reduction, performance improvement, any constraints, alignment to a business need/strategy and associated contract reviews. Detailed planning for the work/product, process requirements, tooling, quality levels, costs, etc., forms the next part. Review and approval of the quality plan provide the authority to implement the plan. The objective of the process is to complete the task within the resources, schedules and efforts. Deviations may occur during the execution of the plan, which should be measured to show the performance of the process to meet the objectives. A full or partial production process verification (first article inspection) may be a desirable method of measuring the product characteristics to meet the requirements at the transferred location of manufacture/work. A review at the end of the process regarding the performance of the new supplier will be beneficial to ensure the continuation or otherwise of the transfer of work process.

What is the difference between outsourcing and work transfer? Outsourcing is when the organisation chooses to have another organisation perform a needed process. Work transfer is when the operation of a function is moved from one location to another. It applies to temporary outsourcing of your regular production work, which might occur when orders exceed your facility's capacity or when your production facilities become temporarily unavailable.

8.1.1 Operational risk management

The scope of this clause (8.1.1) is limited to the risks associated with the operational processes needed to provide products and services.

Risk is an undesirable situation or circumstance with both a likelihood of occurring and a potentially harmful consequence. Risk is a safety hazard. In life, every day, we are likely to encounter many risks. Wearing a helmet when you drive a motorbike is a risk management to protect your most crucial body part, the head. When you go to the market to buy vegetables, you may carry an umbrella to protect you if it is likely to rain when you return. The protective devices help you reduce negative consequences' impacts if and when undesirable situations arise. It is also necessary to evaluate potential circumstances and take adequate precautions and mitigation actions to achieve the desired results and product/service conformity.

Risk management is an iterative process to identify, assess, reduce, accept and control risks systematically, proactively, comprehensively and cost-effectively, considering the business, costs, technical and quality/schedule constraints. Since managers always decide on risk, the term 'risk management' is used. Risk management is a decision-making process designed to identify hazards, systematically assess the degree of risk and impact on quality and determine the best course of action to eliminate or mitigate them.

Risk management is to reduce the chances of something harmful happening to the business. This involves focusing on the risks of meeting customer requirements and preventing the escape of product nonconformities. Risk management processes need to be considered irrespective of the organisation's size. The absence of a risk management programme can result in known or unknown potential problems for the customer and stakeholders concerning the quality and on-time delivery performance of products and services. It can also affect the cost, schedule and technical performance of products/projects.

A risk management programme describes the organisation's attitude and approach towards risks, how it conducts risk management, the risks it is prepared to accept and how it plans on dealing with those it chooses not to accept, thereby defining the main requirements for a risk management plan. An effective risk management process begins with understanding strategic objectives and taking steps to identify, evaluate, report and resolve the various risks.

A risk management programme helps the organisation to:

a. identify risks;
b. reduce occurrences and impacts of risks;
c. understand significance/severity/impact of risks;
d. promote organisational behaviours focused on risk management;
e. increase the effectiveness of product and service delivery to customers;
f. create a process for who, what, when, where, how and how much;
g. maintain information on past experiences and prevent the recurrence of unacceptable issues;
h. bring out hidden risk knowledge so that it can be managed.

Risk introduces potential constraints in achieving overall organisational objectives, objectives of working within the defined schedule, cost or available technical capabilities. Remember, risks are yet to happen. They have future consequences and can be considered taken care of only after successful mitigation through avoiding, controlling, transferring or accepting the risk.

Risks must be planned with various stages of risk identification, assessment, communication, monitoring and planning mitigation, including training personnel and accepting risks beyond the organisation's capabilities. It is necessary to quantify and prioritise the identified risks. The PDCA approach can be applied successfully to the risk management process. Some of the risks associated with the Quality Management System are:

a. the complexity of design;
b. criticality of products for the intended use;
c. new or unproven process or technology;
d. organisation's capability to design or manufacture;
e. specific customer or process requirements are vulnerable to risk, such as short delivery schedules or process/product defects.

Risks within a process are identifying risks inherent in the process/task, communicating with all relevant personnel, understanding how the risks are likely to affect the process, and deciding on implementing the mitigation plan to control the risks or accepting the unavoidable risks.

Some examples of identifying the risks within a process are given here for clarity. Try to figure out what are the possibilities/reviews/risk evaluations that can be done before deciding to accept the order:

a. The customer provides a purchase order, which includes a task the organisation has not done before.
b. The customer wants the organisation to undertake a considerable task involving deploying many skilled personnel with knowledge of using particular software tools. The organisation has no sufficient strength to undertake the mission, and delivery will start within the next two months.
c. The customer has requirements for a very complex sub-assembly of an aircraft system and floated an enquiry to the organisation to send a quote.

The standard prescribes that the risk process should be appropriate to the product/project and the organisation (clauses 6.1 and 8.1.1). 'Appropriate' makes it more convenient for the organisation to assess the risk, which can vary based on the situation of the process, the availability of people for doing the process and the customer, product or service. Therefore, organisations must decide what is appropriate for them and support their decision with evidence of effectively managing the process.

First, the responsibilities for risk management are to be assigned. Generally, the responsibilities concerning the processes carried out are defined in organisations. Particular responsibility for risks associated with those processes can also be identified by the process owners, such as:

a. Sales and marketing: Understanding customer needs/requirements related to the product concerning the organisational capabilities, capacities and constraints.

b. Design and development: Transforming the customer inputs into producible products through suitable documentation (specification, drawings, test schedules, statement of works (SOWs), memorandum of understandings (MOUs), etc.) (clause 8.3).

c. Production planning: Application of appropriate methods through planning for product realisation and associated risks in meeting customer requirements (clause 6.1).

d. Supplier management/purchasing: Supplier evaluation/process capability assessments, communication through purchase information (clauses 8.4.1 and 8.4.2).

e. Manufacturing: Suitable manufacturing methods, production facilities, application of production controls, tools, etc. (clause 8.5).

f. Competency of personnel doing the job related to the product quality (clauses 7.1.2 and 7.2).

g. Quality control: Verification of products (in-process and final), identifying nonconformities to products (clauses 8.4.6 and 10.2).

h. Facilities, monitoring and measuring resources (clauses 7.1.3 and 7.1.5).

Hence, risk management may be carried out by a responsible team consisting of process owners of project management/PPC/engineering, supply chain management/purchase, marketing, design, manufacturing and quality control processes. Respective process owners may identify the risks related to the product, process and people in each process. For example, project management/planning/engineering may identify hazards associated with cost, quality and delivery schedules. The engineering/design department may identify technological capabilities and process/design-related risks. Supply chain management/purchase process may identify risks associated with supplier selection for outsourcing and supplier/external provider's capability/competence.

Within the aviation, space and defence industries, the risk is generally expressed in terms of the impact of the consequences and the likelihood of occurrence. By definition, addressing the risk's impact and probability of occurrence are risk assessment criteria regarding impact (how severe it may be) and likelihood (chance of occurrence). Based on the requirements, every identified risk needs to be assessed to know the level of risk. The risk priority number (RPN) indicates the level of risk. It is evaluated by multiplying the impact of risk by the likelihood of the occurrence of the problem (impact and probability assessed by using the table (Figure 6.1) in Section 6.1.2 in Chapter 6).

It is also necessary to review existing controls to achieve the desired effect. The RPN level will be high if the current controls are ineffective. Additional controls are to be identified and implemented to reduce the risk level from higher to lower levels, where management will be easier.

Acceptance level (RPN), the defined risk acceptance criteria, is a decision the responsible team must make and record in the risk register. Where both the severity and the likelihood of occurrence are high, the risk is significant, and the process requires improvement. Top management must determine what level of risk is acceptable to the organisation to sustain business operations successfully. Higher RPN is an indication that process improvements are mandatory. The risk level cannot be left without mitigation actions, primarily if the RPN level is higher than the acceptable level. The management of actions to mitigate risks that exceed the acceptable level of RPN is mandatory. It is necessary to discuss and identify suitable mitigation actions and plan the implementation of those actions, which are preventive actions to prevent the occurrence of the risks. The mitigation actions may include establishing additional controls to avoid that potential problem. Responsibility and the expected completion date of the mitigation actions can be decided, recorded and communicated through the risk register itself.

Once the mitigation actions are completed and the additional controls have given the desired results, preventive measures effectively reduce the risk's impact. The new risk priority number (RPN) indicates the reassessed/current level of risk. The reassessment of RPN is done by the same multiplication of impact of risk with likelihood using the table (Figure 6.1) in Section 6.1.2 in Chapter 6.

The risk management actions are practical if the RPN is now at acceptable or below acceptable levels. The current risk level is accepted as the remaining risk after implementing mitigating steps. Generally, organisations plan mitigation for identified risks using appropriate tools such as failure mode effect analysis (FMEA) or process failure mode effect analysis (PFMEA). FMEA is a tool that can be used to assess products and processes to improve performance.

Enterprise risk management (ERM) has emerged as a new methodology for managing risk. Instead of relying on a traditional, intra-departmental strategy, where each organisation's process manages its risks, ERM adopts a broader perspective that integrates and coordinates risk management across the entire organisation.

Interdisciplinary teams usually undertake quality risk management activities. When forming teams, they should include experts from the appropriate areas, such as quality units, business development, engineering, regulatory affairs, production operations, sales and marketing, legal cells, etc., and individuals knowledgeable about the quality risk management process. Decision-makers should take responsibility for coordinating quality risk management across various functions and departments of their organisation and ensure that a quality risk management process is defined, deployed and reviewed and adequate resources are made available.

The risk register or log becomes essential as it records identified risks, their severity and the action steps to be taken. It can be a simple document, spreadsheet or a table. The register maintains a great deal of information. The format with necessary data is indicated, and a register is populated with examples of identified risks and their treatment.

Common pitfalls/errors found in the risk register maintained by organisations are as follows:

a. The risk remaining is not indicated in the risk register after the review.

b. Risk (planning/operational) not reviewed for two consecutive years.

c. Risk status as 'closed/stopped' – How can a risk be closed? Potential nonconformities may happen again, for which periodic reviews are carried out to identify and assess. Once identified and assessed, the severity and occurrence level may vary occasionally. Still, it can never be closed, or it has to be eliminated forever by appropriate mitigation actions and kept for the following review. Risk never dies.

d. 'Mitigation is in progress (ongoing for a long time to reduce high risk)'. Working with high-risk for a long time is not desirable. At least some workaround mitigations must be established to control the high risks.

e. Internal defects recur, no reducing trend is seen for the NCs over a long period (zig-zag trend), customer complaints are repeated, corrective action taken is ineffective, etc. This is because preventive actions or mitigation actions taken are not practical. Better mitigation actions must be identified, complied with and monitored to manage the high risks.

f. Verify the current controls existing for the risk. Suppose the risk exposure level is below the anxiety/risk acceptance threshold. Further mitigation actions to push down the risk level are unnecessary. It may not be economical and sometimes becomes infructuous work, as the same risk level will be reviewed soon. They record in the risk register that it is below the acceptance level. However, some mitigation is carried out, and the risk level remains unchanged.

g. The date of completion of mitigation and reassessment of risk exposure using the original assessment methodology needs to be shown in the risk register.

h. A trend chart for monitoring the residual risk is unnecessary (as historical data may not be helpful). Assessment and recording of initial risk and, after every review, indicating that the current level (residual risk) is adequate.

i. Process/product/service/deliverables risks will be identified with additional focus and assessed as high risks. The increased risks must be mitigated vigorously to bring it down to an acceptable lower level.

Various types of risks that can affect ASD business are listed: scientific risk, business risk/sales and marketing, commercial risk, pricing, technological/design and development risk, engineering risk/production planning, supplier assessment, supply/supplier management/purchasing risk/external providers, production risks/manufacturing, process/product/people, unique processes, quality system/quality control, risk during delivery/logistics, risk during use, process trends, planning activities as a consequence to changes in processes, personnel or environment, while monitoring the customer satisfaction, through error proofing or redundancy, while analysing the trends using statistical techniques, while analysing the human factors, events, while examining similar nonconformities, which have occurred in other products/areas.

Considering the appropriate risk, before doing anything/process, and while doing anything/process, ask a simple question: What can go wrong when I do this? Are the existing controls, precautions and monitoring adequate to achieve the desired objective? Is it a new component, a new process, or a new supplier? What additional control and monitoring are necessary to reduce the risk and improve the process? Check the effectiveness of the action; does it work? Learn from experience: How can I avoid, eliminate, or mitigate threats? What is the likelihood (probability) that it will go wrong? What are the consequences (severity)?

Some of the identified risks (examples) are as follows:

a. Key characteristics not identified
b. Critical items/special requirements not identified
c. Configuration not controlled
d. Special conditions cannot be met
e. Regulatory compliance not verified
f. Recurrence of defects reported by customer
g. Special skills/competence not available
h. Process capability Cpk below 1.0
i. Special processes not validated
j. External provider not approved
k. The customer scorecard is below expectations
l. Existing orders get completed in three months
m. Customer inputs are delayed and not likely to arrive on time
n. Unique software tools not available
o. The external provider denies copying validation reports of particular processes.

8.1.2 Configuration management

Configuration management maintains a product's physical and functional features and characteristics/performance during product realisation and throughout the product's service life as the customer requires. The configuration management process consists of the configuration management plan, configuration identification and configuration control during production, including change control, status accounting and delivery at the end of production and periodically auditing the process to ensure that the rules established are in place and are effective. It is a technical and administrative discipline applied over the lifecycle of an aircraft, equipment or product during design, development, production, repeated use and maintenance until obsolescence. The application of the configuration management process depends on the size, complexity and nature of work related to the product/aircraft. It is necessary to have documented information about the product as to the conformity to requirements, design, verification, validation and acceptance records indicating the actual attributes of the product.

The organisation must plan, implement and control the configuration management process appropriate to its product and service. This facilitates proper identification and control of the product's physical features and performance attributes/functional properties throughout its lifecycle. This control process establishes the product identification and traceability to the requirements. Whenever any change happens, the change is also implemented and identified.

When complex design and development, production, maintenance, etc., are carried out on aircraft by industries, there must be detailed planning to incorporate configurable items on the product (including hardware, software, processed material and service).

Configuration identification is a 'top-down' process by dividing the total product into logically related hardware, software, processed material and service aggregates. Illustrated parts catalogue/list, with part numbers, assembly numbers and their drawing numbers and issue/revision status, higher level and critical configured items, interfaces, modifications, changes, concessions and production permits, etc., together make the baseline configuration of an aircraft, which includes aero-engine, as certified by the airworthiness authorities (Regulatory Authority).

During the design and development of an aircraft, aero-engine, or major equipment for aircraft, it is a usual practice to establish a configuration control board consisting of members from the organisation, customer and regulatory authorities to review and approve configuration management plan, procedure, selection of configured items, review and approval of changes to the configuration of equipment/systems/components, etc. Configuration control board may be called the dispositioning authority, which consists of a person or a group of persons assigned responsibility and authority to make decisions on the configuration. All configuration changes are discussed and approved, recorded, classified and distributed to all concerned for implementation in the product and related documents. The configuration control board may authorise deviations from the initially agreed configuration and accept requests for concessions from the organisation/design agency. Activities of the configuration control board may be extended to the production phase and throughout the aircraft's lifecycle. This is because the aircraft's configuration keeps changing for improvements based on service experience. The change may be for incorporating improved features/modifications/functions/performance of aircraft and to keep pace with emerging technology, trends and customer requirements. Whenever any procurement or MRO activities are done on aircraft, the current configuration required by the customer may be identified as the baseline configuration with approved changes since certification.

The aircraft build or maintenance standard is the aircraft's 'Standard of Preparation' (SOP). SOP includes the total list of applicable drawings, specifications, modifications, servicing, technical, special instructions bulletins, etc., for an aircraft under manufacture. For aircraft maintenance, periodic servicing and maintenance schedules are added to the SOP.

SOP is prepared and issued before manufacturing or maintenance activity. It is a document agreed upon among all the interested parties – customers, users, manufacturer/MRO organisation and the concerned regulatory authority. Regulatory authority documents like DDPMAS prescribe annual issue of SOP, modification index, local modification committee/local technical committee procedures, advanced mod information for considering changes/improvements and issue of mod leaflet for compliance, with the point of an embodiment for the lifecycle management of configuration of military aircraft. As per the contract requirements, the production planning and control department or the process owner should identify the configuration while planning the production of the part, assembly, or equipment. This is done by releasing the configuration documentation, such as drawings, specifications, route cards, job cards, work instructions, etc., by the process owner of the production/planning process. Planning is responsible for identifying the configuration as per the contract with customers.

Configuration control ensures compliance with the SOP after the manufacturing or maintenance organisations determine the SOP for a batch of aircraft. A complete set of ground and flight test schedules defines and measures the functional and performance characteristics of the aircraft and meeting configuration.

Changes introduced on the configuration baseline after aircraft certification while the aircraft is in use are documented as upgrades or modifications. This is done after verification, validation and approval of the changes by the customer and regulatory authorities. The configuration management is to identify the required changes to the aircraft and control their implementation at suitable occasions. Changes in components/assembly raise the issue/revision status of the part/assembly drawings; similarly, the specifications and other process/configuration/build/maintenance documents. Minor changes, when introduced in aircraft or aero-engines in the form of an improved component, process change, other improvements and minor modifications, or validation of the changes in configuration, including performance measurements, may not be feasible economically immediately. Therefore, aircraft/aero-engine industries pool together a set of changes on the aircraft or aero-engine and carry out a production quality test/long test for aero-engine and aircraft equipment, which is otherwise carried out to ensure that the production/maintenance activities are being carried out consistently throughout. Periodicity for production quality tests/long tests is mutually agreed upon among the production agencies, regulatory authorities and customers on a calendar or batch quantity basis.

Configuration status accounting ensures that the aircraft's required configuration has been built or the maintenance/servicing has been carried out during production. Status accounting is done while complying with the modifications at customer locations. Status accounting ensures the baseline configuration and all the required changes are identified and traceable. Appropriate records are maintained when the aircraft is delivered after manufacture or maintenance per the customer's requirements. Configuration is verified and evaluated every time before the release of the aircraft to customers.

Configuration management and control of product-related documents are implemented throughout the product realisation processes and subsequently for the entire product lifecycle. Configuration management is widely used in acquiring defence equipment and commercial products and services. A disciplined configuration management process ensures that products conform to their requirements and are identified and documented in sufficient detail to support the product lifecycle management. Configuration management ensures accurate product configuration information, interchangeability, safe operation and maintenance. Authority Holding Sealed Particulars (AHSP) for defence equipment maintains the basic configuration information when procurement is made and is continually updated when changes occur. As for the OEMs, when configuration management principles are applied using effective practices, return on investment is maximised and product lifecycle costs are reduced.

In the small-scale manufacturing sector, where machined components or sub-assemblies of components are made for the aircraft industry, the configuration identification, traceability and control are simple, confined to the part number with its issue/revision status and material traceability as required by the customer. The part is manufactured and supplied, and the sub-assembly is made in accordance with the bill of materials/parts as per the configuration required by the customer. Plan as per customer PO, identify the configuration of parts from drawings, control through PPC/route card/ production documentation, consider changes and incorporate if desired by the customer, verify the configuration status before delivery of the product to the customer as per the configuration management plan (accounting) and periodically audit for ensuring the consistency of the process – that is, configuration management.

In the software design or engineering services industry, the configuration items are a collection of objects related to the specific functionality of a more extensive system, such as requirements, code, documentation, models, assemblies and other files. Altering a particular version of the baseline of a configuration item creates a new version of the same object. Configuration management carried out as appropriate for the software projects is as follows:

Identify the configuration items with information as to how they have been affected. A release, a versioned entity, may consist of several configuration items. The set of changes to each configuration item may appear in the release notes, which may contain specific headings for each configuration item. It is necessary to establish standard vocabulary/name/number convention across all groups connected to the project, preferably through a 'configuration and traceability template'. Configuration items, their versions and their changes form the basis of configuration control in the software design and engineering services sector. The configuration management structure is usually project-related and adapted as necessary to meet the needs of the product's lifecycle stages. Configuration identification involves determining the project structure and selecting configuration items. Configuration control is through the initial release of configuration documents and subsequent management for changes. Configuration status accounting provides information on all configuration identifications and changes from the specified baselines by the customer. Configuration audit is performed for a project, following the trail from the acceptance of a configuration baseline to the project completion, meeting the specified requirements and ensuring that the documentation accurately reflects the project status.

Configuration auditing is a systematic, independent and periodic verification of the current configuration status concerning the contract requirements as the criteria. Both physical and functional/performance configurations of the component/assembly/equipment/project are to be audited. Configuration audits could be done together with the other Quality Management System functions like internal audits or separately before delivery of the product/service to the customer. In many organisations, configuration audits were found to be carried out by quality personnel. The audit does not replace other verifications, reviews, inspections or test activities. Since the audit is not a quality control/review activity, the counterpart of quality personnel /reviewers from the organisation's other projects/product lines must perform a configuration audit. A typical audit checklist is indicated in Annexure J for use.

In some industry sectors, such as small and medium manufacturing units, configuration management is a means by which the identification and traceability of the products are maintained. ISO 10007: 2003 may be referred to for guidance on configuration management.

> *'SMEs, it is generally found, make a few mistakes while managing the configuration of their hardware products:*
> *(a) Part required by the customer as per PO is Xxxxxx Drg Rev C. Production documentation does not indicate the rev status of the drawing.*
> *(b) The part the customer requires as per PO is 9Bxxx-002 Drawing Rev 01. The part was manufactured to Route card 9B xxx-002 Drawing Rev 00.*
> *(c) Configuration audit is not carried out*
> *(d) Part Configuration is not identified in the quality plan; the part No. The indicated difference is in the inspection record of EDM operation. There is no link for the router to the customer drawing.'*

8.1.3 Product safety

Overall analysis of aviation safety (IATA data) shows improvement every decade since the 1950s. The fatal accident rate has decreased from 40 to 0.32 per million departures from 1960 to 2015. The major contributor to this improvement is the Quality Management System followed by all the OEMs, stakeholders and the supply chain (OASIS). Significant improvements in risk management, training and technology have resulted in commendable improvements. Times have changed, and technology has changed for the better. The airline industry's outstanding safety record also comes from decades of intense analysis of errors and crashes – and a determination to learn the lessons. The continual improvement efforts by all the stakeholders and interested parties genuinely made the aircraft safer. Improvements in flight safety are mainly due to the conscious efforts made to increase aircraft systems' reliability. Improvements in the reliability of aircraft systems are due to:

a. use of high-reliability mechanical parts and electronic components;
b. use of more scientific techniques and improved methods for tests for conformity;
c. the relentless drive for improved quality and reliability for equipment on board and ground;
d. improved design techniques, analysed with risk management/FMEA, and finally accepted after validation.

The efforts continue, and flight safety is on the path to improvement.

While designing, developing, producing and marketing a product, one of the primary objectives of an organisation is to make all the efforts and ensure that the product is safe for humans; be it a consumable or durable, be it a toffee or a toy, watches or winches, be it static or dynamic, be it an automotive or aircraft, safety is the paramount consideration.

Governments or regulatory authorities verify and assure the product's safety before it is released for marketing. Certification for airworthiness is one such assurance provided to the public that an aircraft is meeting all the requirements for safety for flight.

Let us have a typical example (not comprehensive) of what planning, implementation and control an organisation does to assure product safety during a military aircraft's design and development phase.

Safety is one of the most fundamental considerations in flight. Safety is paramount, and everyone wants to be safe. Safety is the freedom from those conditions that can cause injury, illness or death and damage to or loss of equipment or property, or harm to the environment. Safety is a state that implies being protected from any risk, danger, damage or cause of injury. In the aeronautical industry, safety means that users and operators do not face any threat or hazard. Safety in the aircraft industry is paramount; therefore, the industry is highly regulated. Aircraft and other airborne equipment, in fact anything that goes up in the air, must comply with certain norms and regulations, whether national or international, to be used.

Rules and regulations are designed to address known or suspected conditions detrimental to safety. Laws are intended to promote safety by eliminating or mitigating conditions that can cause injury, damage or death.

The regulations are comprehensive, but even the strictest compliance with laws may not be sufficient to guarantee safety. Rules and regulations are designed to address known or suspected conditions detrimental to safety. Aircraft are designed for operations within an envelope, which depends mainly on speed and structural load factors. A system may be defined as the combination of people, procedures, equipment, facilities, software, tools and materials that operate in a specific environment to perform a specific task or achieve a specific purpose. In any system, there is a high probability that some new combination of circumstances not contemplated by the regulations will arise.

The aircraft structure is a platform for installing and operating all other systems on the aircraft. The structural integrity of the aircraft is essential for flight safety. Flight safety is established by systematic design analysis, quality planning and control during production and testing against schedules. Stress analysis is verified by measuring the strain and deflections on structural parts. Test results form the basis of acceptance of structural components from your safety point. Airworthiness requirements include strength tests and fatigue (caused by repeated loads, including vibratory loads due to flight operations, handling and service use). The test is done within the flight envelope, environmental conditions and service use.

For a military aircraft, ground testing is done for all structural components or the complete airframe, simulating the aerodynamic load conditions for the entire envelope of the aircraft operations. Theoretical stress analysis is done. The test demonstrates the damage tolerance concept. Other tests are crashworthiness test, rain test, water tightness test, bird strike test, lightning strike test, weapon effect test, nuclear weapon effect test and special tests based on user requirements. Operational tests for all systems, power plant, fuel systems, hydraulic, flight control surface operations, navigation and weapon aiming sub-systems, air-conditioning and landing gear integration/operation, canopy operation and cockpit pressurisation checks.

The total test plan includes load/critical load cases, limit/proof/ultimate load, load distribution, expected strain/deflection and a schedule for loading instrumentation and inspection during the test. Limit loads are the maximum loads expected in service. Ultimate loads are defined as the limit loads × a safety factor (1.2–1.75). Causes of failures are established, and modifications are initiated for improvement. Flight load measurements do load verification. Freedom from instability is tested for the entire envelope of the aircraft. Incremental progress in testing is made by limiting 'g' conditions and speeds of aircraft. Safety critical components, such as ejection seat/canopy operations, parachutes, etc., are accepted based on performance testing.

Similar airworthiness certification tests are done for civil aircraft as well. Civil aircraft programmes undergo a complex, rigorous flight test and certification phase. Once approved and certified, the aircraft is cleared for take-off for the entirety of its lifetime.

Def Stan 00-970 'Design and airworthiness requirements for service aircraft' is the default certification specification for military-registered aircraft in MOD UK and India. Several US military standards, like MIL-STD-1530A: Military Structural Integrity Programme, Airplane Requirements, and FAR- 23 Airworthiness Standards for Normal, Utility, Aerobatic and Commuter Category Aeroplanes are examples of regulatory requirements. Some more examples of standards/regulations are:

a. IEC 61508: Functional Safety of Electrical/Electronic/Programmable Electronic Safety-related Systems (E/E/PE, or E/E/PES) is an international standard published by the International Electro-Technical Commission of rules applied in industry.

b. The standard ISO 26262 is considered one of the best practice frameworks for achieving automotive functional safety.

c. ARP 4761: This document describes guidelines and methods of performing the safety assessment for the certification of civil aircraft. It is primarily associated with showing compliance with FAR/JAR 25.1309.

d. ARP5150: This document describes guidelines, methods and tools used to perform the ongoing safety assessment process for transporting aeroplanes in commercial service (hereafter, aeroplane). The methodology described herein is intended to support an overall safety management program.

e. ARP5151: Safety Assessment of General Aviation Airplanes and Rotorcraft in Commercial Service.

Risk management (more information in Section 8.1.1 in Chapter 8): Assessment of hazards and control of associated risks for product safety is a requirement that was not vigorously pursued until recent times in the Indian aircraft industry. The organisations did not take advantage of risk-based thinking or systematic and elaborate risk assessment for their business operations, which had the objectives of product safety, reliability and availability.

While the regulations offer comprehensive treatment of individual issues, they are not designed to cover the infinite number of possible combinations of hazards that can undermine safety.

Product safety is the 'application of special technical and managerial skills in a systematic, forward-looking manner to identify and control hazards throughout the life cycle of an aircraft'. Influential safety culture includes a commitment to safety, shared perceptions about safety and good communication.

The four fundamental principles of operational risk management are:

a. Accept no unnecessary risk.

b. Make risk decisions at the appropriate level.

c. Accept risk when benefits outweigh dangers (costs).

d. Integrate risk management into planning at all levels.

Every accident or incident in the military aircraft due to technical reasons could have been prevented if there had been efforts to identify the latent/hidden risks and act on time. Every accident or incident that has happened in the military aircraft was due to (a) accepting unnecessary risk, (b) not making risk decisions at the appropriate level, or (c) accepting risk when benefits did not outweigh costs. Most of the time, we have been reactive to the causes identified for the problems for improvement. Perhaps we had very few indigenous projects and relied heavily on our licensors' commitment to safety. We could have been more proactive in bringing out improvements on the aircraft for safety.

Safety critical items: To ensure product safety, the organisation plans, implements and controls the management of critical items. Refer to the examples of special requirements, critical items and key characteristics in Chapter 3. Critical safety items are those systems/items whose failure could result in loss of life, significant property damage or environmental damage. Critical safety item (CSI) may be an aircraft part, assembly or installation containing a critical characteristic whose failure, malfunction or absence may cause a catastrophic failure resulting in loss or severe damage to the aircraft or may cause an un-commanded engine shutdown, resulting in an unsafe condition. The term is synonymous with 'flight safety' and 'flight critical'.

Control of Aviation Critical Safety Items (CSI) is a regulatory requirement. Critical safety items are required to be inspected 100%. Parts are accepted only after production process verification (FAI)/qualification tests. There must be an identified configuration for the part with production lot numbers, serial numbers and stringent record-keeping practices. AS 9017 is an 'Aviation, Space, and Defence' standard issued by SAE Technical Committee G-14 'Americas Aerospace Quality Standards Committee' (AAQSC) designed to harmonise requirements across the supply chain for management and control of Department of Defence Aviation Critical Safety

Items (CSI). CSI identification ensures that proper source approval requirements are applied to potential new suppliers. This is particularly important when evaluating offers from sources with limited knowledge of the item's application, design intent, failure modes, failure effects, critical design characteristics or critical manufacturing, repair or installation processes.

Analysis and reporting of events affecting safety: In aviation, space and defence, there is always a system of investigating and reporting all accidents and incidents that occur during the lifecycle of the equipment/products. This is basically to learn from the mistakes, improve safety and remove the aberrations in the practice of the past. A safety management system recognises the potential for errors and endeavours to establish robust defences to ensure that those errors, if committed, do not result in incidents or accidents. This is a field where organisations contemplate a lot of improvement. The significant areas of product/flight safety concern are:

a. Technical factors
b. Standard operating procedures (SOP)
c. Facilities (e.g., airports and airspace)
d. Flight environment (which includes weather, FOD and bird strike menace)
e. Human factors, including training
f. Organisational factors

Every concern is being addressed regularly by the experts, and there has been successful implementation of the control measures, with which higher standards of aerospace safety could be achieved. Industry concern is mainly on the technical factors (product safety).

> *'Take the example of de-bonding the erosion strip (metal) of the tail rotor blade made of composites. The purpose was to avoid erosion of composite blades due to the impingement of dust and pebbles during ground use. A case of erosion-strip deboning in flight resulted in heavy vibrations on a helicopter. Timely detection helped in landing safely. Consequently, the tail rotor gearbox front lugs cracked due to excessive vibration. This could have resulted in a catastrophe as tail rotor gearbox detachment amounts to the loss of tail rotor control.*
>
> *Bonding of erosion strips has been improved by using mechanical fixtures instead of vacuum bag curing, and a process improvement*

could have been identified as mitigation if a risk assessment was done using the PFMEA tool before production.

An emergency involved a Teleflex cable connecting the cockpit engine throttle to the fuel control unit on an engine. The cord passes through it. There was an incident on a jet trainer when the throttle was moved back from max rpm during flight. There was no corresponding response from the engine because the Teleflex cable outer tube had come out free from the guide during engine slam-acceleration tests, causing a geometric lock, and it got stuck in the new position. The cause-safe distance of the safety check hole (25 mm from the end) was not ensured on the outer tube during servicing/rigging of the engine control run. No confirmation was available regarding the quality/acceptance of the Teleflex cable when brought in for assembly. The process owner did not carefully assess the risk while choosing it for assembly.

One holiday during one of the test flights of a fighter aircraft, the pilot strayed/diverted from his allotted sector and tried to perform aerobatics in a valley (bowl) surrounded by hills, perhaps to impress his friends, who were having a picnic there. The risk could have been prevented or avoided, but what happened was investigated. Obituary reference in the fatal accident inquiry report stated that the pilot, for his mistake, paid the supreme penalty with his life, and the fatality was attributed to human error. Here, the pilot made a mistake in perceiving risk, or the system had not adequately prepared him to perceive and avoid the risk. The pilot has paid the penalty, but what the organisation has done to mitigate the circumstances to prevent or avoid a similar risk is more important. There was no 'HEVOLREP' (Human error voluntary reporting) from people who knew of the flight plan! The honest reporting system that encourages disclosure of errors/aberrations without revealing the informer's identity - a 'Voice Mail Recording' of HEVOLREP to encourage reporting from the field units to capture occurrences and incidents has been introduced!

Learning lessons from past events, the organisations can plan, implement and control the activities that ensure product safety during the product's life cycle '.

The investigation reports of all major events, incidents and accidents are distributed to all the interested parties to derive the benefits of improved safety of aircraft or aviation, space, and defence equipment. Analysis of in-service failures yields useful information involving mechanical, material, functional and environmental factors so that necessary reviews and remedial measures can be undertaken to prevent deficiencies from being revealed and improve quality and reliability. The findings and recommendations in these reports encourage operational risk management, establishing mitigations/preventive measures such as introducing modifications on the equipment or establishing additional verifications and controls for the existing configuration. Training the personnel concerned is necessary to optimise the controls and get the desired and safe performance results. Since many incidents and accidents originate in humans, efforts to reduce or control 'human error' begin with training. That should make a huge difference. If any deficiency in the equipment or personal capability exists, it must be improved by appropriate training. Training must be focused and appropriate to ensure safety.

Organisations may issue a 'Quality Instruction' as indicated below (as an example) for implementing the requirements of 'Product Safety':

a. 'Product safety' is defined as the state in which a product can perform to its designed or intended purpose without causing unacceptable risk of harm to persons or property damage. AS 9100 D/EN 9100:2018 requires that we plan, implement and control the processes needed to assure product safety during the entire product lifecycle, vis-a-vis design, manufacturing, maintenance, repair and overhaul of the product, as appropriate to the business and the product we deliver to our customers. The following plan is implemented to control the process to ensure product safety in our organisation:

    1. Assessment of hazards and management of associated risks: Refer to risk registers, where operational process level risks, which are likely to occur, are identified and mitigated to ensure safety. Design risk with respect to product safety is managed through the 'Failure Mode Effect Analysis' or Failure Mode Effect and Criticality Analysis, a factor of safety for the component and introduction of redundancy in the functional system. The risk register is reviewed periodically to review the existing risks, identify any new risks and address all the risks related to the organisation (enterprise-level risks) and the products/process/people during manufacturing and MRO activities (refer to Annexure H-risk register).

    2. Management of critical items: Safety critical items have any deviation in functions, parts, software, quality characteristics, processes, fracture critical items, mission-critical items and key

characteristics. Variations/defects in characteristics/items may significantly affect the function/performance, service life, interchangeability, maintainability, safety and strength (FLIMSS) or form, fit, function and manufacturability. These items require specific actions to ensure they conform and any nonconformity risk is adequately managed. Refer to AAQG/SAE AS 9017 Control of Aviation Critical Safety Items (CSI).

3. Suppliers of reliable/safety critical items should be assessed for exercising the necessary operational/process controls and ensuring process capabilities before producing these parts or components by following the quality system. Production process verification (First Article Inspection Report) is mandatory for a part being produced for the first time. Further control guidelines on particular requirements and critical items (SRCI) analysis and reporting of occurred events affecting safety, etc., are available in the Supply Chain Management Handbook Section 2.4.3 (IAQG). Select and retain those suppliers, contractors, outsourced manufacturers and service providers that comply with all applicable laws and regulations and meet our product safety, quality and environmental performance requirements.

4. Ensure that all nonconformities and deviations detected in the past have been identified and practical corrective actions were taken. Records of nonconformity reports are available for reference. Lessons learned are collected and included in the 'knowledge bank'/knowledge repository for all personnel information. Communication of all such events and training of persons is the best practice. Effective communication can be achieved through safety posters displayed on the shop floor.

5. Safety has to be an integral part of the entire operation. All procedures, training programmes, operating manuals and work instructions may include aspects of product safety. All personnel are to be involved in product safety. It is a matter of applying safety in the conduct of the overall business. All employees working in an organisation (including contract employees) must be aware of their contribution to product conformity, product safety, and ethical behaviour. Management and supervisory training programmes plan training sessions that should cover product safety.

6. Employees may identify their requirements for safety procedures, selection of personnel protective equipment (PPE) and other

decisions that affect their safety. The basics of workplace safety should be in place. The overall culture should encourage employee participation. Involve the employees in problem-solving, safety observations, workplace inspections and training. Communicate the results widely and cultivate a culture that values and rewards these efforts. Treat employees with respect, honesty and fairness. Managers get employees involved in safety like they get anyone to do anything. Explore what they want and then respond to their requests, meeting their needs and addressing their concerns. When people begin to see positive results by eliminating hazards, they all want to get on board, and you end up with a waiting list of employees ready to contribute to safety committees and other activities.

7. Other practical points for ensuring product safety are the following:

b. They ensure proper ergonomics of the products (free from sharp edges and burr), safe handling and storage (avoiding metal-to-metal contact, adequate storage, protection from dust/moisture, avoiding fall or spillage, etc.). Personnel working in this organisation are aware of their contribution to product safety and quality. We machine metallic components. We do sheet metal work. All parts have sharp edges and corners. Handle them carefully so that your fingers or hands are not injured. All sharp edges are to be smoothened out.

c. Requirements for the products and services should consider personal and product safety. Employees use knowledge of their environment to understand what things are putting them in risky situations. Workers' ideas and suggestions have support from management to enable employees to feel they can genuinely make a difference in the lives of their co-workers. Bringing out any issues for others' safety is more important. Always be alert and report all product safety and quality concerns, including concerns raised by fellow employees, customers, product users or other interested parties. Safety is the concern of every employee – employees' motivation results in employee ownership of the safety process. Performance is always better when it comes out of commitment.

d. Cleaning: Prevention, detection and removal of foreign objects (like dust, moisture, broken pieces of rivets, washers and unwanted metallic and non-metallic debris) from the products. Establish a system of FOD checks.

e. Exceptional handling and storage for sensitive products, marking and labelling, including safety warnings and cautions. Be aware of sensitive products and give clear instructions on handling them. There should be special safe handling and storage for hazardous materials.

f. Shelf-life control and stock rotation of items having specified shelf life are to be established. 'First in, First Out' methodology should be adopted for storing and issuing items received for production work, with periodic checks for conditions/deterioration during storage.

g. Safety precautions are to be taken by the customers on the products after the same has been delivered to them. These precautions are communicated along with the delivery of products and documents. Examples are 'Leaflet on post-delivery, operation and maintenance manual, and safe handling/storage of the products' at the user's/customer's premises, all of which must accompany the product when delivered.

h. All personnel must comply with the above instructions to ensure product safety.'

8.1.4 Prevention of counterfeit parts

Using the words 'appropriate to the organisation and the product' in the Standard provides adequate freedom for organisations to prevent counterfeit or suspect counterfeit parts and their use in the product being delivered to the customer. In the case of organisations offering hardware products like machined components, the use of counterfeit or suspected counterfeit parts does not arise. Similarly, counterfeit or suspect parts are not used for organisations delivering design and engineering software services. Therefore, the requirement applies to only organisations that assemble two or more hardware parts into a sub-assembly, out of which at least one part is received from an external source. Counterfeit parts or suspected counterfeit parts come from external sources. Suppose an adequate control/verification mechanism does not exist. In that case, there is a chance that the counterfeit or suspected counterfeit part can be used in the product/equipment/assembly being delivered to the customer. Therefore, it is mandatory that the organisation plan, implement and control the process to prevent the entry of counterfeit or suspected counterfeit parts into the production line.

'Counterfeit part' is a fraudulent product. It is an unauthorised copy, imitation, substitute or modified part (e.g., material, part, component), which is knowingly misrepresented as a specified genuine part of an original, authorised manufacturer. It may be an altered part made by a source without legal right, with intent to mislead, deceive or defraud. Examples of counterfeit work include, but are not limited to, used components represented as new. Counterfeit parts can cause personal injury, mission failure, reduced reliability and product recall, potential loss of contracts, shutdown of manufacturing lines, negative cost and schedule impacts, and penalties for companies and individuals. Government and industry investigations have shown that the risks of receiving counterfeit parts vary based on the supplier entity providing the parts. Procuring directly from the original component or equipment manufacturer (OCM/OEM) reduces counterfeit risk. Procurement requires a documented, unbroken chain of custody from the source of manufacture for all components. Purchase order requirements typically require that the suppliers provide original component or equipment manufacturer (OCM/OEM) traceability documentation, which could include authorised supplier certificates of conformance, supplier shipping and receiving documents and authorised supplier packing slips. Part manufacturer's approved (PMA) parts are not counterfeit.

The organisation should plan, implement and control the process that prevents entry of counterfeit or suspected counterfeit parts into the production line by providing adequate training to persons. Training relates to the awareness and prevention of counterfeit parts. Designers/manufacturers of electronic assemblies must establish an obsolescence management and monitoring program for the parts used in the assembly/equipment. Suppose the parts are being procured from external sources. In that case, arrangements like direct procurement from the original manufacturers, authorised distributors or other approved sources will be arranged. Ensure processes are in place to quarantine parts that require testing and verification until they are verified as authentic. The detected counterfeit parts are to be reported to all concerned to prevent the entry of such parts into the supply chain. During the purchased product's verification and test, methods to detect and isolate counterfeit parts will be used. Suppose it is suspected that a counterfeit item may have infiltrated the supply chain. In that case, action must be taken to locate any parts in stock and on assemblies and ensure they are quarantined and marked as nonconforming. If suspect items are left at the customer facility, a recall procedure must be adopted. It is necessary to keep monitoring counterfeit parts reporting from external sources.

To safeguard against the purchase and use of spurious spares/equipment/modifications, it is stipulated that the organisation's purchasing process shall have to satisfy respective regulatory authority requirements with regard to the use of non-certificated suppliers. The organisation shall take appropriate measures to prevent purchasing counterfeit/unapproved products. It makes it mandatory that maintenance organisations purchase spares/any additional equipment/modification for fitment on aircraft only from authorised suppliers duly authorised by the regulatory authorities.

Organisations may issue a 'Quality Instruction' (indicated below as an example) for implementing the requirements of 'Prevention of Counterfeit Parts':

'Counterfeit part is defined as an unauthorised copy, imitation, substitute or modified part (e.g., material, part, component), which is knowingly misrepresented as a specified genuine part of an original or authorised manufacturer. Examples of a counterfeit part include, but are not limited to, the false identification of marking or labelling, grade, serial number, lot number, date code, documentation or performance characteristics. Counterfeit part means the part that contains unlawful or unauthorised reproduction alterations that have been knowingly mismarked and misidentified to be an authentic, unmodified part from the original manufacturer or a source without the express written authority of the original manufacturer. They include suspected parts and fraudulent parts.

Counterfeit parts can jeopardise the performance, reliability and safety of Aviation, Space and Defense (ASD) products. ASD products are susceptible to introducing counterfeit components into the supply chain because the systems are intended for use over a long lifecycle. As a result of the potential risks associated with counterfeits in ASD products, several regulatory agencies have established a regulatory basis for addressing false and misleading statements, including assessing potential civil remedies and deterrents (fines, certificate forfeiture, etc.) for persons engaged in such activity. Governments have also enacted laws that penalise trafficking unapproved and counterfeit parts. SAE guideline standard AS 5553: 'Counterfeit Electronic Parts; Avoidance, Detection, Mitigation, and Disposition' helps end users detect counterfeit components in their inventory (refer to Supply Chain Management Handbook, IAQG). Although most of the material and examples are from the electronic sector of the industry, the principles and methods apply to other commodity types.

AS 9100 D/EN 9100:2018 requires that organisations plan, implement and control the processes needed to prevent the use of counterfeit or suspect counterfeit parts and their inclusion in the product(s) they deliver to the customers. The best way of avoiding receipt or use of counterfeit parts is to be aware and prepared at all levels and functions of the organisation. The following control plan is implemented to control the processes needed for the prevention of counterfeit or suspect counterfeit part use in the organisation:

1. Personnel training: Purchase and quality control personnel are trained in the awareness and prevention of counterfeit parts. Training of

appropriate persons in the understanding and prevention of counterfeit parts; written instructions are given on identifying, avoiding and correctly handling and disposing of counterfeit products.

2. Parts obsolescence management: Diminishing manufacturing sources and material shortages (DMSMS) is defined as: 'The loss or impending loss of manufacturers of items or suppliers of items or raw materials'. Monitor shelf life of items like rubber, seals, araldite, sealant, paint and other consumables and monitor parts obsolescence programme (applicable for electronic parts, which become life expired or obsolescent during the product's useful life). Shelf-life monitoring of life items is carried out in stores. There must be the application of a 'parts obsolescence monitoring programme'.

3. Risk evaluation is carried out before approving and using a supplier concerning the risk of supplying nonconforming/counterfeit parts from any supplier and identifying and controlling other risks.

4. Purchasing process and control of source of supply: It is essential to flow down purchase orders and contractual requirements to ensure sub-tier suppliers implement an appropriate strategy to ensure that products delivered or furnished are not counterfeit/unapproved parts. The purchase order issued includes a communication to the suppliers about the need to prevent the use/supply of counterfeit parts in the parts/items they supply to us. Purchase of items with shelf life, which are likely to expire during production or post-delivery life of the product, and items likely to become obsolescent are purchased with a specified minimum life remaining at the time of receipt of supply. Terms and conditions included in purchase orders mention the particular life requirements. The minimum life required is specified in the PO. Life available should be marked on the container/certificate of the item. Purchase orders shall ensure that all received items are inspected/verified before acceptance. Parts are to be bought from approved suppliers only. Most counterfeit parts enter the supply chain through independent distribution, i.e., through brokers or agents. Purchase orders shall include a condition that if any item is found or detected/suspected to be a counterfeit part due to verification at the receiving inspection stage, legal action will be initiated with jurisdiction '(your place)', and the concerned government enforcement authorities will be informed. Suspected counterfeit parts should not be returned to suppliers or recycled but sent to proper legal officers. Mentioning the same in the PO will deter suppliers from making a mistake.

5. Organisations should assess the long-term availability of authentic parts and part sources for production and support systems during the proposal stage and program design and planning efforts. When assessments indicate availability risk, consideration shall be given to steps such as lifetime buy, system redesign, alternate/multiple sources, substitutions and planning for adequate procurement lead times.

6. Verification of purchased/returned products: A conformity certificate/test certificate/warranty shall be obtained from the original manufacturer/OEM for all metallic raw materials and other supplies in the airborne products delivered to customers. The items shall have their parts number/specification with revision status and traceability/batch number identified on the part and the conformity certificate/test certificate/warranty. To meet the requirements for assuring traceability traceability of parts and components to their original or authorised manufacturers; identification of batch number/heat number/lot number of the item is insisted upon and ensured before acceptance. Authorised distributors who supply OEM products, raw materials and other airborne items shall get the original conformity certificate/test certificate/warranty from the original manufacturer or other approved sources. A copy of the original conformity certificate/test certificate/warranty shall be supplied to us per the purchase order's conditions. The items shall have their parts number/specification with revision status and traceability/batch number identified on the part and the conformity certificate/test certificate/warranty. Test reports of critical raw materials of airborne products shall be validated by testing a sample of the raw material received in an approved laboratory. However, this inspection or periodic testing is undertaken when there is a high risk of nonconformities, including counterfeit parts from the supplier, as determined by the organisation/customer.

7. In-process investigation and reporting: Any instance of supply of a non-conforming part coming to light is notified as 'flow down of corrective action requirement' to suppliers. Any received counterfeit part or detected counterfeit part should be quarantined and shall not be returned to the suppliers since it is the evidence of receipt of counterfeit parts from the supplier. Inspection or periodic testing, as applicable, should be planned and verified when there is a high risk of receipt of non-conformities, including counterfeit parts from suppliers. Once rejected, counterfeit or suspected counterfeit parts shall be made unsuitable for the intended use, broken, bent, damaged and permanently marked to

prevent re-entry into the supply chain with permission from enforcement agencies.

8. Material and parts control: Ensure lifecycle planning and counterfeit avoidance plan are negotiated with the customer. Ensure a common understanding of the prevention of counterfeit requirements. Ensure all customer requirements are shared with the affected functions. Consideration should be given to options that include parts obtained from OEMs, OEM-authorised distributors and other authorised sources.
9. Software organisations must verify and validate the OEM software/open software tools they buy from authorised/approved sources and obtain a warranty for their performance characteristics.

All personnel are requested to comply with the above instructions to prevent counterfeit parts in products delivered to customers'.

8.2 Requirements for Products and Services

8.2.1 Customer communication

Get closer than ever to your customers; so close that you tell them what they need well before they realize it themselves

– Steve Jobs

The organisation needs to know much more than its customers. As customers, when we want to buy something, we only look at the source, perhaps inquire about it from the customers, and based on the communication, we decide to buy it before we get to know more about the product. What was communicated, when found to be accurate, we were satisfied. If the product meets our requirements and enhances our satisfaction, we may pursue it or look elsewhere.

Practical communication skills are fundamental to success in many aspects of life, and effective communication with customers is essential for business success.

Even after designing and producing the best product that meets all the general requirements of the customers, if the product's goodness is not appropriately communicated to the customers by any media, the customer may not come to know about it. The business process can commence only when the customer knows about the product and interest is communicated to the organisation. This is an inquiry or a request for a quote (RFQ). The investigation or RFQ may not give all the features and characteristics of the product, may not indicate the performance results, and perhaps provide only a few product/technical requirements. The customer wants it for use. He does not know how it is designed, developed or manufactured, or the technical details of making it suitable for use. Still, he wants all the comfort/benefit of its use.

So, when the customer requirements are determined, the organisation further communicates with the customer, explaining what the organisation can provide to meet the customer requirements and enhance customer satisfaction. This is the review of the needs related to the product. Let the customer be an individual or an institution.

Once the contract review is completed, the marketing department or the authorised person in the organisation has to make arrangements to send a quotation to the customer within the due date. This is the quotation in business language, including terms and conditions; the price the customer has to pay is indicated.

In this communication to the customer, details of the product being offered with a full description, cost and the terms and conditions of supply, including the organisation's business practices, are included. The communication should also clearly answer specific customer inquiries, indicate the organisation's requirements and offer any additional product features that the organisation can supply. The product details being offered must be communicated in the quotations. Most of the hassles in understanding the customers' and products' requirements are related to communication. As an example of packing for export, if the products must be packed in wooden packing cases, pallets, boxes, etc., it must comply with the guidelines for regulating wood packaging material in international trade for fumigation. The quotation must clearly state and include compliance with the regulatory requirements, such as taxes, customs duties, etc. Quotation is the formal handshake towards the contract between the customer and the organisation, and it is a commitment to supply. In the quotation, it is desirable to indicate how subsequent changes to the product, if any, and the changes in terms and conditions that the customer may request will be addressed.

The quotation is the first formal communication the organisation makes with the customer. Therefore, the customer expects a comprehensive answer to all the potential questions the customer may have. Suppose the organisation has listened to the customer and understood precisely what the customer's needs are. In that case, most of the questions are answered or implied in the quotation or the clarifications sought from the customer. The customer indicates his views or seeks further clarifications until the communication is effective and beneficial to both the customer and the organisation. Acceptance of the quote when received by the organisation, the contract is signed, and matters are mutually agreed upon.

Suppose the acceptance of the customer order is acknowledged. In that case, the communication accepts the contract between the customer and the organisation, and the commitment to supply has become legal.

While the above sequence and communication methodologies are accurate in small- and medium-sized enterprises (SMEs), manufacturing components or small assemblies of any engineering discipline, more prominent organisations designing and producing aviation, space and defence equipment have long-standing business relationships with their customers. It might have been started by building capabilities over time and entering into a memorandum of understanding (MoU). It might be the designer organisation issuing a license to a manufacturing organisation based on a mutually beneficial relationship like a partnership or joint venture. Large organisations like public sector undertakings, private corporations and software product design and engineering service providers in India have MoUs or service-level agreements (SLAs) signed with their customers. Purchase orders may be on an annual contract basis or for equipment/projects to be delivered over a few months. Regular, periodic, run-of-the-mill inquiries, contract reviews, quotations, order acceptance, etc., may not happen in such cases.

MOUs are not legally binding but carry a degree of seriousness and mutual respect stronger than a gentlemen's agreement. Often, MOUs are the first steps towards a legal contract between two organisations. MOUs are usually followed by purchase orders indicating the statement of works (SOW). A service-level agreement is an agreement between two parties, where one is the customer and the other is the engineering service provider. This can be a legally binding formal or an informal 'contract'. SOWs are framed after mutual consultations, feasibility studies and agreements on service, schedule, efforts and other terms and conditions.

The department or section that enters into the contract with the customer has the obligation and responsibility to obtain feedback from the customer and initiate actions concerning any complaints that the customer may make during the contract period.

The business depends on effective communication between the organisation and their customers on all necessary occasions. While establishing resources, the means of contact are also decided upon, and arrangements are made. Customer communication is intended in the Standard at several places such as customer focus, contract review, quotation and acceptance of order, design and development stages, notification of nonconformity to product, if any, etc. Customer communication usually takes place in any of the following methods:

a. Communication through advertisements or marketing information to existing or potential customers, identification of customer needs, product details, drawings and specifications.

b. Specific information in response to a customer enquiry, request or instruction such as quotation, customer order, order acceptance, amendment to order, delivery documentation, invoices, release notes, conformity certificates, e-mail and general correspondence.

c. Communication in reply to customer feedback, suggestions and complaints.

Customer enquiries, contracts or order handling, including amendments, quotations, order forms, confirmation of order, amendment to order, e-mail and general correspondence, visit reports or notes to/from customers, delivery documentation, invoices, credit notes, etc., demonstrate compliance with the customer communication process. Response to complaints and complaint handling process (timely response and adequate actions), customer feedback, customer satisfaction determination, etc., demonstrate communication effectiveness with the customer. More objective evidence of the customer communication process is found when the order acceptance process is verified. Communication with respect to the acceptance of concessions by customers should be available.

> '*Organisations usually respond to customers on time. However, it is observed that the customers do not respond in time to the queries/clarifications from the organisations, especially when there is an error, and it is determined that the customer is responsible for the error. For example, the component drawing number and rev status quoted in the PO do not match the current drawings available with the manufacturer, or there is undue delay in responding to the customer satisfaction survey sheet sent by the organisation. In such cases, an organisation can use all their communication skills effectively to not derive any dissatisfaction from the customers. It is business, and "the purpose of business is to create and keep customers"!*'

Care and custody of property belonging to the customers or external providers are addressed in para 8.5.3.

At the same time, if the response from the organisation is delayed for any feedback or suggestions made by the customer, even if the customer is not asking, the effectiveness of the Quality Management System is affected. Any undue delay, say more than a month, without even an interim reply to the customer through any media, may attract even nonconformity from the external auditor!

8.2.2 Determining the requirements for products and services

The quotation is the culmination of all the interactions and discussions about the products and services the organisation can offer. The quote indicates the products and services the organisation offers, the expected time frame for completion of all the products and services and post-delivery support, if any. In this, the organisation is committing to the customer to fulfil the customer's needs as to the quality of the product or service with full description, cost and the terms and conditions of supply. The quotation stage in more significant projects and organisations comes after several clarifications and correspondence, teleconferences and meetings. This results from a feasibility study where all the features and characteristics of the product/service that fulfil the customer's requirements are mutually agreed to. The requirements for the products and services are defined with drawings, parts list, materials and process specifications or standards, and tools like jigs, fixtures, moulds, software programs, etc., as appropriate. Process flow charts required production documents such as manufacturing plans/schedules, travellers, routers, process cards, quality plans and inspection documents may also be indicated. This information may act as a confidence-building measure for the customer. Positive actions and steps, care, considerations and verifications being carried out by the organisation to ensure consistency in production may help develop and sustain a healthy relationship with the customer. Reiterate those actions and steps such as production process verification, control of production process changes and control of production equipment, tools and software programs.

Determining the requirements for the products and services is an important activity that must be done in the early stages of the business process. If the organisation is known, the customer sends a request for quotation (RFQ) or identifies their need and asks the organisation to conduct a feasibility study and make a proposal. Customers may float a tender enquiry through any media if they do not identify a capable organisation. The customer may even communicate their needs through a formal contract, RFQ, an e-mail, a phone call, etc. The customer may specify the requirements in general or specific if details are known to him, including product drawing, technical specifications of the product/equipment, brand names, quantity and performance requirements as known to the customer.

> '*Yes, as known to them, because most customers do not know precisely what they want. They may have a fair idea of their need and hear about the products manufactured by various organisations. Your family felt the necessity of a refrigerator, and you want to buy one. What requirements will you specify while doing the market search for the product? As a customer, you do not know about the refrigerator you want - except perhaps the size, colour, etc. You may not know whether the refrigerator you want should be working on a vapour compression system or a vapour absorption system, what is the refrigerant used in the system, what is the size of the capillary tube, what is the coefficient of performance (COP) of the refrigerator? Is the product RoHS compliant, i.e. meeting the statutory and regulatory requirements? Is the refrigerant environmentally friendly, and will it not act to deplete the ozone layer?*
>
> *Almost all customers are in the same boat when identifying the requirements for the product they want to buy and use. They get consoled by the fact that the designer and the manufacturer of the product know what the user wants, and they have ensured compliance with the requirements and the desired qualities - characteristics that meet the requirements*'.

The customer may see a demonstration and compare the costs of various components to check if the product will likely meet his needs. That is common for consumer durables. For aerospace products, generally, designs are made by one organisation, manufactured by another and used by some other significant organisations. Therefore, each type of aviation, space and defence equipment has to be clearly defined by the customers/users based on their requirements to the designer organisation. Suppose the design has been verified, validated, and released for use by the design organisation. In that case, the manufacturing technology must be transferred to the manufacturing organisation, or the design, drawings, and specifications must be given to the manufacturer. When it is unclear what exactly will satisfy the customer's needs, the organisation undertaking design and development has to complete a task involving modern or futuristic design and 'state of the art' technology with the available resources. When the design specifications are met, and technology is transferred, the manufacturer's job becomes more accessible. Maintenance and repair schemes are arrived at during the development stages so that when the equipment is available after manufacture, the necessary resources and methods are laid down for upkeep and servicing.

The organisation, with the involvement of various functions like marketing, customer services, sales and finance, etc., may initiate a process of identifying the requirements specified by the customer, including the requirements for delivery and post-delivery activities. Responding to the enquiry quickly with a quotation is possible only if the products are to be delivered off the shelf. But aviation, space and defence organisations generally sell no products off the shelf except OEMs, which stock spares for sale. Before production, most airborne products or services are designed and approved by the customer/regulatory authority. Otherwise, the organisation must initially carry out design and development activities or associated engineering services. Even for the fabrication of ground support/handling equipment, tooling, etc., detailed design, development and production processes will evolve according to the aircraft-related specific applications. The purpose is to find out precisely what the customer wants, and it may be a product already available off the shelf in the organisation or ultimately designing, developing and manufacturing a new product. Therefore, identifying the customer requirements and requirements related to the product is an involved and time-consuming activity that must be done effectively and meticulously. In the case of significant equipment, aero-engine, aircraft, rotorcraft or any associated equipment, therefore, when the customer initiates the requirements, a detailed feasibility study is carried out in consultation with all the related agencies to finalise or propose the requirements related to the product and its intended use. The feasibility study identifies all the requirements associated with the product.

In the case of established products or manufacturing components and sub-assemblies, determining requirements related to the product can be done by scrutinising the tender requirements, presenting product information, visiting the customer/site and discussing with the customer. Where there is a need for clarity, the customer should be asked to confirm all that is required and identified by the customer for the product. Indicating, explaining and justifying additional or alternate requirements that the customer did not indicate will add value to the process. But organisations should do that with a prime motive/principle of enhancing customer satisfaction and not increasing the cost or complexities.

Organisations sometimes face the dilemma; they express it sometimes: how do they confirm customer requirements when the customer provides no documented statement of need?

> '*It was only a telephone call, and we responded quickly, thinking it was a good business from a good customer. They do not accept the product,*

> *saying it was not what they wanted. Who will bear the cost of transportation? And we lost the customer forever*'.

It is always better to communicate through the simplest documented way of communication – an e-mail to the customer or his representative – and request confirmation about what was understood about the requirement conveyed by the customer.

Organisations revolve around their customers' needs, enabling success for both. Therefore, organisations know and listen to their customers, anticipate needs and provide solutions. Organisations must be proactive, innovative and responsive, provide value and inspire customer loyalty. It is, therefore, essential that requirements related to the products are clearly understood by the organisation and described in as many details as necessary in a document rather than just meeting the requirements stated by the customer.

> '*Someone says- "Is it insufficient to meet the requirements specified by the customer? Why indicate better alternatives and additional features to the product, which adds to the cost of the product? After all, the customer is always right."*
>
> *Many debates occur on the adage "the customer is always right" and more recent findings that "the customers need not be right always"*'.

Robust and long-lasting business relationships thrive only when both parties identify and nurture complementary, long-term profit objectives, respectable relationships and satisfaction. Suppose the organisation and customer relationship is described as symbiotic or creating synergy. In that case, it may meet the country's overall economic objectives. It may help achieve the objective of creating a pleasant industrial environment in the government. Suppose it is a mutually beneficial relationship between the organisation and the customer. In that case, that is the model for business excellence, and it is this relationship principle embedded in the Quality Management System. It is not a question of who is correct but what can be done to enhance customer satisfaction. (Symbiotic means living together in more or less intimate association or close union of two dissimilar organisms or species. Synergy is when different entities cooperate advantageously for an outcome.) If used in a business application, teamwork will produce better overall results than working towards the same goal individually or having individual pockets of excellence. My boss used to say, 'We don't want pockets of excellence; we want excellence all over, spread across the organisation'.

Everyone talks about the organisation adding 'value' to the product or services when trying to meet the requirements considered necessary by the organisation but not explicitly demanded by the customer. The customer has not demanded it because the customer is not aware of it. Therefore, indicating alternative or additional features that add to the product's usefulness, if known, while confirming the product information to the customers along with the quotation meets the objective of enhancing customer satisfaction. The international standard requires that organisations identify product requirements not stated by the customer but necessary for specified or intended use, where known. The conditions not stated by the customer but necessary for the product to reach the customer in good shape can be as simple as the packaging for the product and conveying the safety requirements for handling the same.

External auditors specifically ask the project managers to indicate the actions or features considered necessary and added to the organisation's project/product/service. Still, the customer did not explicitly demand those actions or features. Most project managers could not indicate or respond in a manner that would expose their competency, process excellence or reveal their goodness, even when providing some of those values to the customers and projects.

'*Good auditors persist by asking them –*

- *What was the value added to the engineering service?*
- *What is the brand value you have provided to the customers?*
- *Is there any unique way of doing this service which others may not?*
- *Why do you consider that the customer should return to you for more business?*
- *What prevents the customers from returning to your competitors? Some organisations may do the same or better service at a lower cost than you provide?*

Now, the project managers understand and reveal a long list of value additions they usually provide to the customers- Unique processes, automation, Lessons Learned, Value creations, "Zero Distance" (remember the quote from Steve Jobs) and Best Practices, Knowledge Management, etc. '.

In many instances, it was observed that some of the organisations providing engineering services for ASD customers do not explicitly indicate the value additions they provide. It is a prerogative for the organisation to indicate the same in their feasibility study reports or proposals sent to the customers so that they take note of that as a reason to derive enhanced satisfaction. The organisation complies with the Quality Management System standard and shows excellence and goodness. It will enable new customers to open gates for excellent business with the organisation!

All customers, whether individuals or organisations large or small, do not know everything about the product, especially hardware products they want to use. How much do we know about the gadgets we use? We may know how to use them. But the organisation that produces the product should know more and convince the customer with more technical and other information about the process of realising the product, its performance characteristics, operation and maintenance features, benefits and ill-effects the products can provide to the customer during its use and maintenance – safety, reliability and maintainability.

Such value additions can include information on actions being taken by the organisation with regard to:

a. Prevent, detect, and remove foreign objects from the products.

b. Control of utilities and supplies during the production of the products.

c. Maintain good workmanship for the product.

Customers must be convinced about the additional value they can derive from the product. Besides the product's quality and on-time delivery performance, that aspect enhances customer satisfaction. It makes the customer hold the organisation that provides the product in high esteem and honour. The customer returns when the products do not!

Knowledge about the equipment is incomplete when organisations place an enquiry or order on aviation, space and defence organisations, where complex equipment is designed, manufactured and serviced. When requests are placed to go ahead with a feasibility study and define the configuration of a modern aircraft with its various features for the intended use, it takes years of effort and resources to bring out a viable proposal for project sanction. State-of-the-art technology, which the customers generally desire for the project configuration, might have become obsolete, and the customers want improvements when the product is realised. Therefore, some broad consensus is necessary for the identified and stated customer requirements and the requirements not stated by the customer but necessary for the specified or intended use.

Design, development, manufacturing, maintenance and use of aviation, space and defence equipment, including critical items used in these, are always under the scrutiny, surveillance and regulation of statutory and regulatory requirements. Therefore, compliance with the requirements when organisations engage in business is mandatory. If the scope includes any product that ultimately goes up in the air, the statutory and regulatory requirements must be met mandatorily for the product/organisation. The standard requires organisations to identify the statutory and regulatory requirements associated with the Quality Management System/business or applicable to the product (Clauses 1.0, 4.2, 5.1.2, 8.2.2, 8.3.3, 8.4.2, 8.5.4, 8.5.5 and 8.2.3.1, respectively). This is integral to the organisation's marketing or business development activities. It is only advantageous if the customers identify the requirements before the contract. Still, organisations must identify and act appropriately to coordinate with the authorities for the applicable statutory and regulatory requirements. Even when the customer has not indicated the requirement, it cannot be taken as an excuse for evasion. It is the organisation's responsibility to identify and comply with the statutory requirements. The organisation should also understand other requirements for the product's intended or specified use. The customer may identify some of the regulatory requirements if, ultimately, the customer has to obtain regulatory clearances like an 'Airworthiness Certificate' for the equipment. Therefore, it is always beneficial to get it clarified from the customer whether there are any regulatory requirements to be met by the organisation. Suppose the customer obtains certification, which is indicated in the contract. In that case, the organisation can be absolved of complying with the regulatory requirement. It is not true regarding the statutory requirements as the law applies to all citizens and organisations, and no one is above the law.

Director General of Civil Aviation (DGCA) and Centre for Military Airworthiness and Certification (CEMILAC) are the Civil and Military regulatory authorities respectively in India; Regulatory authority for Europe is the European Aviation Safety Agency (EASA) and for the United States of America it is the Federal Aviation Agency (FAA). Technical Standard Order (TSO) and Parts Manufacturer Approval (PMA) Procedure/Regulations; DO-178C, Software Considerations in Airborne Systems and Equipment Certification (FAA, EASA), etc., are some of the regulatory requirements related to aviation, space and defence products to be complied by the organisation, as applicable, in consultation with customers.

Any additional requirements considered necessary by the organisation for the successful conclusion of the contract and the supply of the products and services are also to be identified by the organisation. This can include identifying critical items, key characteristics and special requirements related to the product. It can be unique features, like in special processes while producing hardware components or assemblies.

The proposals cannot be valid for an indefinite period. Since everything changes over time, including incorporating improvements in the product characteristics, process, technology and cost, drawing a line and indicating a cut-off date in the quotation is necessary for the proposals to be valid. The customer thoroughly evaluates the quotations/suggestions before responding to the organisation.

They indicate their views or seek further clarifications until the communication is effective – matters are mutually agreed upon when the customer issues an order.

Some software organisations usually document customer-related processes as follows: The inputs received from the customer are – statement of works (SOW) or statement of requirements (SOR), work instructions, guidelines, specifications, customer standards along with relevant policies, standards and regulations used to prepare the statement of works (SOW). Unspecified/unstated requirements for product realisation and intended use are also identified and included in the SOW. Input requirement checklist is used to ensure that the product requirements, including delivery, post-delivery, statutory, regulatory and customer QMS requirements, have been considered. Deliverables are identified. Estimates for the efforts and milestones with deliverables are determined. Day-to-day project work is monitored on dashboards visible to all the team members and the delivery head. Process measurements may be the KPIs about the quality of the deliverables, effort variance, schedule variance, etc. Some organisations use customised process measurement metrics/tools such as 'On Time and Above Customer Expectations' (OTACE).

It is mandatory to reiterate or recall the organisational-wide activity of identifying the operational risks and review the actions taken to mitigate them when responding to the enquiry or RFQ to the customer. Determining if any potential issues can affect the business because of the customer's requirements is necessary. It may be regarding a new technology or the constraints concerning the ability and capacity of the organisation to provide or the short delivery time frame demanded by the customer, etc. The corresponding risks are to be assessed with the same criteria of impact and likelihood of occurrence (refer to Risk Register and Figure 6.1 Risk Assessment (RPN)). Suppose the risk (RPN) level is found to be more than the acceptance level decided by the organisation. In that case, appropriate mitigation actions are to be taken and ensure that the remaining risk is at or below the acceptance level.

Every organisation maintains an enquiry register to record all the customer requests related to the marketing/sale of products and requests for quotations (RFQ). After registration, a team of personnel will deal with the enquiry, at least from the marketing, production planning and control, production and quality departments/processes. The group will identify all the requirements, usually in a checklist, prepared specifically for each project/product/service delivery.

The checklist also verifies that the organisation can meet the claims for the products and services it offers in the quotation. Let the checklist be a comprehensive list of:

- characteristics and requirements for the product/service that the customer demands, along with other requirements like logistics, packing, delivery, post-delivery activities, etc.;
- corresponding offers that the organisation is providing;
- related statutory and regulatory requirements identified and being complied with;
- value additions the organisation is offering (already provided or additionally being provided) to the specific customer;
- special provisions (including crucial characteristics and critical items) of the products and services determined by the organisation;
- a confirmation that the operational risks (e.g., new technology, ability and capacity to provide, short delivery time frame) have been identified and addressed/are being addressed;
- pending issues, if any, with the customer or any contract or order requirements differing from the previous occasion are recalled and resolved;
- any other problems, requirements or related matters.

The checklist can be used for all subsequent references and contract reviews. Once completed correctly, the list becomes the documented information (a record) maintained for future reference.

It must be ensured that contract reviews are completed for every enquiry or order from the customer. Once the customer receives the order and the review is completed, the customer requirements are communicated to all concerned. When deviations from original conditions have been agreed upon with the customer, the appropriate record is maintained. The records retained will become a valuable reference in the future.

If, at the end of the review, the organisation determines that some customer requirements cannot be met or can only partially be met, communication must be given to the customer. The organisation is free to negotiate a mutually acceptable requirement with the customer.

8.2.3 Review of the requirements for products and services

8.2.3.1 Requirements

The requirements related to the product have been determined as above. It includes requirements specified by the customer, including the requirements for delivery and post-delivery activities; requirements not specified by the customer but necessary for specified or intended use, i.e., the value additions offered by the organisation; statutory and regulatory requirements applicable to the product and service and any additional requirements considered necessary or specified by the organisation.

Sending out a quotation to supply something, a product or service, and acknowledging an order from the customer are both deemed a commitment by the organisation to supply. Therefore, it is a contract with all the legal rights and remedies.

When customer requirements, especially requirements related to the product/service, are identified, a contract opportunity is at the doorstep. The organisation has the intention to do something good, some good business.

Therefore, there is a need to review all the arrangements to receive the guest, i.e., confirm the organisational readiness, whether the organisation can meet these requirements or not, with an objective of customer satisfaction. First, the review relates to the organisation's capability to address customer requirements.

1. What are we aiming at by doing it? What is the purpose? Objectives.
2. Do we have the entire wherewithal (money, energy, support) for doing it? Resources, men, machines, tools, software programme, measuring devices, logistics, time to do and deliver.
3. What are the processes required to do it? List out internal processes and externally provided processes, if any.
   (For each process, can we identify the following: Inputs, Steps, Activities, Interactions, Output?
   a. Who will do it? Competency/knowledge.
   b. How do we do it? Steps, activities, work instruction/ procedure.
   c. What are the game's rules? Controls.
   d. Where does the result of the process (product) go? External customer or internal customer).

e. How are the measurements made for the effectiveness of the process?

i. For conformity to requirements: quality, on-time delivery.

ii. For effectiveness: measure objectives /KPIs against the targets set.

4. What are the quality system requirements/criteria for conformity of the process Planning, Operation, Check and Act (PDCA): verifications, monitoring, measurement, analysis, review, improvements?

5. What documented information is necessary? As per the system requirements.

6. Can we do it? Capability decided by the cross-functional team indicated in the contract review record.

In other words, it is the feasibility review as to whether the organisation can do this within the time frame given, at the cost and quality levels specified by the customer. The assessment must be conducted before the organisation commits to supply a product to the customer. Commitment to supply becomes a contractual agreement. This is effective when the organisation conveys a message or agrees to undertake the production or supply of the item to the customer. The communication may be a quotation against a tender enquiry or a response to an interest indicated by the customer. Submission of tenders, acceptance of contracts or orders and acceptance of changes to existing contracts or orders are commitments to supply. The requirements in the Standard are specific to the reviews that the organisation has to complete mandatorily before communicating the commitment to supply.

This mainly points to the folly, sometimes, when some enthusiastic marketing manager or the chief telephonically agrees to supply something very soon because the customer is significant to the organisation. When the chief conveys the importance of meeting the customer requirements, product manufacture or delivery down the line to the planning or production/quality departments, he gets the stumbling blocks such as – 'Sir, we don't have the specific raw material, our specific machine is out of order, or the specialist is on leave'. The organisation is embarrassed when faced with the situation that 'the customer needs to be advised to wait or to accept the delay in delivery' because the internal resources/planning/production constraints do not reflect the organisation's capability to meet the customer's requirements. The contract review before communicating the commitment to supply is the best solution to the potentially embarrassing situation. When systematically done, the reviews reiterate the capability of the organisation to meet the customer requirements. The chief will henceforth hold back for review/support before committing to the customers!

It must be understood that verbal agreements or instructions shall not be construed as approval or authorisation to proceed.

Organisations need to specifically identify and document the stated and implied requirements of the customer. This should include relevant product-related documents like drawings, specifications, etc., given by the customer. Standards to be followed during manufacture, compliance to be ensured for the applicable statutory and regulatory requirements, special requirements such as First Article Inspection Reports (FAIR), technologies available and additionally needed, short-term delivery schedule, arrangements to interact with offshore customer reps, introduction of new processes or machines, requirement of new competencies, etc. (identified by the customer or the organisation) are the additional requirements related to the product. Some details, such as standard parts, commercial off-the-shelf (COTS) parts, standard hardware, etc., may not have been identified by the customer. Organisations should specifically identify them, including the current capability, the agreement reached with the customer on product specifications, etc., and document and communicate it before proceeding with the production planning process.

A systematic review should involve all the affected functions/processes in the organisation, such as planning/process engineering, manufacturing, quality, sales, etc. There is a need to include the respective departments/sections/processes to confirm their readiness to meet the customer requirements. It may be planning for considering the production schedules, production department for considering the resources, facilities and availability of competent human resources, purchase department for procurement and sources, stores for availability of raw materials, quality department for inspection/testing and associated requirements and marketing or finance for costing, etc.

When you know your strengths, capabilities and resources, you can estimate the time and cost required to produce and deliver the products or services and evaluate for post-delivery activities like operation, servicing and maintenance. This is done after a review of the requirements jointly by the affected functions in the organisation. The review process aims to identify if the organisation can meet such conditions.

Risks and opportunities are identified, and adequate actions are taken to mitigate identified and potential risks. Risk management activities must be linked to the contract reviews. Controller process sheets for planning production exist in the organisation for all types of contracts and customer requirements. The measurement of results must show that the targets are achieved. One point in the checklist concerns the risks identified and actions taken. Reference can be made to the risk register, where risks identified during planning and operations are listed. This register may have covered all the risks. When this register is linked to the contract review, any additional risks are identified because of the new inquiry or order from the customer, which is only to be assessed and acted upon.

Activity-based costing (ABC) is a method that measures the cost of a product based on the activities performed to produce the product. The technique involves the assumption that the actions drive the price. The product or customer goes the cost. This radically differs from the conventional costing systems, built on the assumption that the product drives the cost directly. The ABC system operates indirectly and supports expenses, first for activities and processes and then for products, services and customers, giving managers a clearer picture of the economics of their operations and services.

Well, what is the cost of an aircraft, say Airbus A320? Ninety-eight million US dollars (approximately). Everyone wonders why an aircraft should cost so much. What would be the cost of designing and developing such an aircraft? Development costs include engineering, tooling, recurring and non-recurring manufacturing labour, recurring and non-recurring manufacturing material, ground test specimen, cost of prototype aircraft and technology demonstrators, flight test, quality control, items purchased from external sources and financing costs. This development cost is to be recovered from each production aircraft. Production costs include all the above costs except ground test specimens, prototype aircraft, and technology demonstrators.

There are two approaches for cost prediction of an aircraft during the design and development stages: the 'top-down' approach and the 'bottom-up' approach. The 'top-down' approach is convenient for the preliminary stages of design. However, the bottom-up approach uses a detailed analysis of the cost of every aircraft component, material, production and overhead. All this information is combined to produce parts, sub-assemblies and assemblies of systems and integrated into the whole aircraft cost.

Organisations may use any specific format to facilitate the detailed review, where details can be identified through a planned checklist prepared by the concerned process owners together. The list, when filled, can become a record of the reviews conducted.

The review's conclusion should identify whether the organisation can meet the defined requirements. Documented information (records) of the results of the assessments must be maintained. The record should also indicate the issues identified and actions completed for the problems.

In the earlier example of the enthusiastic marketing manager or the chief of the organisation telephonically agreeing to supply something, the embarrassment could have been avoided if the organisation, after a contract review, identified the customer requirements, confirmed what they can provide and what they cannot, and when it is possible to supply, etc., through a recorded communication. The customer's requirements are more important than the individual needs in the customer organisation. It means that when working with a customer organization, the specific needs and priorities expressed by the customer as a whole (their official "requirements") should always take precedence over the individual needs of any single person within that organization. The standard brings out the solution that the organisation must confirm customer requirements before acceptance when the customer does not provide a documented statement of their needs.

The customer is always on the lookout for continual improvements in his needs. Therefore, the customer enjoys the privilege of changing or improving his requirements related to the product. Customers pay, and the organisations exist because of the customers and not vice versa. Customers may indicate a change in the already placed order, which the organisation has committed to supply. The customer may change supply conditions during the contract's validity. Significant changes, which require detailed design and development study, development of technology for manufacture, additional resources for realisation processes, changes or modifications in the existing aircraft/equipment, etc., may be required to address the customer needs. Simple changes in the component drawings or configuration, supply conditions, etc., will need a contract review to evaluate the changes and make appropriate decisions.

Organisations that deal with significant aircraft, aero-engine or accessories projects generally complain that the project delays are partly due to the changed requirements from the customer. State-of-the-art technology, scientific improvements and optimisations, which continue to happen worldwide, especially in materials, processes and equipment, add to the lead time required to reach various milestones in these projects. The Standard intends to ensure that the business always fulfils customer expectations, even though all the customer expectations cannot always be met. Organisations should review the contract for the impact it brings, technically and financially, due to the changes suggested by the customer. Organisations that plan for these issues pragmatically, communicate with customers honestly and forthrightly and manage the associated risks will be able to improve their business sustainably.

> *'A newspaper recently reported that the top management of a private software giant stated that his organisation suffered huge losses*

on account of effort and time schedules due to periodic changes suggested by the government for a government-funded project. Some works were re-done after 30 to 40 % of the work was progressed. Perhaps the contract reviews, if done, were ineffective in identifying the impact of 'the changes' suggested by the customer! The willingness to accept customers' opinions should not be at the cost of the organisation's/stakeholders' interest, even if the changes/opinions are from bureaucracy/authority. The organisation must have evaluated the impact of the suggested changes and then convinced the customer to share the consequences.

Those bureaucrats from the North Block (government office) would have found that a flying visit and happy vacation in the beautiful sprawling campus with a floating restaurant beneficial to deliver the piecemeal requirements before the bi-weekly review by the ministry!

It was also possible for the organisation to adopt a more flexible methodology, such as Agile, for the software service realisation project. With a high priority to satisfy the customer, Agile methodology can accommodate changes in requirements later in developing the product/service. (Read more on Agile methodology in Para 10.1)'.

In large projects, the organisation's decision-making concerning these changed requirements from the customer is critical concerning the progress and timely completion of the projects. Therefore, organisations should immediately review the requirements and proudly convey their capability or politely decline to incorporate the changed conditions into the current contract. The resultant slippages in delivery schedules and changes in cost, if any, as a result of implementing the changes in the requirements of the customer should also be communicated while agreeing to comply with the changes in the product. Without fail, the organisation has to ensure that the related and relevant documents, such as drawings, specifications, test schedules, estimates, delivery schedules, quality plans, etc., are amended according to the customer's demands. Customer satisfaction or the organisational reputation will be at stake if the agreed changes are not documented, communicated and complied with once the customer has indicated the changes. There should be a process to identify and take appropriate actions of review, communication and documentation, including the changes in the realisation processes and verification for the configuration status accounting at the time of final acceptance or delivery of the product. The requirement is a point in the contract review checklist.

When the customer provides no documented statement of requirement, the organisation must have a system to confirm these requirements, changes to PO, etc., before the organisation accepts the order. However, the same should be confirmed in writing to the customer with a default statement that if no communication is received within 15 days, it will be deemed accepted by the customer. Sending mail indicating the oral acceptance of the changes is adequate.

The note that is added in the Standard points to some situations like internet sales, where a formal contract review is impractical for every order. In such cases, it is advised that the review can cover relevant product information such as catalogues or advertising material for off-the-shelf items/products such as hardware, electronic components, etc.

In software service provider organisations, a review of requirements related to the products is done when a 'request for proposal'/quotation (RFQ) is received for any project. An input checklist is used to verify the requirements related to the organisation's engineering services, including identifying and using relevant technology resources, including software tools. Contract review is performed as a 'Project Management Meeting' involving relevant functions. Decisions are recorded in the form of minutes of the meeting and stored in the project folder. Schedules and efforts are estimated. Costing is done. Identified deliverables to the customer are the process output monitored and measured for the quality requirements and meeting the pre-determined organisational objectives of controlled effort variance and schedule variance.

'Should costing' is a process by which one can estimate the cost of a part or component with reasonable accuracy. It is based on the raw material, manufacturing and overhead production costs. Designers estimate the cost by analysing the engineering models to understand the required raw material, defining the manufacturing processes required to deliver the required configuration and calculating the total costs using rate data related to material and processing costs. Generic or customised software tools are available for 'should costing'. SMEs can use such tools to respond to the RFQs from organisations that look for competitive quotes from SMEs for economical purchases.

Effort variance is the difference between planned and actual effort spent on the activities. Schedule variance is the difference between the planned milestone dates and the actual milestone dates. It is estimated that a job can be done by taking eight person-days. You have only two persons to do the job and fix a schedule for completion of the work/scheduled delivery before closing hours after four days. If two persons do the assignment in four days, there is no variance in effort, and the schedule is met (no variance in schedule). If four people are employed, it takes two days to complete it. The effort is not varied (effort variance is zero). Still, the schedule gets shortened by employing more people (schedule variance is −50%). If two people are used on the job, and it takes five days to complete it, effort variance is +25%, and schedule variance is +25%. The target in most software engineering service organisations for effort variance and schedule variance is +3%, and defect density is less than 3%. (You can imagine why some software engineers return home late after the day's work!)

Software organisations engaged in engineering services usually document the review of requirements related to their service. Statement of works (SOW) is reviewed for product requirements, planning schedule, estimates of efforts, unspecified needs and organisational capability. Issues arising out of the review are resolved. Approval from the customer is sought for the identified service requirements, mutually agreed upon. Records of reviews and resultant actions are retained. Customer approval is taken on the agreed SOW. Guideline documents such as domain checklists, project planning, control documents, etc., are reviewed. In case changes to product/service requirements are indicated subsequently, the same is communicated to the relevant personnel. The risk management template, part of the review documentation/domain checklist, is used to identify, evaluate and prepare mitigation plans for all the identified risks. The risk register will have identified threats, assessments and actions as necessary for high risks for the planning and operations stages of the service delivery.

Maintenance organisations have to maintain the contracts with the OEM for the scope of work defined and maintain the data required for servicing, which may include subcontracting parts of works. There may be a standard list of works to be carried out for each type of aircraft. However, additional works, repairs or replacements may not be pre-determined but discovered during pre-survey, dismantling and maintenance. The organisation's commitment to supply a maintenance service to the customer should ensure the above aspects as part of the contract, and reviews should consider it.

8.2.3.2 Review records

It is mandatory that the organisation retain the records of the review results. A typical example of the contract review record is indicated in Annexure I for information. The contract review must be redone when any new requirement for the product or service is added. The new requirement can be a change in the product characteristics, physical or performance characteristics, or product configuration. The same cross-functional team must review the changes or new requirements for impact on cost, quality, schedule and objectives. The negative impacts must also be studied. Record the decisions taken appropriately.

LOGO	CONTRACT REVIEW	NAME OE ORGANISATION	
CUSTOMER		PO #	
DRAWING		REV	
PROCESS DEFINED			
SPECIAL REQUIREMENTS			
SURFACE TREATMENT/JIG PT			
RAW MATERIAL/ GRADE			
RISK EVALUATION REFER REGISTER (ANY ADDITIONAL RISK?)			
PRODUCTION PROCESS CAPABILITY			
LEAD TIME/ SPECIAL PROCESSES			
WORK ORDER			
SIGN / DATE			

Figure 8.3 The 'chop' or stamp.

Well, the contract review record checklist indicated is quite exhaustive. Most points concern components, tools and hardware assembly manufacturers, partly for software engineering services organisations. You can customise it for your products and services. Incidentally, a miniature contract review record was found 'chopped' on the POs received by some Southeast Asian SMEs (Refer Figure 8.3).

A stamp is generated with minimum information because everything is controlled by the head of the organisation, who has no time for elaborate review and record-keeping because of small parts, batch quantities, and several orders in a year. Every order is similar and stands alone. So, he takes only a cursory look at the POs and, with a smile, hands over the contract to the shop floor for chop and execution. The documented information 'chopped'/ stamped and retained at the back of every PO is as follows:

'The advantage of the implemented QMS is that when these kinds of brief records are seen, the auditor can go into many other details to understand and evaluate the process for effectiveness. The maintained documented information showed that the company had contract review planned/restricted to only POs of more than 37000 USD. About 200 odd POs from about 45 customers were received during the year—the auditor requested to show another chopped PO, another sample, for further verification.

The CEO says, "We don't have any, as we have not done any".

"Very well," said the auditor, "then you can take action on this nonconformity".

"Why it's nonconformity?"

"The standard requirements do not allow any organisational freedom or discretion concerning the retention of contract review records, which is the result of the review of requirements related to the product."

As they wake up from the nightmare, the CEO says, "The power of Consultants over our QMS is terrifying!"

You know very well when similar kinds of 'chops' were seen in more than one organisation in the country. Still, you don't reveal the influence of consultants on the organisational processes and documented information!

The time when the slogan was "We document what we practice, we practice what we document" has gone! What we document and practice must conform to the Standard'.

Other mistakes found in the contract review process of some organisations are:

a) The configuration of the product was not captured correctly
b) No operational risks have been identified for the contract/PO and
c) The organisation did not ensure that relevant documents were amended when product requirements were changed'.

8.2.4 Changes to requirements for products and services

It is the organisation's prerogative to change the internal documents. Product- or service-related documents are held by the designer or the manufacturer of the product/service. However, the customer requirements may vary according to the needs or expectations of the customer. Customers can request a change in the product or service that the organisation is delivering. The request has to be studied, impact analysed and contract reviewed before accepting the change request from the customer. Where product or service requirements are changed by the customer, management or process owner has to ensure that relevant personnel in the organisation are made aware of the changed requirements once the organisation accepts the changes after review.

8.3 Design and Development of Products and Services

8.3.1 General

Design and development of products or services is a significant process. The scope of the main activity in an organisation can be the design and development of a product or service. Ensuring the proper design and development activity leads to the product realisation process, which is necessary. That

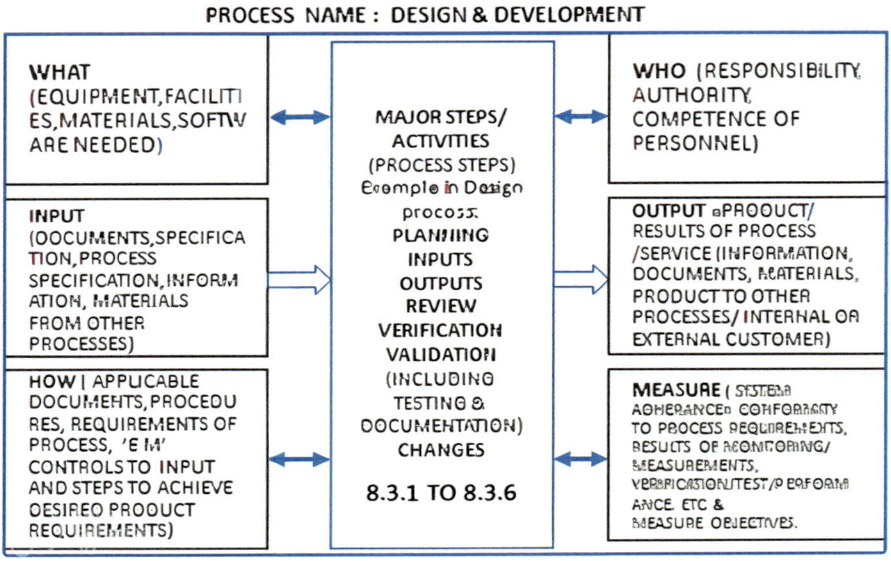

Figure 8.4 Typical process map/planet diagram for 'design and development process'.

means what is designed and developed shall be produced, whether a product or service. Independent organisations design and develop aviation, space and defence equipment. Software engineering service providers design and develop engineering services or software services for various applications.

8.3.2 Design and development planning

Design and development are product realisation activities, and the organisation has to meticulously plan and control the activity related to the realisation of the product design and development. A Process Map is shown in Figure 8.4.

Design and development of aircraft, aero-engines, space vehicles and defence equipment involve appropriate use of engineering, technology and processes, long cycle time and a high level of competency in various engineering and management disciplines. Therefore, the design and development of aircraft, military or civil, small or large, are undertaken by prime organisations. Here is a brief description about how an organisation plans for this massive task of designing, developing and manufacturing aircraft or any other airborne equipment.

The design process starts with the intended purpose of the aircraft. Civil aircraft are designed for carrying passengers or cargo long-range and have greater fuel efficiency. In contrast, Air Force fighter jets are designed to perform high-speed manoeuvres, aerobatics and tree-top flying ability; some provide close support to ground troops, and some have profound penetration strike ability. Naval fighters have VTOL (vertical take-off and landing) ability, helicopters can hover over an area for a time, and some have combat ability. A feasibility study is carried out based on the requirements projected by the customers or features and requirements identified for an aircraft for the intended purpose proposed by the organisation.

A feasibility study lays down all the estimated capabilities, possibilities and planned aerofoil for the wings, control surfaces, propulsion system, aircraft structure, required hardware, aircraft systems and other details for the aircraft or similar equipment. This feasibility study may compare the proposals with different design concepts or proven contemporary designs of similar aircraft or equipment worldwide. The requirements identified in the feasibility study may undergo several revisions and changes as the preliminary design progresses. The iterative process progressively determines more accurate and acceptable output, improving the design. The feasibility study clarifies and lays down the specification upon which all later stages of the design process will be based. Technical requirements include the capability of the aircraft to reach a certain all-up-weight, wing span, length and conditions of runway for take-off and landing, airspeed, forward speed, range, ceiling height, manoeuvrability, rate of climb (time to reach a certain altitude from the ground), maintenance and servicing characteristics, etc. The customer often indicates these requirements at the beginning of the feasibility study and is used as guidelines in each design project phase. Even if the customer does not indicate the requirements of meeting the safety standards and statutory and regulatory requirements, it is mandatory that these are to be considered and appropriate features are included during the design and development stages.

The design and development project is for the whole aircraft for the organisation. For many teams of engineers in the organisation, the design and development activity occurs at aircraft sub-system levels – the power plant, the fuel system, the hydraulic system, the flight control system, the air-conditioning system, navigation and communication systems, etc. There has to be close coordination and interaction among all sub-systems to integrate into the aircraft project. The industry has to manufacture a fully integrated, tested and certified airworthy product that is safe to fly and fit for the purpose.

In the early years of aircraft design, designers used analytical theory to do the various engineering calculations that went into the design process, along with a lot of experimentation. With the invention of the computer, most calculations could be automated. With the arrival of programming languages, engineers could write programs meant for aircraft design. With the introduction of personal computers, designers used more user-friendly programs.

The first design step is to sketch aircraft configurations that meet the design specifications. Designers seek a design configuration that satisfactorily meets all requirements, including performance characteristics such as aerodynamics, structural, propulsion, flight performance and control systems. This is design optimisation. Basic configuration features such as fuselage shape, aerofoil sections, wing contour and location, engine size and type, etc., are determined at this stage.

The design configuration arrived at in the conceptual phase is then tested in the wind tunnel, and computational fluid dynamic calculations of the airflow around the aircraft are done. Significant structural and control analyses are carried out. Aerodynamic flaws and structural instabilities, if any, are corrected, and the final design is drawn and finalised.

The features and characteristics proposed in the feasibility study are subjected to detailed analysis and evaluations, which may lead to more precise definitions of the aircraft or equipment requirements. Some of the extensive studies include wind tunnel testing of aircraft wings and fuselage structure, the investigation of characteristics of different materials and shapes, lift-to-weight ratio, thrust-to-drag ratio, aerodynamic features, etc. When the exercise and experiments are completed, the report may reflect a complete description of the proposed aircraft or equipment. Design of the central systems, sub-systems, hardware, software, and other requirements are generated through engineering data for further design activities. Aerodynamic features of the aircraft structure and control surfaces are also studied in detail, and detailed designs are carried out. The layout or selection of the power plant and its installation on the aircraft are considered at this stage.

Detailed structural design and selection of materials are to meet the requirements of quality without compromising safety and depending on the estimated capabilities of the aircraft. The ability of the materials to withstand various static and aero-dynamic loads, which the aircraft will be subjected to during use, is evaluated.

There are two different design concepts used for aircraft designs:

a. Fail-safe design: Design failure-tolerant aircraft, building systems that can withstand failures. It is a rugged design considering the effects of a single potential failure, and combinations of several similar failures are used in defining a safe design. The objective is to maintain a safe environment for aircraft passengers and crew when a failure occurs.

b. Safe-life design: It is the life within which fatigue failures of the aircraft structure may not happen. The aircraft's structural life requirements are usually stated in terms of the total flight hours possible.

Flight is achieved by the aerodynamic forces created on the aircraft. All military and civil aircraft are built with several complex systems that enable sustained flight. The major systems are Power plant/ propulsion system, fuel system, wings, fuselage, structures and control surfaces, flight controls, hydraulic system (fluid power), air-conditioning and cabin pressurisation system, undercarriage, nose wheel steering and brake system, oxygen system (for high altitude breathing), electrical system (AC/DC power), avionics (for communication, radio navigation, inertial navigation), lighting system, etc. Military aircraft may also have armament and weapon systems, electronic countermeasures, search and identification, escape systems, arresting hooks, brake parachutes, etc.

Installing various systems, routing electrical wires and tubes and connecting structural and movable members is complex. Computer-assisted fly-by-wire controls bring significant benefits through improved flight safety, reduced pilot workload, a reduction of mechanical parts, and real-time monitoring of all aircraft systems.

Fabrication of a partial or complete aircraft mock-up (iron-bird) helps identify the locations of structural and system components. Iron-bird is a full-scale aircraft model made of commercial and inexpensive materials.

One of the essential aspects of the feasibility study is to plan the entire activity of design and development of the aircraft from the conceptual stage to the completion of design to obtain the airworthiness clearance, production clearance and technology transfer for manufacturing the aircraft. The last stage of the design of defence equipment may be the user trials to validate the design and technology transfer for manufacture. Planning involves identifying the milestones for achieving various project progress stage objectives. It is also necessary to identify the different stages of design and development.

While determining the stages and controls for the design and development activity, the organisation has to consider the nature, duration and complexity of the actions associated with the project. One complexity concerns the designers' professional skill, competence, and experience level. For the design and development project, it is the whole aircraft. For many teams of engineers in the organisation, the design and development activity occurs at aircraft sub-system levels – the power plant, the fuel system, the hydraulic system, the flight control system, the air-conditioning system, navigation and communication systems, etc. Another complexity is due to enhancements like high lift devices on aircraft structures. Increased use of avionics and flight management system (FMS), navigation and weapon aiming sub-systems, 'fly by wire' flight control systems in the aircraft, different types of radars used in aircraft (radar is an electronic system used to detect, range (determine the distance of) and map various types of targets) and FADEC – (a full authority digital engine (or electronics) control (FADEC) is a system consisting of a digital computer, called an 'electronic engine controller' (EEC) or 'engine control unit' (ECU), and its related accessories that control all aspects of aircraft engine performance), etc., increase the complexities of design of aircraft for the intended purpose.

There has to be close coordination and interaction among all sub-systems to integrate into the aircraft project. The industry has to manufacture a fully integrated, tested and certified airworthy product that is safe to fly and fit for the purpose.

While determining the stages and controls for the design and development activity, the organisation must identify the process stages for smooth progress and associated reviews at appropriate design and development stages.

So, project management tools or Gantt charts are prepared to monitor the achievement of milestones and meet the delivery schedules. The Gantt chart will identify the integration and production of technology development full-scale test models. It can be further extended to flight test phases.

The standard in subsequent paragraphs/sub-clauses identifies the main stages of the progress of the design of aircraft/aerospace equipment. Different steps, activities in each phase, tasks, resources, responsibilities, design content, input and output data planning constraints, etc., are decided appropriately by the organisation, which has the specific skills to understand how best the processes are established. Adopting the known best practices would be an advantage.

The processes are the inputs for the design, outputs, various reviews that take place concurrently, verification of the design at multiple stages, validation and validation testing and documentation necessary to prove the design and subsequent changes in design or design improvements that take place on the completed design. It is also necessary to identify and document the responsibilities and authorities in the organisation for progressing the design and development of various systems and sub-systems for the aircraft.

In the case of aircraft projects, the organisation undertaking the project has to subdivide the entire design and development effort into specific and smaller projects. Organisation divides the design and development effort into distinct activities. The task is defined for each activity, necessary resources are provided, responsibilities are assigned and the process is clearly defined with inputs, design contents, actions, analyses and outputs.

For example, the aircraft structure design may be subdivided into wings, front fuselage, centre fuselage, rear fuselage, vertical fin, horizontal stabilisers, etc. The most common high-lift devices that go on to the wings, such as flaps, slats, etc., may be designed by smaller groups within the design organisation. More groups may be involved in the design of control surfaces such as tailplanes, rudders and ailerons. In each type of aircraft, depending on the end use and other flying characteristics, the class of structures, high lift devices and control surfaces may vary in size, shape, material, location and the type of actuation for the moving surface. For each design sub-group, the task is clearly defined, necessary resources are provided, responsibilities are to be fixed and the deliverables are to be identified. The data used for design as inputs, steps involved in the detailed design, including the assumptions made during design stages and expected design outputs such as drawings and exact specifications are to be laid down. No design is going to be perfect. Certain constraints may limit man's ability to achieve the desired objectives. Therefore, all design and development activities compromise the needs, expectations, possibilities and requirements. Planning should be done to achieve the options that safely and satisfactorily meet the conditions. The constraints should be known while planning to meet the requirements so that necessary safeguards can evolve while moving the design forward. Any design effort should be based on the fundamental principles and guidelines of safety and aimed at meeting the functional objectives of the product. Design should be progressed through the available and proven technology so that the designed product meets the requirements of the customer and regulatory authorities, apart from meeting the law of the land. Concurrent engineering is the methodology used for reducing the lead time for many successful projects.

Design and development planning should consider the ability to manufacture/produce, inspect, test and maintain the product. Why put such a massive effort if the design will not be used for production? Product design should be such that the quality of the product should be measurable through inspection and testing. No equipment can be used for long periods without maintenance; therefore, the equipment should be designed and produced to be maintainable and inspectable. To achieve high quality in a product, the desired quality standards should be incorporated into the design phase itself, then monitored and checked for consistency throughout the production processes through inspections and tests, and ultimately validated by the product's reliable performance when used by the customer. Design and development planning consider the ability to provide, verify, test and maintain products and services.

Efficient project management and coordination among various design groups is an essential management function to ensure effective communication. The project chief will define the assignment and associated responsibilities of each group. Appropriate interaction among the various groups is necessary for flawless coordination, and effective communication is a method of achieving this coordination.

The design and development plan documented and circulated to all concerned helps establish adequate controls to achieve the milestones without delay. However, the identified and indicated plan may require revision, and fresh targets and milestones may have to be identified and updated as the design and development progress. Constraints and unforeseen events cause the modification of plans. It is always desirable to indicate the changed plan and targets to the customer and agree upon the revised plans.

Design milestones such as reviews, verification and validation have very distinct purposes. Review, verification and validation have some differences and similarities. These may be separate or combined depending upon the needs of the product or the organisation. However, these events may be conducted, and the records may be retained in any combination, in accordance with the product's complexity and the organisation's convenience. The choice is for the organisation to decide.

Organisations must consider involving customers, production agencies and users in the design and development process at the appropriate stages, especially during reviews, verifications and validation. The association of regulatory authority, which ultimately decides on aircraft certification for safety, quality and airworthiness, is mandatory.

Organisations that carry out engineering services in connection with the design and development of aircraft/automobiles may treat the activity as a project management function. Design and analysis software programs and their adoption in the manufacturing industry have helped prime organisations to outsource the design engineering services by giving aircraft/equipment geometry and continuing to work on the design concurrently. This reduces the waiting time necessary to analyse and progress the design, compared to the older methods, which relied on sequencing the different stages of activities.

Modern facilities and software programs help the designers in modelling the external surfaces of the aircraft in 'MATLAB' (matrix laboratory), which is a multi-paradigm numerical computing environment and fourth-generation programming language. To carry out static, dynamic, and thermal analyses using 'NASTRAN', which is a finite element analysis (FEA) programme designed to perform a specific task of processing model geometry, assembling matrices, applying constraints, solving matrix problems, calculating output quantities, conversing with the database, printing the solution, etc. Essential design tools (e.g., CAD/CAM/Ada (programming language) used and resulting data are correlated with manufacturing. Formal design reviews (e.g., preliminary design review, critical design review, production readiness review, etc.) are regularly conducted.

The standard engineering services provided by these organisations are related to computer-aided design (CAD), computer aided-manufacture (CAM), computer-aided engineering (CAE) (simulation and modelling technologies), finite element method and finite element analysis (FEM/FEA), computational fluid dynamics (CFD) and design of airborne items/fixtures/dies, etc., as an outsourced process from prime manufacturers of aircraft. The analysis reports and results are the products delivered by these organisations. In addition, techniques such as the computer-aided three-dimensional interactive approach (CATIA) and simulation-based design enable engineers to capture design errors early in the development cycle.

In these design engineering services, specific design and development activity milestones, such as design validation, may not be possible since the verification must be done on the final product, not the analysis report.

While prime organisations carry out the design and development of aviation, space and defence equipment and their spare parts, other organisations develop alternate parts for fitment on original equipment designed and manufactured by prime organisations. This is another business model, designing and developing non-critical parts through reverse engineering and making them available to customers for use on original equipment as replacement spares. Longer lead times required by the original equipment manufacturers (OEM) and their higher costs force customers to go for such alternate parts for replacements, though there may be conditions and risks in doing so specified by the OEMs to their customers. When designed and manufactured by non-OEM organisations through reverse engineering, parts of aviation, space and defence equipment have the scope of activities as 'engineering design and documentation service of mechanical and electro-mechanical components'. The parts they designed and manufactured are known as part manufacturer approval (PMA) parts. The parts have to undergo the Quality Management System routes of planning, input, output, review, verification and validation of design and production process verification of first batch parts (first article inspection) successfully to obtain airworthiness clearance from the regulatory authorities before use on aviation, space and defence (ASD) equipment. Regulatory authority periodically publishes the approval status of PMA parts.

Parts manufacturer approval (PMA) is not only for design and production approval for replacement parts but also for modifications on aircraft/equipment. It allows a manufacturer to produce and sell these parts for installation on type-approved aviation, space and defence products. Some regulatory authorities specify PMA for non-critical aircraft parts only, and some others do not allow even that. Some regulatory authorities allow only locally registered aircraft to fulfil international legal obligations.

Through the study of an OEM spare part, details of materials, physical and functional characteristics, etc., are laid down in drawings and specifications. Prototype/first production batch parts are manufactured to the re-drawn specifications. Along with the design/drawings and specifications, inspection and test reports and other documentation prescribed by the regulatory authority, the manufactured parts are submitted for airworthiness clearance to regulatory authority approval of PMA parts for design and production.

Indigenisation of imported spares/parts of a large number of ASD equipment is also approved following the above procedure/methodology. It is reiterated that the Quality Management System requirements for organisations dealing with the design and production of PMA parts are the same as for prime organisations.

The following issues should be considered during the planning function for design and development. The overall plan should depict the entire design and development process. Various stages should be described. Resources and competencies required for the processes should be identified. If any part of the design and development activity is planned for outsourcing, that plan and details must be available. Stages of design, responsibilities and authorities for each functional group should be identified, and the internal and external interfaces between various groups and between the organisation and customers/ regulatory authorities should be determined. Stages at which reviews, verifications and validations are carried out for each of the design groups/subsystems/systems and aircraft as a whole product should be identified in the plan – milestones and deliverables with dates of delivery defined. Implementation and effectiveness of the plan monitoring should be effective. Verify whether the plan is updated and communicated to all relevant functions as necessary.

The process for FAA-PMA part, briefly, is as follows:

- Identify the proposed FAA-PMA part, identify the OEM part that the proposed FAA-PMA part will replace and provide a statement that the organisation is applying for FAA-PMA based on test and computations as per regulatory authority requirement 14CFR Part 21.303 (c).
- Add a description of the OEM part and its function, identify the proposed FAA-PMA part installation to the next higher assembly and, if applicable, add a description of the part's design features, materials and any necessary pictures, schematics, figures or diagrams to support and clarify the above points.

- List out similar and related PMAs obtained and currently being produced. Include a safety and criticality (FMECA) assessment to fulfil the requirements of the CFRs applicable to the proposed FAA-PMA part.
- Describe the reverse engineering design methodology, process and test methods and part marking.
- Compliance is certified for airworthiness standards, inspection standards (FAI), continued operational safety of design and subsequent design change management.

Any aviation, space and defence equipment-related indigenisation activity can be carried out similarly.

The modification/fitment of equipment on aircraft is possible only when the regulatory authority approves the same with a certificate of airworthiness. The corresponding civil or military aircraft certification regulatory authority is the 'Authority'. The maintenance organisations must get the necessary airworthiness certificate for the fitment of modification/equipment unless the equipment manufacturer has obtained the same. If so, when the equipment/modification kit is demanded by the maintenance organisation and supplied by the manufacturer, it should always be accompanied by the certificate of airworthiness issued by the relevant regulatory authority.

Design and development tasks are to create inherently safer aviation, space or defence equipment and systems. The focus is on safety in the design of aircraft and aircraft systems and the procedures used to operate them. This requires understanding how and why accidents occur, including continuously reinforcing lessons learned.

8.3.3 Design and development inputs

Design-related or project-related inputs are required to be determined by the organisation. Inputs for the design or project are necessary information related to the present status, expected output and functional and performance requirements concerning the final product. While determining the requirements for the specific types of products and services being designed, the organisation first considers applicable and performance requirements.

Suppose the organisation did a similar design project earlier for the new design project. In that case, the organisation can use information from the previous design and development activities. In an aircraft-related design or an engineering project connected with aircraft design analysis, data collected from similar aircraft designs or projects would be of immense help.

Design, planned to meet the product's safety, functional and performance requirements, should also meet the statutory and regulatory requirements. Regulatory authorities must be involved right from the beginning of the project so that the certification path is hassle-free.

The organisation has a quality policy and relevant objectives. Every organisation has its vision, mission and values proclaimed in its business ethics. They may have specific standards and codes of practice that the organisation has committed to implementing. Codes of practice are principles and procedures to produce trustworthy, high-quality statistics of public value. The reputation of the organisation and its relevance to society depend on the Standards and codes of practices followed by the organisation.

Before proceeding with the design and development project, at the initial planning stage itself, it is prudent to deliberate on the potential consequences and probability of failure of the efforts because the design and development will be a long and strenuous journey. The results of it will be available after a few years. If the results are not going to satisfy, first of all, the organisation itself and, secondly, the customers and other stakeholders, it would be a failure with all the losses to the country due to the nature of the product. It is always necessary to define the success of the project stage by stage. If any potential issues may lead to a failure, appropriate mitigation can be taken.

It is a desirable practice to collect all the input information in the form of a checklist so that an accounting of whether all the required information has been received from various sources can be monitored. Once the list is completed, it can be maintained as the record of received inputs for the design and development.

Since the design and development project involves different specific materials, processes, components, equipment and products, which (if applicable) may have some life constraints and chances of becoming obsolescent, ageing or going out of use, it is necessary that the organisation considers it and solves the issue by appropriate decision keeping 'safety first'. There are several such life-limited parts and materials in aircraft. Some examples include rubber/elastomeric seals, hoses, cartridges for canopy, door openings, ejection seats for emergency use, personal survival packs in military aircraft, etc.

As in the case of a review of requirements related to the product (contract reviews), design inputs must be reviewed for adequacy and completeness. Any requirements should not conflict with each other, and the gathered inputs should be verified for clarity concerning the design processes and the final output of the design. No information should be left ambiguous.

If there is any conflict in the inputs, it must be resolved before proceeding.

The final input data used for the design or development or the engineering analysis must be agreed upon or concurred with by the customer before the same is used to work out the output. It can be maintained as documented information and updated when any change happens, initiated from the customer side or proactively introduced by the organisation. Organisations can demonstrate compliance by identifying the necessary inputs for the processes and products, including financial, environmental, health and safety issues, organisational risks and impacts, customers and expectations, statutory and regulatory requirements (airworthiness requirements) applicable to the product, etc. Risk management and contract review records demonstrate how the organisation systematically undertakes design and development tasks.

Depending on the product or service and according to the organisational policies and vision, the organisation can also consider other information such as benchmarking, external provider feedback, internally generated data and appropriate in-service data for use as inputs for the new design project. By following this methodology, organisations can bring out different aircraft variants with better designs and performance results. If the inputs are controlled, getting a desirable output is most likely.

8.3.4 Design and development controls

Design and development are a process; therefore, if appropriate controls are applied, the result of the approach will be helpful. In the function, the results are planned and defined. All the controls are to get the described outcomes. Controls include reviews, verifications and validations.

Reviews: At different stages of design and development, systematic reviews are necessary, and the reviews are conducted.

These reviews are required to be carried out by the organisation as per a defined plan. The reviews are essential to monitor the progress of the design projects and ensure that the final product will meet the specifications of the product and objectives of the organisation. This review plan may be part of the design and development plan (Gantt chart), which is duly updated occasionally.

Participants in these reviews are multi-disciplinary and from the design department of the organisation (persons dealing with various stages/systems and integration groups), manufacturing agency, users, customers and regulatory authorities. Reviews mainly deal with or evaluate the ability of the design and development results to meet input requirements, agreed customer requirements, design results/outputs, method of achieving the results and verification of the results. In this, it is confirmed that all the original input/customer requirements and subsequently changed needs, if any, from the customers have been considered. If the original requirements are altered or modified, whether the relevant output/results have been worked out needs to be checked. The results should demonstrate the product's suitability, adequacy and effectiveness.

Reviews are to identify any deficiency in the design, any environmental or other factors not considered during the design and development stages, review any identified problems/nonconformities and evaluate potential problems and their mitigations, producibility and verification methodology.

In aircraft design or aeronautical development activities, initial requirements are reviewed at an early stage of design after the conceptual stages. The next stage is system design review, where the hardware and software requirements and their configurations are identified. System design review is a stage where the overall requirements are analysed. A preliminary system design is used to review hardware and software requirements. While the preliminary stages progress, major reviews are planned as preliminary design reviews (PDR) and critical design reviews (CDR). PDR is reviewing the initial design before the detailed design progresses. Detailed hardware and software design follow the decisions taken in the PDR. CDR takes place almost at the final stages of hardware and software design. CDR is highly critical since further changes in the design and layout would be time-consuming and costly. CDR also includes evaluating potential deficiencies or defects in the design and exploring the ways and means of overcoming them. Risk evaluation, failure mode and effects analysis (FMEA), failure mode, effects and criticality analysis (FMECA), fault-tree analysis (FTA), etc., are essential steps towards solving potential deficiencies and finalisation of the design. Software designs also undergo similar studies and analyses. Reliability analysis and predictions of significant aircraft systems' performance and mean time between failures are undertaken. Aircraft structural integrity programmes, reliability studies and predictions of hydraulic/fuel/electrical/electronic systems, etc., are such analyses. The aircraft structural integrity programme is envisaged in MIL-STD-1530C: Aircraft Structural Integrity Programme (ASIP) (03 AUG 2010), which includes design and development tests of full-scale static and full-scale durability tests.

Internal reviews should also evaluate and measure the organisational objectives achieved concerning the design in each functional group's completeness, correctness, timeliness, number of iterations exceeding the plan and other targets the organisation sets while planning the controls.

Authorisation to progress the design and development activity to the following stages will depend on the results of the reviews and decisions taken to carry out the improvements suggested by the review team. Reviews are, in fact, gates through, allowing further progress of the designs. Design organisations have to plan appropriate reviews and take decisions as the authorisation to progress the designs to the following stages. The organisation must define the authorisation method for each design and development stage.

Design reviews are an ideal environment to discuss and challenge tolerances in specifications. It has been seen that organisations resort to several engineering change requests or raise product concessions later to adjust the tolerance of a specification, thereby reducing the design process's effectiveness. There is a tendency for designers to specify tighter tolerances than are strictly needed for the stated application. The rationale for this is easy enough to understand, as designers want the product to function flawlessly every time it is called upon. They plan for every eventuality: thermal expansion, manufacturing errors, presence of foreign objects, unfavourable environmental conditions, etc. They reduce the potential variation by tightening the tolerances for the parts, which may not seem like a bad idea. But is it achievable through manufacturing processes?

The best design can be consistently produced to meet customer expectations cost-effectively. Informed tolerance selection is just one aspect that can be used to meet this goal.

All design reviews at the planned stages should be recorded with decisions, responsibilities and planned dates of completion of the improvements. Follow-up actions on all points are to be recorded and maintained.

There may be problems determined during the reviews. As a result of the reviews, if any actions are necessary, all such steps have to be completed. Issues may be resolved during the verification and validation activities. Solutions to these problems must also be arrived at and implemented for improvements.

Participants in design and development reviews include representatives of all respective organisational functions with the design and development stage being reviewed. The reviews can be internal and include the participation of customers, users, regulatory authority personnel, external experts, intellectuals, professionals and academicians, who can contribute to the design of the equipment at the review stage.

'It was always a vulnerable issue for the regulatory authorities to put across critical comments when reputed organisations, more reputed scientists, and much more experienced engineers reviewed their work together.

No one dares to get clarifications from the designers on the performance results expected on the aircraft after the design and development activities. The performance results are the results of calculations/ simulations. No one dares to get clarifications from the designers or engineers on the details of operations/controls to get the aircraft's performance results. The details have not yet been worked out or designed/ decided. Still, they will undoubtedly be included in the detailed design documents /drawings. The ground simulation or test is often neither practical nor easy to perform without investments. Hence, the matter is closed with a statement - "will be checked in flight test".

Some clarifications get submerged in the flood of respect for the institution, scientists and engineers. Finally, if the requirements are met, everyone depends on the design validations and validation tests for the certification of the equipment/aircraft. We always learn our lessons from the results of catastrophes or the resulting negative consequences of doing better.

For example, India's most ambitious civilian aircraft project started in 1999 to build an indigenous 14-seater aircraft, which came to the stage of its maiden flight in 2004. Still, the project has always been plagued with glitches. Weight was a significant concern as the aircraft was overweight by about 1,000 kg. In fact, during the maiden flight, all seats except three — for two pilots and a flight test engineer — were removed to bring down the weight!

The aircraft was on a test flight in 2009 when it crashed, killing two pilots and a flight test engineer from the IAF while carrying out an "engine relight" procedure.

The court of inquiry, constituted by the regulatory authority to investigate the crash, also highlighted the deficiencies in the design of suitable control surfaces to enable the crew to fly the aircraft manually. The project never fully recovered from that crash.

No one dares to get clarifications from the designers or engineers on the details of operations/controls for getting the desired performance results on the aero-engine/aircraft.

When the indigenous experimental gas turbine engine was run on the test bed, 96 Nos of High-Pressure Turbine Rotor (HPTR) blades had Tip rub, over-temp on the tip, erosion of tip, trailing edge, burn on the tip and trailing edge, puncture of the top brazed cap (other than cooling holes), chip off of Platinum-Aluminide coating layer and hair line crack at the trailing edge. During design, creep characteristics were studied, and the optimum clearance between the stator and rotor was arrived at.

However, the failures/tip rub could not be simulated before the physical run of the engine.

The inability to design an optimum clearance between the stator and rotor was also probably one of the reasons that the engine could only produce an 80kN thrust class engine instead of the desired 85kN to 98kN thrust class engine for the light combat aircraft'.

Verifications: Every design output, such as drawings, specifications, calculations, analysis reports and data brought out as a result of the progress in design and development activity, must be verified for conformity to the design requirements concerning the inputs and customer requirements.

Verification includes activities such as methods and results of calculations, using alternate methods of evaluating the results, comparing the design/drawings/specification with that of a similar and proven design/drawing/specification, etc., if available. Fabrication of prototype samples, simulation, actual tests on the specimens, etc., are all verification methods that can be employed. The verification activity should be such that it is a planned arrangement and completed during the design and development activity process. All verifications should be against acceptance criteria, which may be based on the design input data or the agreed customer requirements. The sequence of events in the design process, inter-relation with other design functions, use of adequate resources and competency for arriving at the results, etc., are all points that need to be verified. The verification should also ensure that the design results meet safety requirements, regulatory requirements and customer input. All related documents and records will be made available to the interested persons and organisations, viz., customers, users and regulatory authorities, only after adequate review and approval by the process owner/designer/organisation. The verified and approved documents and records of the design output (component/assembly/sub-system/system) should be available as inputs to the next stage of the design and development activity.

The organisation demonstrates design verification by completing planned arrangements, which shows acceptable and consistent verification results with reference to the initial design input requirements and any improvements made after not achieving the expected/planned results. Records must indicate that the verified design outputs were only cleared to the next design stage.

Design and development validation: Verification and validation are different means of ensuring conformity with the customer requirements. 'Proof of the pudding is in the eating'. Tasting for quality characteristics and performance during the design and development process may be termed verification. Eating and enjoying, earning and sustaining the benefits and having no adverse effects later are validation. Customers should validate it as they use it, with allowance and limits for abuse and misuse as much as design-maintained documented information permits.

Validation is subjecting the systems and product to the rigours of testing or use (including abuse, if intended) to the designed values under the environmental conditions at which the systems and product are likely to be used by the customer so that the customer gets an assurance that the requirements are met. Validation in the contest is confirmation through the provision of objective evidence that the requirements for a specified or intended use or application have been fulfilled. Validation is a design requirement for aerospace and defence products to ensure compliance with design specifications and satisfy the user's needs. Certification follows for airborne equipment.

In aviation, space and defence equipment, after satisfactory design verification by simulation, prototype manufacture, inspection and local testing, the equipment may be taken to the user's premises for user trials and performance evaluation. This evaluation subjects the equipment to all the design loads and stresses and the working environment to which the equipment is exposed during normal usage. That is, checking that the equipment will meet or does meet the customer's needs when it is in use. Local flight tests of aircraft can only verify and validate some of the performance characteristics per the flight envelope. Aircraft or helicopters are flown at high altitudes and low-temperature terrain to validate the designs before they can be declared fit for use per the customer's requirements. User trials carried out on equipment throughout the year at different terrains and peak seasons/adverse conditions (as designed and allowable) are also for validating the designed equipment. Validation is the confirmation by examination and providing evidence that the specified requirements for the specified/intended purpose or use of the product are fulfilled.

All validation and validation testing are planned activities during the design and development of any aviation, space and defence equipment. Validation methods should be specified in the design and development planning process. Planning should include the details of tests, test methods, performance results expected (acceptance criteria) and contingency plans for any unforeseen results (risk assessment and mitigation actions). Strategies can always be modified according to the prevailing situations, with the concurrence of the customer and regulatory authorities. Test schedules must be issued indicating the test setups, test procedures, duration and list of personnel involved in witnessing the tests, including customer and regulatory authorities who typically coordinate the test/verification/evaluation results. Validation ensures that the product, which passes the tests, can meet the customer's requirements for the specified application or intended use. When tests are conducted to verify and validate the design and development activity, there is a need to meticulously plan them, control and document all the methods and results to prove that the verification and validation activities fully meet design requirements. It is always advantageous for the designer to know the ultimate use of the designed equipment. Equipment validation is challenging in space applications since the intended use is known. Still, the designer cannot take the equipment there to test and validate the design! In such cases, simulations and adequate testing validate the design by creating the user's environment at the designer's premises as much as possible! Design validation may be outsourced or performed by external providers for non-aerospace equipment per the customers' requirements.

The standard stipulates that wherever practicable, the validation and validation testing should be completed before the delivery for use or implementation of the project. In major aerospace, aircraft or aviation projects, it generally takes a very long time to complete the design; even with the concurrent engineering methodology, validation may be completed after the equipment is used. However, the Standard intends that validation should have been completed before the equipment is subjected to the rigours of use. Records of the results of validation and follow-up actions, such as correction/corrective action/improvements, etc., arising out of the results obtained during validation must be retained.

In software, validation declares that the documented solution or process is correct, complete and current in the indicated/identified configuration and is suitable for the intended use. The validity period is until the next/revised version of the software!

Only validated design outputs are submitted to the customer unless the validations are not completed before the delivery due to the reasons agreed to by the customer.

Records retained by the organisation demonstrate that the validation activities have been carried out as planned. Validation indicates that the product, designed and verified, can meet the customer's requirements and specifications and performs as designed/expected. Records include correction/corrective actions/improvements made to meet the customer requirements, and action taken to overcome deficiencies, if any, in the results. Concessions/acceptance of deviations from the customer/regulatory authority are also retained as records.

Even though the design and development reviews, verification and validation stages have distinct purposes, it is for the organisation to decide whether such reviews/verifications and validations can be conducted separately or in any combination, as is suitable for the products and services of the organisation.

Controls include validation activities to ensure the final product meets the customer's intended use or application requirements. Validation is a design requirement for aerospace and defence products to ensure compliance with design specifications and satisfy the user's needs. It must be proved during validation that the product meets the customer's specifications, including the environment in which the final equipment is used.

Documented information (records) for completing stages of reviews, verifications and validations are to be retained. Documented information (records) for the issues identified during reviews, verification and validations and the actions taken to solve the issues are to be included in the above records. Progression to the next activity stage during design and development is authorised only on completion of the review, verification and validation activities.

Organisations may demonstrate compliance with these requirements by showing records of reviews completed at the planned stages throughout the design and development process, where reviews are carried out with the involvement of personnel from various functions/processes concerned along with others. Records should indicate the action taken to the observations made during reviews, verifications and validations. Changes to the inputs from the customer and corresponding revised design outputs, reviewed and approved by the relevant authority, should be produced. The design review, verification and validation records should include the designed product's suitability, adequacy and effectiveness.

8.3.4.1 Verification and validation testing

When tests are necessary for verification and validation, the organisation shall plan these tests, control, review and document them to ensure and prove the result of the process meets the requirements.

Most of the verification and validation activities in design involve testing. Therefore, when tests are conducted to verify and validate the design, all the tests are planned and executed in a controlled manner, and the test results are to be reviewed and recorded. Correct configuration of the part is required to be submitted for the test. All tests are conducted on identified products or components/assemblies/systems. There needs to be a test plan, the requirements of the test plan, and the test procedures are to be observed. The description of methods of operation mentions the resources being used for the performance of the tests and defining test objectives and conditions. The test results and parameters to be recorded are pre-determined. Actual measurements will be taken and compared with the acceptance criteria. If the test results meet the acceptance criteria, it meets the verification and validation testing requirements. The test equipment and monitoring and measuring devices used for testing will be calibrated as defined in clause 7.1.5.

After the validation testing, the organisation confirms that the reports, calculations, test results, etc., demonstrate the product design results, which meet the specification requirements for all identified operational conditions. The relevant documentation (a record) will be the evidence of completing the tests.

In any design and development activity connected with space, aviation and defence, verification and validation of the product's design are mandatory. There are various means of verification and validation, as explained above. The designer's prerogative is to choose the appropriate method of verifying and validating the product's design and take the customer into confidence. While doing so, the designer may undertake some testing methods and get the test results in a safe, faster and cost-effective manner so that they can be readily compared to the design values and proven to the customer more convincingly. However, the necessary tests undertaken will also have the customer's concurrence. Whatever the tests are undertaken, the plan and details are available. After conducting the examinations, the results are documented to compare with the acceptance criteria.

The organisation demonstrates the compliance status of verification and validation testing through test records in accordance with the accepted test plan. The test reports indicate the responsibilities, product configuration, test procedure, resources used and the acceptance criteria for all the tests. In case of failures during the test, actions taken to correct the nonconformities and the reasons for the nonconformities should be recorded.

Verification and validation documentation: Comprehensive documentation with regard to various design and development activities is required when completed. This documentation includes related review, verification and validation activities, test reports, calculations and results. This documentation (records) finally proves that the design and development activities have been completed. The final product definition meets the specification requirements for all the identified operational conditions, the conditions the customer specified and the conditions required for satisfactory operations.

The organisation prepares a document of compliance matrix between all specification/customer requirements and the verification and validation results obtained. If discrepancies are noticed, the customer acceptance/concession for the deviation needs to be recorded.

To demonstrate successful design and certification of aviation, space or defence equipment for production clearances, the organisation has to show the following documented information:

a. Airworthiness/provisional clearance for parts/assemblies/systems/equipment issued by the regulatory authority.

b. The regulatory authority issued the type approval/airworthiness certificate for all aviation, space and defence equipment/assemblies/systems/aircraft/airborne stores.

c. Final documents for product lifecycle management:

   i. Standard of preparation for manufacturing of aviation, space and defence equipment/assemblies/systems/aircraft/airborne stores.
   ii. Repair and servicing manuals, including maintenance schedules, illustrated parts catalogue, modifications list, list of servicing instructions, inspection instructions, etc.

8.3.5 Design and development outputs

Upon completion of the design and development activity, the organisation ensures that the process outputs meet the input requirements – customer requirements, value additions the organisation has provided, statutory and regulatory requirements and other stakeholders' requirements.

Organisations should ensure that the design output documents are adequate for production. The organisation must extend all clarifications and design support to the production agency.

Design output should include the monitoring and measuring requirements as appropriate for the product and relevant acceptance criteria.

Design drawings, specifications and other information the designer provides should include the product's essential characteristics for its safe and proper use.

The ideal way of specifying those essential characteristics is to indicate them as key characteristics, critical items or special items. The very purpose of identifying them is to control them and specify what actions to take while manufacturing them so that no deviation happens to them. Therefore, the aerospace standard explicitly calls for identifying critical items and key characteristics. To ensure that these essential items and key characteristics do not deviate from the specified requirements during manufacture, assembly, testing, operation and maintenance, specific actions to be taken are to be indicated in the design documents. Design documents will be inadequate and incomplete without this information. Customer requirements for the designation, documentation and control of unique characteristics must be identified in the early stages of design and indicated in the design outputs. Customers would expect them to be available in the applicable documents such as drawings, process documents, control plans, etc. The output documents contain or refer to product acceptance criteria and product characteristics essential for its function, life, interchangeability, maintainability, safety and strength (FLIMSS) and proper use.

Suppose a report is generated based on the results obtained after the analysis. In that case, the input data used for getting the results must also be referred to in the report. It means that if the data used were different, the outcome of the study/analysis would have been further from the one reported.

Usually, the project managers or their peers approve the verified output before delivery of the analysis reports. If drawings and specifications are the output documents, they are also verified. The documents must be released after approval by an authorised senior management person from design.

The design and development output documents are verified before being issued to internal or external parties. Like the authorised person releasing the final product after necessary verifications and inspections from a production unit, a designated authority/person in the organisation must approve the design documents before they are issued or released to external parties.

The organisation should deliver the output/results/reports of the final product/deliverable in a form that enables verification against the design and development input, and the organisation must approve the design before release. The ability to verify the design as a requirement is emphasised here rather than providing a simple possibility of substantiating the design.

If the design of an aircraft part is carried out using the agreed inputs, the output data in the form of drawings or specifications must reflect the data used for the design. Design outputs data, which facilitate manufacture, identification, verification, use and maintenance, include:

a. The drawings, part lists and specifications necessary to define the configuration (physical features and performance results) and the design features of the product.

b. The material, process (including unique processes), manufacturing, assembly, handling, packaging and preservation data needed to provide and maintain a conforming product.

c. The technical data include the operating manual, illustrated parts list/catalogue and their source if needed, and service and repair schemes for operating and maintaining the product.

The organisation shall retain all the documented information on design and development outputs. A total list of product/project configuration master documents – a 'bottom-up approach', as in the case of an aircraft costing – is preferable (read about the cost of an aircraft under para 8.2.3.1).

Individual components/parts/hardware (identified with part number and revision status) together form the bill of materials for a sub-assembly, several such sub-assemblies and linkages/connections form an assembly/system, and several such systems together form an aircraft/aero-engine; for that matter, any equipment in aviation, space or defence or any other sector. Rubber or elastomeric seals have a limited shelf life, prescribed by the designer based on the property of the material being used.

Preparing the design and development output documents should ensure that the information in the documents meets the input requirements for the design and development. For this, a compliance statement for each particular input would suffice to indicate the results obtained in the output after the design. Even if specific results are not received, the same has to be communicated to the customer and/or regulatory authorities for seeking concessions or accepting it as is. The design outputs, like drawings and specifications, should provide appropriate information for purchasing, production and maintenance/service operations. This is particularly important for all products produced through drawings and specifications so that the raw materials or unique materials procured and/or processed are indicated in the design documents. It is desirable to display alternate materials that meet the design output requirements for rare materials or are not indigenously available. If a raw material is machined and further processed to achieve the final design characteristics/properties of the component/product, the process specifications must also be indicated in the drawings. It helps the production process engineering or methods department establish suitable processes to achieve the end characteristics for the component/product during manufacturing. Process engineers shall refer the process specifications to the latest revision status, which may not be indicated in the design/drawing/specification.

The drawings of components or assemblies indicate the final acceptable dimensions and characteristics. No process can give out an absolute output, result or characteristic like no two things are identical in their genuine sense/characteristics, which is most negligible for artificial or manufactured objects. Therefore, it is always necessary to specify the permissible tolerances for all measurements in the design output documents. The tolerances are decided by the designer, considering the technological and environmental constraints and producibility of the designed part or product. Specific dimensions/characteristics may be given open tolerances and others close or unilateral tolerances.

What is the close and open tolerance in drawings? What is the right tolerance for a part feature?

Open tolerance: This tolerance needs only minor concentration from the operator and quality control. For example, per customer drawing, there may be some tolerance with more than + or −1 mm variation. This can be easily achieved.

Close tolerance: These are very crucial/critical dimensions that need the utmost effort and care. Usually, tolerances below + or −1 mm may be considered close tolerance.

No rule book or programme is available to dictate open or close tolerances. Tight tolerances are expensive to fabricate and lead to higher scrap rates. 'International Geometric and Dimensional Tolerance Chart' decides on unspecified tolerances. The permissible deviations are defined as fine, medium, coarse and very coarse tolerance classes (angular and linear) in DIN 7168/ISO 2768 – General Tolerances Package. This package includes ISO 2768-1:1989, ISO 2768-2:1989 and IS 2102 part 1-1980. Acceptability or the product acceptance criteria of a dimension/characteristic/measurement is decided by the fact that the measured result falls within the specified tolerance and limits indicated in the drawings.

The design and development documents should indicate how the product will be identified, manufactured, inspected, used and maintained when the parts/assemblies/products are produced per the drawings and specifications. Drawings and specifications should indicate adequate data/information, like a regular test schedule describes how the testing is done. The data/information is usually in drawings, part lists and specifications. The documents also identify the items necessary to define the product's configuration and design features. Configuration identification of a part is by the drawing/part number accompanied by its issue/revision status if a drawing is for a specific part only. It should show the relationship with assemblies. Part numbers should also be identified with their part issue/revision status, which indicates the chronological changes incorporated in the drawings. Sometimes, drawings can show different part numbers in one drawing. It may be wrong to assume that the drawing and part numbers are identical. Therefore, the drawing numbers with its issue/revision status and the part number with its issue/revision status should be brought out. Design office practices of different organisations are different. With subsequent changes in drawings or parts, the drawing/part issue/revision status gets updated every time. 'British Standard 8888:2013 Technical product documentation (TPD), Specification for defining, specifying and graphically representing products' is a reference for engineering design office procedures. NASA Engineering Drawing Standards Manual (GSFC X-673-64-1F) is another useful reference.

Configuration identification in software designs, computer-aided drafting and design (CAD) and design analysis of aircraft components are usually done by the numbering system adopted for the document/analysis results. Once a change is authorised, the version identification and control establish the configuration for the above output documents and reports.

The material, process, manufacturing and assembly data needed to ensure product conformity are all to be described in the design documents.

In established organisations, where more than one activity of the design, manufacture, assembly, testing and final delivery of the products take place in various divisions or departments of the same organisation, some designers may feel that:

'The planning knows what to do; production engineering knows how to produce. Quality control knows how to inspect and test (and they have been doing it for a long time now (!). The drawings do not need to indicate all production and inspection requirements. It is a waste of time and effort. If done, it will be interference in their work'.

The standard indicates a solution to this kind of issue. Design documents will be inadequate and incomplete without this information. Without this information, the design and development process may not be effective. As a result, 'the designer' may become a design liaison engineer, fully engaged in giving clarifications and information to the rest of the organisation, day in and day out, and the chief will have a dozen assistants doing the same, less critical and unending consultancy!

Design output documents should include qualification requirements of the constituent parts/product. Suppose the parts/assemblies/equipment are used in an airborne vehicle. In that case, when the qualification test requirements (ground testing and flight testing) are satisfied only, the parts/assemblies/equipment are given an airworthiness certificate/type certificate or a certificate as fit for flying.

The designer earns appreciation from the user for such information. For the designer to deserve thanks and appreciation, it is necessary to associate other departmental personnel (e.g., planning, engineering, tooling, production and quality control), customers/users and regulatory authorities before finalising the design output documents so that the inputs/contributions from these departments and personnel will help the designer to indicate the relevant, appropriate and feasible information in the drawings, specifications, test schedules, qualification and type certification/type approval requirements, scheduled servicing at periodic intervals, servicing procedures, etc., for the product lifecycle management. The designer is responsible for indicating how and when the aircraft/aeronautical products/defence equipment will be disposed of after its useful life.

Organisations can demonstrate compliance by providing information concerning the completion of design and development stages, which may be indicated in a compliance matrix with respect to the input/customer requirements and indicated critical items and key characteristics in the output documents demonstrate the care taken by the organisation with respect to the needs. The retained documented information (all design documents, test/review records, reports, etc.) on design and development outputs demonstrates compliance with the requirements.

8.3.6 Design and development changes

Once the design and development activity is completed and the airworthiness clearance is obtained for the aerospace part or assembly and/or manufacture or use, there will be several occasions where changes may be necessary for the design, manufacture or maintenance of the aviation, space and defence equipment. The changes may be initiated by the customer, user or manufacturing agency based on the experience of use or manufacture, or it may be a proactive continual improvement process initiated by the designer for upgrading technology, manufacturing process or material or an improved design. All these improvements are termed changes, including corrective actions to correct any design deficiency or deviation in design documentation and preventive actions for the potential deficiencies determined later. Whenever a change is required in the design output/documents, the needed change must be identified, and appropriate documentation, 'advance modification information' (AMI), must be created to propose the change. Even before creating the document for change, it is necessary to study the impact of the change on the resulting product, including products already delivered to the customer and the change in the manufacturing process. Examine if the change must be reviewed for the total population of the product. The proposed changes are to be studied for their impact on the quality/conformity of the product, verified and validated. This is to ensure no adverse impact on the conformity to requirements. The organisation informs the customer if there are any changes in the requirements before accepting and implementing the changes. Verification and validation are done by trying out one or more equipment. Each of these changes is finalised in consultation with the customers or regulatory authorities after careful study for the impact of the changes/improvements on the configuration of the part or assembly or its performance, without any adverse effect on safety, quality, reliability and maintainability. Any change should be for the benefit of the product and its use.

Once the change is accepted and approved for implementation, the change is recorded. The change primarily affects the configuration of the part/assembly/product. The design office practice must indicate what type of change must be recorded. In the aircraft industry, the drawing/part/assembly number change may be indicated by an increment in the issue/revision status. A more detailed change is usually initiated or introduced by advanced modification information. After the aircraft has been given an airworthiness certificate or production clearance, several such change requests/advance modification information may be introduced based on the field use/maintenance experiences or the need for improvements in the manufacturing or overhaul/servicing requirements as the aircraft use continues. These improvements are all identified as modifications, technical instructions, servicing bulletins, inspection instructions, etc., which require final approval by all stakeholders and regulatory authorities. All records of the reviews, verifications and validations are to be retained. The accepted and implemented changes are to be maintained as documented information. The documented information includes the actions taken to prevent any adverse impacts due to the changes. All design and development changes are to be controlled in accordance with the configuration management process, as explained earlier (para 8.1.2). Records should indicate where the change has been made effective in the product's lifecycle and at what stage of the contract with the customer. All design changes, whether made during the development stage, production stage or maintenance stage, are to be controlled in line with the configuration management process throughout the aircraft's lifecycle or any other relevant aviation, space or defence equipment.

In military aviation, minor changes that do not affect the configuration of the equipment are introduced as servicing technical instructions, servicing bulletins, inspection instructions, etc., with the customer's and regulatory authorities' approval. All modifications are proposed as 'advance modification information' (AMI) and discussed among the organisation, customers and regulatory officers. Approved changes are categorised based on the criticality and convenience of embodiment on production aircraft and the aircraft already delivered to the customer.

When a change is proposed and implemented, the organisation demonstrates compliance with the design and development changes through the following:

a. advanced modification information;

b. impact study of the changes, including trial compliance, additional testing, etc., and their records;

c. regulatory approval of the changes, including documented information on the modifications, configuration changes, compliance with the equipment;

d. changes affected the delivered product (retro-compliance), including documented information and configuration changes.

8.4 Control of Externally Provided Processes, Products and Services

8.4.1 General

The purchasing process is an important function in any organisation since all required for product realisation can seldom be produced. The supply chain is 'supplier–organisation–customer', and the organisation always purchases products or services from suppliers. The suppliers are now known as external providers. How do you buy materials, products or services of the right quality, quantity and price from the right external source?

Control of externally provided processes, i.e. outsourced processes, is achieved through the application of requirements of both clause 8.4 (Control of externally provided processes, products and services) and clause 4.4 (Quality Management System and its processes).

In some situations, the organisation might not 'purchase' the outsourced process; it might, for example, receive the service from a corporate head office or another division within the organisation without any financial transaction. Under these circumstances, clause 8.4 (Control of externally provided processes, products and services) and clause 4.4 are still applicable.

Two situations frequently need to be considered when deciding the appropriate level of control for an outsourced process:

a. When an organisation has the competence and ability to carry out a process but chooses to outsource that process (for commercial or other reasons), in this situation, the process control criteria should already have been defined and can be included in the requirements for the supplier of the outsourced process.

b. When the organisation does not have the competence to carry out the process and chooses to outsource it, it has to ensure that the controls proposed by the supplier of the outsourced operation are adequate. In some cases, it may be necessary to involve external specialists or the ultimate customer in making this evaluation, selection and approval.

Defining some methods to control the outsourced processes in a contract between the organisation and the supplier may be necessary. The potential impact of the outsourced process is based on the supplier's capability to provide a product that conforms to requirements. There is no room for the supplier to experiment or propose innovations to the outsourced process without the permission of the ultimate customer. The organisation's control of the outsourced process has to be based on the need for product conformity to the customer requirements.

Ensuring control over outsourced processes does not absolve the organisation of the responsibility of conformity to all – customer, statutory and regulatory requirements. When unique processes are outsourced, the organisation needs to ensure that the control over the outsourced process includes process validation in accordance with clause 8.5.1.2 (Validation and management of unique processes).

The organisation must ensure that the purchased or externally provided product, process and service meet specified purchase requirements. Every purchased product, process or service impacts the final quality of the deliverable to the customer. Therefore, the purchasing process assumes equal importance in the production operation, like all other production or quality control processes. Purchase requirements are to be specified by the user/indenter of the product, process or service. The purchasing process owner orders the supplier/external provider to supply the product, process or service. Therefore, the organisation is responsible for ensuring that the product/process or service conforms to specified purchase requirements. The organisation shall know fully what they buy. Therefore, the organisation is responsible for confirming all externally provided processes, products and services, including from customer-defined sources. That means the organisation should have already evaluated the product or process they propose to buy, made arrangements to ensure that the product/process they buy from an external source would meet the quality requirements even before placing the order and also verified that the organisation has indeed received the conforming product or process they wanted after placing the order.

It is usual for the customer to designate or specify the source of some of the constituent parts or processes that go into the final product. In such cases, the organisation has to communicate with the external providers and ensure that the customer-designated source for the constituent part/process is used. It is to be ensured that the organisation does not use verification by the customer as evidence of effective control of quality by the supplier (it does not absolve the organisation of the responsibility to provide an acceptable product, nor shall it preclude subsequent rejection by the customer – AS 9100B).

> *'Abex Hydraulic Pump went out of production when the customer wanted more trainer versions of a fighter aircraft fitted with the original Abex pump. The customer spotted an equivalent "Messier Bugatti Pump" and requested the manufacturer to purchase them in place of Abex. Based on the operational compulsion, a decision was taken by all the stakeholders together to go for it. However, the responsible manufacturing agency did not evaluate the suitability of the substitute pump for the intended use in the local/operational terrain and environment. Not aware of the risk and consequences of not matching the Messier Bugatti Pump's performance characteristics on the aircraft under tropical conditions, the manufacturer purchased 34 Nos to speed up the aircraft's production and supply. No comparative study was made before procurement; the manufacturing agency did no inspection. "A pump is, after all, a pump, will work like a pump", someone pointed out the need to ensure intended quality before procurement, though the esteemed customer designated it. The standard requirement, which indicated the best practice in such circumstances, was known (AS 9100 Rev B). We always learn things in quality the hard way. Correct fitting on the aircraft itself was not possible because the input connectors of the pumps were not interchangeable. Modifications were carried out. Loss of the hydraulic power system was experienced in the first test flight on the new aircraft fitted with the new pump, similar to the failures noticed on the hydraulic power system on Eurocopter helicopters (Refer French Airworthiness Directive No F-2007-010 dated September 12, 2007). Not doing the required things at the right time means we spend enormous resources and time learning the hard way'.*

External providers are identified, evaluated and selected by the organisation. They may be direct or sub-tier suppliers or suppliers designated by the customers. The organisation determines, identifies and manages the risks associated with the external provision of processes, products and services, as well as the selection and use of all external providers/suppliers. While dealing with suppliers, some risks are monopoly/proprietary supply, restricted technology stoppage of business, etc. The organisation-wide risk register contains risks identified, assessed and managed concerning the selection and use of external product/process/service providers.

The organisation shall make it mandatory, through contract, for the suppliers to apply necessary controls to ensure that the requirements for the product/process are met when they get their supplies directly or when they get their supplies from sub-tier suppliers.

The organisations must determine the controls when they obtain the processes, products and services from suppliers/external providers. Usually, the controls are through the purchase process. This is when we have the following:

a. The products and services from suppliers are intended for incorporation into the organisation's products and services, for example, raw materials for machined parts, components bought for assembly, etc.

b. Products and services are provided directly to the customer(s) by suppliers on behalf of the organisation. Example: The organisation arranges a supply of lifed items/critical items directly to the user from the approved suppliers for which the organisation agrees on purchase/inspection controls and payment.

c. A process, or part of a process, is provided by an external provider as a result of a decision by the organisation. Example: The organisation is arranging to supply a process or part of the process (like calibration of measuring equipment), heat treatment or similar process on customer-owned part/equipment or logistics, which the organisation has agreed. The parts need not have to come to the organisational premises to get this external process from the service provider directly to the customer parts when this process or service has been agreed by or based on a decision by the organisation.

The organisation has to determine and apply criteria for evaluating, selecting, monitoring performance and re-evaluating all suppliers based on their ability to provide processes, products and services per requirements.

That highlights the importance of evaluating the supplier and their capabilities, including quality system and scope of supply, before placing the order. Specify in the purchase order exactly what product, process or service the organisation wants and ensure that the product/process/service received is exactly what the organisation wishes and meets the quality requirements. That is precisely the purchasing process.

How should the organisation evaluate and select the suppliers? Selection of suppliers should be based on the supplier's ability to supply products in accordance with the organisation's requirements. The organisation should evaluate the capacity of the supplier. For that, the organisation must establish criteria for evaluating, selecting, monitoring performance and re-evaluating all suppliers. Examples of different types of suppliers/external providers are:

a. Raw materials, casting/forgings, standard hardware, bought-out standard parts, finished equipment, accessories, consumables, coolant oil, DM water, packing materials, spare parts for existing machinery, tooling, jigs, fixtures, imports, OEM software, computer and computer peripherals, cutting tools, stationery.

b. Unique process sources: Heat treatment, plating, anodising, non-deliverable software programs, chemical processes, NDT, painting, machining, and other services.

c. Training providers, calibration agency services, annual maintenance contractors, servicing agents for computers and peripherals, taxi operators, truck service providers, and logistics providers.

The criteria can be in the form of a checklist, which may contain questions or points for evaluation and grading, which, when added up, will cross a level, say 75%, which is directly proportional to the ability of the supplier to supply products/process/service that meets the requirements of the organisation. Evaluating all suppliers must use appropriate criteria (it can be a standard criterion or unique criteria for every supplier/external provider). The checklist/criteria should be completed, and the supplier's approval should be recorded by the designated authority in the organisation.

This is to register, evaluate, select and approve a supplier/external provider to begin a mutually beneficial relationship.

This is a pre-condition for purchases by any industry in general and aviation, space and defence organisations in particular. The supplier is evaluated further based on the supplier's performance for each purchase order. The supplier's performance means the supplier's ability to provide materials/parts/products/processes/services that meet the specified requirements and timely delivery of the product/process or services.

One of the many factors that can be used as criteria during supplier selection and evaluation is the supplier quality data from objective and reliable external sources. The data evaluated by the organisation, information from accredited Quality Management System certification bodies, and approvals received from government authorities could all become part of the evaluation criteria. There is a tendency for organisations to use approval certifications or even recommendations given by other external parties or organisations to claim their ability. Using such data would be only one of the several items in the criteria for an organisation's supplier evaluation. Approval and control processes for the supplier are after evaluation. To say that 'the supplier is an ISO 9001:2008/2015 certified organisation and therefore the organisation need not institute any selection process or controls on the supplier for the quality of the supplies' would be incorrect. First, the certification is not for any product/process/service supplied by the supplier. Even if the supplier follows the best Quality Management System for their business operations, the organisation that purchases the product/process/service continues to remain responsible for its quality. They are responsible for the verification at the time of receipt to ensure that it meets specified purchase requirements. It is also necessary that the organisation considers appropriate risks while selecting and using the suppliers for the purchase process. The organisation should take action in such a way as to mitigate the risk factors in favour of the customer to whom the finished product is delivered.

Supplier's evaluation, selection using criteria, monitoring of their performance, etc., are to be retained as documented information (records). Any actions arising from the assessments must also be retained as records.

8.4.1.1 Supplier controls

It is essential that the organisation define the process of approving suppliers. It is necessary to determine who is responsible for evaluation and selection and who holds the authority for the organisation's 'approval status of suppliers' decisions. What the conditions are by which the approval status may change, what the ways of using the external providers are when the approval status changes, etc., are to be determined.

'If you ask exactly what you want, you may get what you want. If you don't ask what you want, you will get exactly what you don't want. With the additional responsibility of specifying precisely what you want, the organisation has to apply controls to the supplier. The type and extent of control used by the supplier are dependent upon several aspects, such as the following:

a. How does the purchased product affect the production operation or final product?

b. What are the qualities required for the purchased product so that the final product also will meet the requirements?

c. What is the proven capability of the supplier to supply a product exactly as the organisation wants it? What is the credibility of the supplier? How does the supplier follow the quality system? What guarantees the supplier will consistently provide a product meeting organisational requirements? How do the suppliers provide evidence of the product's quality? Several such questions can be asked.

'The type and extent of control exercised on the supplier and the purchased product depends on the supplier's answers and the supplied product's quality. The standard shows the light towards the best practices in the world, where there are buyers and suppliers of quite different kinds.

You talk to one of the purchase managers. He will say- "We have been buying things from them for a long time. It is often assumed that the supplier will supply the usual one even if we do not specify things in so much detail, as he was supplying always. There is no need to check the quality of his supply since the supplier is dependable. There is no necessity to check the quality on the receipt because the supplier has given a certificate, etc. He may even replace the entire lot free of cost if found not acceptable to us or when we have a complaint about it!"

With all this system "requirements" of the purchase process, anticipating verification of approved suppliers list, formal RFQ to approved suppliers, vetting, comparative statement, purchase order with adequate information, verification of the adequacy of the information, verification of purchased materials, conformity and test certificate etc., when the auditor asked the Managing Director, who was also the process owner for purchase, in an SME in Komatsu-Shi, Ishikawa Prefecture, Narita, he showed the auditor nothing but the trade/telephone directory and said they got the correct material with the right dimensions as requested and made the payment also electronically, on the same day. (He added – "the purchased raw material was not related to products deliverable to Aviation, Space or Defence customers" – big relief for the safety-conscious auditor!). The auditor from India was awed and amazed by the honesty of the people in that country but requested through a "nonconformity" to establish the required documentation before certification'.

To ensure that an approved supplier list is always used for purchases, the organisation needs to maintain a register of its current suppliers that includes approval status (type of approval such as temporary/trial/permanent/conditional/ approved/disapproved, etc.). The scope of the approval should be indicated as manufacturers, suppliers of raw materials, bought-out spares, consumables, software, tools/fixtures/templates, special process service providers, calibration agencies/training providers/annual maintenance contractors, product type, critical items, process family, type of service etc., as the case may be. Ensure that the approved supplier list is up-to-date (at least quarterly) and that orders have been placed only with approved suppliers and that activities necessary for ensuring conformity and on-time delivery and with the specified purchase requirements have been carried out. The approved suppliers list may be reviewed and updated once a year.

Organisations should periodically review supplier performance based on the conformity of the product, process or service, i.e., quality of supply, on-time delivery schedule adherence and response to their unique needs expressed to the suppliers, if any, etc. The organisation shall use the results of these reviews to establish the level of controls to be implemented. Levels of management mean a scalable control based on the effect of the product/service supplied by the supplier on the final product intended for the customer. Perhaps the effect of tooling supplied by a supplier will be different from the effect of calibration services provided by another supplier on the quality of the final product. A poor calibration may have a high risk of product nonconformity. The organisation has to define the necessary actions to take when dealing with suppliers that do not meet requirements or when their performance falls below an expected benchmark.

Performance evaluation of suppliers differs from the evaluation and selection of suppliers carried out initially for their approval, including them in the approved supplier's list. Supplier performance must be evaluated and reviewed periodically based on the supplier's performance against purchase orders placed on them each time throughout the year. There could be suitable evaluation criteria used for the review, such as quality issues identified by the organisation at the time of receipt of the product from the suppliers or during use by the customers, on-time delivery or delays in adhering to the schedules that are scheduled variance, communication/response of the supplier to special requests/requirements of the organisation, etc. The evaluation should finally give out a message, such as passing a benchmark, whether the supplier can be retained for business, whether the supplier is a good partner for progress, is dependable, and whether the relationship benefits the company. Whatever the evaluation result is, it should be recorded and used as the basis for establishing and maintaining the level of control for the suppliers.

The definition of criteria for evaluating suppliers initially and after their performance after every supply reduces the risks of the vulnerability of organisations associated with the activity of suppliers.

The organisation expects the score/rating of a supplier to be above a certain percentage to continue the business/retain the suppliers for future trade. Suppose any supplier fails to achieve the benchmark. In that case, suitable advice/guidelines/instructions/warnings are to be communicated to the suppliers, providing them another chance for improvement. When the organisation finds that some suppliers are consistently failing to meet the benchmark, the organisation may take appropriate actions, like removing the supplier from the approved list and having no purchases from the supplier. Communication of a supplier's performance evaluation results, even when the supplier meets the benchmark, will encourage them to improve and continue the mutually beneficial relationship with the organisation.

Sometimes, a few organisations evaluate the suppliers' scores. Still, no benchmark is determined so that all the suppliers are retained for the future, and no communication is sent to anyone to appreciate or advise improvements of their role in the business. That does not help in improving the quality and industry. Organisations need suppliers to perform to their advantage. Organisations can prefer a supplier to others, and the preference is established over some time based on the supplier's performance. Business always aims at long-term relationships and nobler objectives rather than shorter financial gains. It is not a 'one-time procurement' type. If so, the suppliers occasionally need to be encouraged and persuaded to do better and help them show the results of their efforts to be a preferred partner in business. Effective communication helps in doing so. Suppose a supplier consistently performs poorly, not meeting the set benchmark. In that case, politely declining to do business with such suppliers may encourage other enthusiastic suppliers to do better and help healthy competition among suppliers.

Suppose the customer has indicated any preference for using special process sources/suppliers of their choice through direct approvals. In that case, the organisation must use the customer-approved sources. The organisation is bound by the customer's obligation to select such sources. However, (apart from the selection of the source), there is no obligation on the part of the customer to bear responsibility for the quality or performance of the supplier, for which the organisation has to take responsibility. It does not absolve the obligation of the organisation to establish adequate controls on the supplier, ensure conformity for the product, process or service and ensure on-time delivery.

The organisation shall ensure that the external providers control and retain the documented information (records) they create. It can also include documentation (documents and records) supplied to the external providers by the organisation for their use and return after use.

8.4.2 Type and extent of control

The organisation's main concern is that the suppliers' products, processes or services do not adversely affect their business – delivering conforming products and services to customers. The organisation establishes controls to ensure everything goes well with the externally provided processes, products and services.

Controls include initial evaluation of the supplier using a defined criterion, including risk evaluation, selection and approval if the supplier adequately meets the criteria, contractually communicating the exact requirements with sufficient description (PO) and monitoring the supplier's performance every time.

The first thing the organisation does is to ensure that the externally provided processes remain within the established and implemented Quality Management System. The organisation then takes into consideration the following before deciding on the type and extent of control:

a. the potential impact of the supplier-given processes, products and services on the business of delivering conforming products and services to its customers;

b. the potential impact of the supplier-given processes, products and services on the applicable statutory and regulatory requirements;

c. the effectiveness of the controls applied by the supplier towards the supply of conforming products, processes or services;

d. The results of the periodic evaluation of the supplier's performance (see 8.4.1.1).

Then, the organisation defines the controls that:

i. it intends to apply to an external provider/supplier and

ii. it intends to apply to the resulting output – the product/process or service delivered by the supplier.

The organisation determines the verification or other activities necessary to ensure that the supplier-supplied processes, products and services meet the specified requirements.

When the purchased products are received, the organisation has to inspect or verify the product to ensure that the purchased product meets the specified requirements. For this, the organisation has to make necessary arrangements, establish the facilities and implement the activities.

Verification or inspection activities of processes, products and services are carried out concerning the risks identified by the organisation. If there is a likelihood of receiving nonconforming parts/processes from the supplier, goods receiving inspection is strengthened. Periodic testing may also be introduced on the products/processes being received from the suppliers. If there is a likelihood of receiving counterfeit parts from the supplier, goods receiving inspection is vigorously implemented. Verification or inspection activities of processes, products and services established by the organisation can include an assessment to verify conformity, verifying the documentation submitted by the supplier, certificate of conformity, test documentation, statistical documentation, process control documentation and results of production process verification. Changes to the production process and its verification subsequently can also be included. Getting the confidence that the supplied product, process or service fully meets the requirements is more important than what verifications have been done.

The organisation may also carry out the following as part of the verifications, but none of them is mandatory:

a. inspection and audit at the supplier premises for the system, process, product or service;

b. review of the required documentation;

c. review of production part approval documentation, including process data;

d. inspection of products or verification of services upon receipt at the organisation;

e. review of delegations carried out for the product verification to the supplier himself.

In the meantime, customers may establish verification activities in the organisation or at the supplier premises. The customer carrying out this verification or inspection does not absolve the organisation of its responsibility to provide products, processes or services to the customer meeting the requirements.

When the organisation performs inspection or verification activity, it can include getting appropriate documentation (means documents and records) from the suppliers. The documentation should indicate objective evidence of conformity to the product. Corresponding evidence may be a certificate of conformity and records like test reports, statistical documents or process control records. Review of the required documentation like verifying the test certificate with reference to the specification, inspection and audit conducted at the supplier's premises and examination of the product upon receipt like testing a sample of the raw material in an approved laboratory and verifying conformity to specification, etc., are all part of the verification activities.

Additional verification arrangements are prescribed in the aerospace standard because of the risk of nonconforming material finding its way into the final product if the organisation receiving the part accepts it without any verification. The component may get integrated into the final product, where the individual quality of the part may further escape verification. The undetected/inherent flaws, if any, may surface during use and affect the quality/performance of the delivered product. Similar lapses and defects detected during the use and maintenance of equipment/aircraft justify the additional verification arrangements for purchased products.

Suppose the purchased material is urgently required at the production shops, and the necessary verification activities are not completed. In that case, management can decide to release the material for production use pending completion of all the required verification activities. However, management has no intention of skipping the verifications since there will be a loss of time and effort if production operation is not given the material in time for use. It is a risk taken by the management. There will be no concerns if the material is found acceptable after verification. But if the materials are found unacceptable after verification, the entire production run using the material will have to be withdrawn, scrapped or rejected. That is the risk the management takes. When material is released in this manner, it is known as the 'release of material under positive recall' (refer to Annexure D for positive recall procedure).

When the material is released under a positive recall procedure, the material should be identified as such. Production records should identify this situation so that, if necessary, recall and replacement would be possible if the material does not meet the requirements. If the organisation is not willing to take the risk/responsibility of a positive recall procedure, exclusion of the condition from the scope/application of the Standard must be resorted to claiming that the received materials/products are not released for use without completing all the required verification activities, despite any compelling situations.

Where the organisation delegates verification activities to the supplier, the details of the delegation need to be defined. For this, the organisation may maintain a register of delegations. What type of verification is to be carried out, and to whom have the verification activities been delegated, who has authorised it, etc., and could they be recorded in this delegation register, in addition to purchase information? The organisation establishes periodic monitoring of the supplier's delegated verification activities.

The organisation has to communicate with external providers and ensure they use customer-designated sources if the customer has done so.

Sometimes, the organisation and the customer want to verify the purchased products/materials at the supplier's premises. It is necessary to indicate the requirement to the suppliers well in advance through the purchase order or an amendment in the purchase order. Details of verification arrangements and the product's release method should also be intimated.

Verification of quality of critical items like raw materials may be included, even when the quality/test certifications are received along with such materials when purchased. Though the requirement of validating the test reports for the raw materials by independent tests when test reports are received from suppliers is removed from the latest revision of the Standard, validation of the report for some of the important raw materials at specified intervals by the organisation may encourage suppliers to strengthen their Quality Management System. Alternatively, a representative or a third party appointed by the organisation to witness the tests at the supplier suffices the need to validate the test reports for critical raw materials.

An organisation should invariably assess the raw material, which is critical to quality and impacts the final product performance. Therefore, validation of test reports of raw materials is an excellent method to mitigate the potential risks and their negative consequences.

The decision to select, approve and disapprove suppliers is effectively based on inputs from all functions in the organisation, previous performances of the suppliers and risk analysis. Initial evaluation of suppliers based on criteria, their approval by top management and maintaining a register of all such suppliers indicating their scope of supply are the first steps. When an indent/request for purchase (RFP) is received, the purchase manager should scrutinise the current register of approved suppliers to identify the scope of supply and the probable suppliers to float a tender enquiry. Against every purchase order placed, the supplier's performance is evaluated as per performance evaluation criteria. The data is compiled annually to arrive at the score achieved by the suppliers. The organisation communicates the performance evaluation result to all the suppliers, indicating the need for sustenance or improvement. The suppliers may be recommended for removal from the approved list based on consistent poor performance. Additional suppliers may be approved and added to the register. Removal from or addition to the supplier register will be done only with the approval of top management. Responsibilities are defined and communicated among relevant functions (production planning, manufacturing, design, quality, post-delivery customer support, etc.) The checklist is used to evaluate and select suppliers. Records concerning selection, approval and contracting results (e.g., audit reports, performance data, action/consequences of deviations, etc.) are maintained. A list of suppliers is available and is regularly updated. Supplier assessment results during the selection and approval process (e.g., supplier rating) are used effectively for the purchase process. The measurement results of defined performance metrics show that the targets are achieved.

Supplier monitoring and on-site surveillance activities are planned (incoming/receiving inspection, sampling plans, supplier audits, etc.) regularly based on product criticality assessment, risk of using the supplier and being a special process source, etc. Performance results such as on-time delivery and quality of supplies are assessed. It must be ensured that a process is in place to demonstrate that the supplier can consistently fulfil given requirements. The supplier should have a system to detect, document, and control nonconforming products. The supplier should prepare and maintain the records required to support the conformity of the products delivered. There should be objective evidence such as procedures, test reports, certificates of conformity (C of C), records of dealing with nonconformities, etc., at the supplier's operations, which the organisation periodically audits.

Checklists and forms are used when performing on-site reviews and supplier audits. The organisation's performance measurements against targets are communicated to the suppliers to trigger improvement actions on the supplier's end.

Organisations would do well to demonstrate compliance with the control of externally provided processes, products and services by the records of supplier evaluation using some criteria, supplier performance evaluations, actions arising from the assessments, type and extent of control implemented on the external providers, purchase process and verifications carried out for the purchased product, process and service, etc. Refer to the associated documentation (drawings, specifications, contract documents, etc.), identify configuration, verify inspection/test reports and conformity certificates, carry out visual checks, take physical measurements using calibrated equipment and make decisions with respect to the acceptance criteria. The availability of a delegation register, even when delegations are not given to suppliers, indicates the organisation has established a contingency plan to act when necessary. Suppliers must provide raw materials test reports/certification results/laboratory analysis requirements (e.g., tensile strength, hardness, chemical composition, etc.) as defined by the product definition and/or the purchase order.

8.4.3 Information for external providers

Verification of adequacy of the requirements and information indicated in the purchase order before their communication to the external provider is the responsibility of the process owner in the organisation. The organisation should ensure the adequacy of specified purchase requirements before communicating with the supplier. It means there must be a supervisory check or peer review of the requirements and information specified in the purchase order for adequacy, completeness and correctness before the PO is released to the supplier. The indenter or the purchasing committee generally verify the adequacy of specified purchase requirements, if constituted in the organisation. The indenter asks the purchase manager to include all that information in the purchase order, detailed product description, specification mentioned, drawing, documents attached and terms and conditions enclosed with the PO. All are purchase information. The purchase order describes the configuration of the product being purchased.

'One production chief telephoned the purchase manager and asked him to buy 8x4 Aluminium sheets, 20 No's, as he ran short of the material for the next day's production task. The purchase manager telephoned and placed the order, adding that it was urgent. It would be appreciated if it could be delivered the next day, and the cheque could be collected immediately after that. The production manager was informed accordingly and came to the factory gate, waiting for the aluminium sheets truck to arrive.

A courier delivery man came on a motorbike and delivered an envelope to the production manager! On opening the envelope, he was furious at the purchase manager, finding that it contained 8mmX4mm Aluminium foils, 20 microns thick, 20 No's, with an "Urgent Delivery to Person" marked on the envelope! Moral of the story: "If you ask for what you want, you probably may get what you want. If you don't ask exactly what you want, You will get what you don't want"!

Don't just go by the telephonic description of the product the indenter gives. Get it in writing. If you find it wanting additional information, return the indent and request to provide more details of the product, process or service indented. "You know it. They will supply right, they know it, they do it, they have done it correctly last time, I know not about it; Some-how does it, it is urgent, MD wants it ..." are all excuses you hear, my dear Purchase Manager. It is your responsibility, and we shall get it right'.

When documents such as process specifications, drawings, process requirements, inspection/verification instructions and other relevant technical data are attached to the PO, their identification number/name and current revision status should be indicated in the purchase order.

Ensure that the specification quoted in a purchase order is the same as that in the design document/drawing or the specification received from the customer. It is necessary to ensure that there were discussions inside the organisation and between the organisation and potential suppliers regarding the design/drawing/specification of critical components during the design process or before the placement of an order.

It is a widespread mistake that organisations send out POs to raw material suppliers or special process suppliers, indicating only the material specification or process specification numbers. That is what is available in the drawing. The revision status is not generally shown in the product drawings/specifications. Therefore, the revision status is not indicated in the PO. Owners of the specifications issue updates periodically whenever any change is introduced for improvement. Material specifications (example) such as AMS 4027 6061-T6 Sheet and T651 Plate and special process specifications (example) such as MIL-DTL-5541 F Class 3 Chemical Conversion Coatings on Aluminium and Aluminium Alloys, and MIL-A-8625 F Type II Class 2 Anodic Coatings for Aluminium and Aluminium Alloys, all carry current revision status after it has been updated after improvement. Suppose the latest revision status of the specifications is not indicated in the POs to suppliers. In that case, the organisation is not asking for improvements on the material or process. Suppliers may push materials and processes of lower/earlier versions of the specifications. If not indicated in the contract/PO documents, an organisation cannot accept the lower/earlier versions of the material/process. PO issued to suppliers can also request the supply of material or process to the latest revised status after quoting the specification number.

Other information that could be included in the purchase information, where appropriate, are requirements for approval of product/services, methods, procedures, processes and equipment, requirements for competence, including any required qualification of persons, compliance of Quality Management System requirements by the supplier, etc. Specifying the personnel qualification in a purchase order may appear weird to an ordinary purchase manager who may not know that their purchased unique processes, like welding or NDT, should only be carried out by qualified and approved personnel to ensure process quality. Similarly, all other purchase information requirements to be described in the purchase order must be considered from the point of 'where appropriate' and included. It is necessary to justify why specific requirements are not indicated in the PO. The reasons could be neither that there is any benefit to the purchasing process nor that the requirement does reduce the risk of not meeting the customer requirements for the product.

Where statistical techniques could be used for product acceptance, such as statistically valid sampling plans, necessary information and related instructions are to be given through PO to the supplier for approval by the organisation. If it is identified and applicable, special requirements and critical items, including key characteristics, should be indicated.

Purchase orders should describe the requirements for test specimens (e.g., production method, controller process sheets, number of samples and identification/handling/storage conditions) for design approval, inspection/verification, investigation or auditing. It should be specified in the purchase information if there are requirements for the supplier to act on the following (which are known as the flow-down requirements):

a. Notify the organisation of a nonconforming process/product or service and obtain (purchasing) the organisation's approval of nonconforming product disposition. This ensures that the organisation expects the supplier to have a system to address all nonconformities within their processes, including a review of nonconformities that may affect the supplied product. In such cases, the decision to accept any nonconformity is taken jointly by the appropriate authority within the supplier, the purchasing organisation and the concerned regulatory authority. While making such decisions, if re-grading or downgrading of a product is necessary, it is done only with the purchaser's specific permission.

b. Prevent the use of counterfeit parts (see 8.1.4).

c. Notify the (purchasing) organisation of changes in process, product and services, including changes of (sub-tier) suppliers, changes of manufacturing facility/location, and obtain (purchasing) organisation's approval.

d. Flow down to the suppliers (and sub-tier suppliers) the applicable requirements, including specific customer requirements.

e. Notify the organisation of the requirement for retaining documented information (records), retention periods and disposition requirements after the retention periods.

f. Right of access by the organisation, their customer and regulatory authorities to the applicable areas of all facilities, at any level of the supply chain, involved in the order and to all relevant records (related to the supply). This is applicable only if the government/statutory and regulatory authorities are the interested parties in the business.

g. Ensuring that persons (external parties) are aware of their contribution to conformity of the process/product, their contribution to product safety and the importance of ethical behaviour.

Purchase orders should contain or refer to any applicable statutory or regulatory requirements to be complied with by the supplier. All the above additions in the purchase information are to act as a deterrent for some of the past problems faced by the industries.

Usually, PO to suppliers for the machining of components may indicate the essential configuration of the part with its part/drawing number and issue/revision status. When a copy of the drawing is attached to the PO, it should be a controlled copy issued by the organisation and linked to the PO number against which the drawing is issued. It is a good practice to indicate the PO reference and date issued at the backside of drawings/specifications, etc., when issued (controlled/hard copies) along with the PO. The documents are valid for the particular PO only unless otherwise advised later. PO for raw materials should indicate the material specification. It is mandatory to request a test certificate or conformity certificate (for physical properties, chemistry heat-treated conditions, etc.) against the specification of the material, heat number/melt number or batch number of the material for traceability identification. Suppose the raw material, proprietary or critical part is procured for aviation, space and defence applications, or airborne use. In that case, the material/part should have airworthiness clearance from the airworthiness authorities of the country of origin. Such components are authorised for release by the manufacturer through a release note signatory authorised by the regulatory rules of that country.

The organisation should ensure the products are ordered according to the proper specification or supplier documentation. Objective evidence such as purchase orders, contracts or other procurement instruments should exist.

Every purchase order generated should contain adequate, comprehensive and complete information with respect to the product being purchased, including proper configuration so that it meets the customer's requirements when used in the final product. Ensure the rest of the purchase information is attached to the purchase orders as standard terms and conditions.

8.5 Production and Service Provision

8.5.1 Control of production and service provision

Clause 8.5 of the Standard deals with production and service provision, which is the core business activity of any production organisation. The production planning process/department has information about the controlled conditions for all the processes in the organisation. Therefore, all necessary

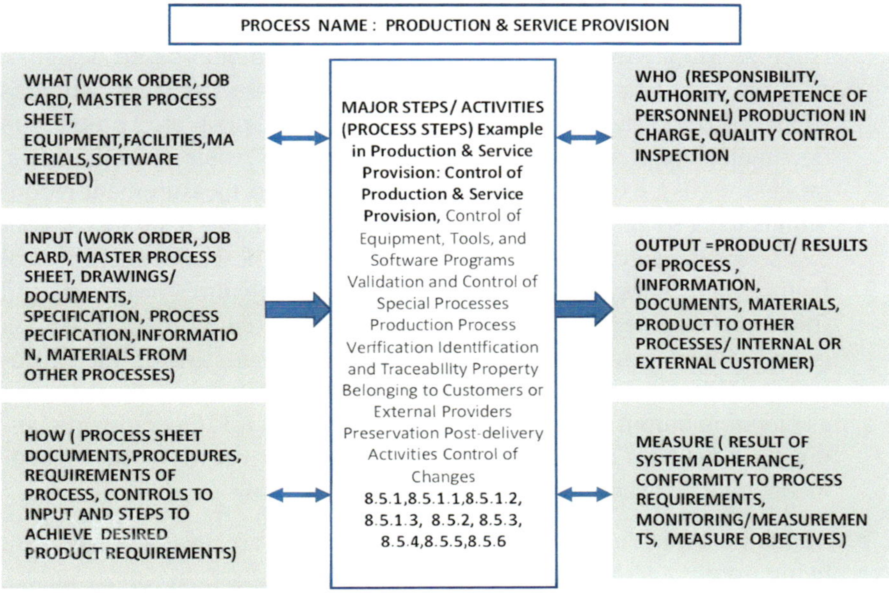

Figure 8.5 Process map/planet diagram for 'production and service provision'.

for production can be identified, planned and documented by that process/ department efficiently and cost-effectively.

Clause 8.5.1 (Control of production and service provision) deals with the production planning activity, which is an extension of the planning for product realisation, dealt with under clauses 8.1 (Operation planning and control) and 8.2 (Requirements for products and services) earlier. The Standard requires the organisation to plan and carry out production and service provision under controlled conditions. Controlled conditions are further defined as applicable. Each production item has specific inputs, steps and activities to complete the production process to get the output, the result of the process and the product itself. The controls established and used for the production process, conforming to the requirements of the Standard, result in the desired product, a beneficial product. A Process Map is indicated in Figure 8.5

What are the controlled conditions for production?

a. All the information about the product, documented information that includes product characteristics -quality- to be produced, processes to be complied with, activities to be performed, services to be provided and the expected results on the product. In the case of a hardware product, they are the documented information such as drawings and other design documents, parts list, including digital product definition data (such as 3D CAD data, DXF or 2D drawings, model-based designs), material specifications, process specifications, process steps/activities (master process sheet, process flow charts, control plans etc.) production documents, (e.g., manufacturing plans, travellers, routers, work orders, process cards), expected performance results and measurement record with respect to verification documents like the quality plan, inspection/test schedules, etc. The product's configuration describes the physical features and characteristics/properties, technical features/properties and functional and performance requirements expected on the product (as given by the design outputs) and the measured results against the expectations. When production is completed, these documents are the same as planning output documents (mentioned in para 8.1, Operational planning and control) or production documentation (records). These documents may be used individually or in combination.

b. Suitable monitoring and measuring resources are available and are being used.

c. Monitoring and measurement activities are carried out at appropriate stages of production to verify that the criteria for control of processes or outputs and acceptance criteria for products have been met. It is to do stage inspections and final inspections. Do these inspections during production; measurements are done per the criteria/reference dimensions/characteristics in the drawings/specifications and recorded in the inspection record sheet. When recording the results of measurement in the inspection record sheets, make sure that:

   i. Criteria for acceptance and rejection is indicated (example indicate drawing dimension 25 (+/–) 0.1 mm, i.e., dimensions with tolerance).
   ii. Record it in the same sequence as inspections are carried out.
   iii. Record the results of measurement. If the measured reading with a Vernier calliper indicates 25.05, record it as 25.05. (There is no need to say it is acceptable or not acceptable, as the reading recorded can be understood as a desired result within the tolerance.) In case using an attribute gauge (example go/no go gauge) (data has only two possible outcomes, such as pass or fail), record the result as pass or fail.
   iv. Retain the measurement results as an 'Inspection Record' with remarks that the results are acceptable or unacceptable (objective evidence of unacceptable results has to be highlighted).

v. Identify the monitoring and measurement equipment used, such as (instrument ID), and confirm its valid calibration status. Follow any instructions associated with its use.
vi. When sampling is used for product acceptance, the sampling plan is justified based on recognised statistical principles and appropriate for use (with respect to the product's criticality and process capability). When large quantities of the same component or hardware store are manufactured, and the parts' features and characteristics are not critical, sampling plans may be used to inspect a batch of products to save time. It is necessary to establish that the sampling plan used for inspection is based on recognised statistical principles. The easiest way is to adopt sampling plans established through national or international standards or prescribed customer standards while receiving the customer requirements. Mil-Std-105E, Mil-Std-414, ANSI/ASQC Z1.4-2003, ANSI/ASQC Z1.9-2003, IS 2500-Part 1 Sampling Procedure for Inspection by Attributes, ISO 2859-Part 1, 2, 3, etc., indicate statistically valid sampling plans. Sampling plans must match the part's criticality and process capability while in use. Risk analysis may provide an answer to whether the plan is suitable for sampling inspection and evaluate if there is a likely chance for developing deviation/defect while the part is in use, etc. The statistical validity of the sampling plan is ensured when the sampling plans selected meet the sample size in a lot as recommended in the national standards unless the customer has indicated a specific requirement with respect to sampling plans. Customer approval of sampling plans insisted earlier has been deleted from the Standard since it is contractual. Customers can insist on this if they desire to. Periodically verifying and maintaining the process capability index for stable processes is a method of validating the sampling plans.

d. Controlled production conditions include suitable infrastructure (para 7.1.3) and environment (para 7.1.4) for the operation of processes. Suitable infrastructure can consist of product-specific tools (e.g., jigs, fixtures, moulds) and software programs (para 8.5.1.1), as in the case of CNC machine operations.

e. Controlled conditions for production include the appointment of competent persons having the necessary qualifications (para 7.2).

f. Controlled conditions established by the organisation for production may include the validation and periodic revalidation of the special processes (read more on the special procedures and their validation in para 8.5.1.2). Validation is establishing that the method can achieve the planned results of a process, where due to the physical or chemical changes in the final product, the resulting output cannot be verified by subsequent monitoring or measurement except by destructive means. It is not significant if the measurement involves dissections that cannot be usefully joined or mended later. However, the microstructure and physical property changes benefit the material/product for engineering uses.

g. Controlled conditions established by the organisation include implementing actions to prevent human error. A common finding recorded in the nonconformity report after root cause analysis in SMEs, especially machine shops, is that the dimensional deviation or tool mark has happened due to 'operator error'. The investigation concludes that the root cause is operator error. The root cause is something else. Further analysis may reveal the reasons for operator error, but management seldom goes beyond true root cause analysis. People make mistakes. But why they make mistakes is important to know. With that objective in mind, human error is when a person makes a mistake. The auditor becomes helpless and has to label the practice as 'conforming' but observes that the root cause analysis could be improved since the operator probably had no intention of making an error. Design deficiencies are often misdiagnosed as human error. The law attributed to 'Murphy' states, 'Anything that can go wrong will go wrong'. It is trying to propagate a key principle of designing parts for assembly. If something can be assembled wrongly, human error will ensure it is done someday. 'Poka-yoke' refers to the Japanese technique of making mistakes/human errors impossible. This is now commonly referred to as 'mistake proofing'. Limiting the possibility of human error is a fundamental part of the pursuit of excellence in quality.

h. Controlled conditions for production include the implementation of release, delivery and post-delivery activities.

i. Establishing criteria for workmanship is a controlled condition for production. Workmanship can be in the form of written standards, representative samples or illustrations. Defining some workmanship criteria in production processes, like blending sharp edges, surface finishes at locations, edge distance from holes and structural damages/minor deviations, may be impossible. It is seldom defined in drawings/CAD models or detailed descriptions. Keeping a sample of 'acceptable deviation' for reference is a better way to judge the acceptability of workmanship. When you are in doubt, compare. Representative examples (etalons are visual comparators) or written standards, illustrations, etc., may help specify the requirement most practically. Plenty of them are available for each stage of production of aero-engine components in a public sector unit. They are standards of weights and measures and true guides for dimensions, surface finish, special features and even acceptable defects. Radii at edges and transitions are measured using etalon. Surface finish determines the fatigue strength of a component. Steps and transitions on diameter, holes, slots, mating surfaces, etc., should have adequate surface finish to avoid stress concentration. MiG technology calls for smooth transitions and necessary surface finish on all components.

Aero-engine compressor discs are manufactured with dovetail slots to fit the blades. When manufactured with BT3-1 titanium alloy, the discs will likely have reduced durability if stress concentrators like nicks, dents, scratches, etc., are present at the dovetail slots. Longitudinal notches, score marks, machining tool marks, etc., are stress raisers.

'*We have passed through this gruelling phase of aero-engine manufacturing. Low-pressure compressor Rotor, IV stage disc of aero-engines, failed on two MiG aircraft, setting both the aircraft on fire within a week in 1990. On investigation of the failures, it was concluded that both the discs had failed due to low cycle fatigue. There were no metallurgical defects observed. Scanning electron microscope revealed fracture striations of about 75mm along the circumference near the rim to the web transition radius. The cracks have probably originated from the tool marks/deep machining marks under the effect of enhanced variable dynamic stresses. The discs machined indigenously had no such defects, and no failure has happened since then*'.

Another speciality is that 75%–80 % of the dimension checks are done by special gauges/jigs/tools. Human error likely to occur during reading is minimised, so checking dimensions with close tolerance becomes easy. Proper tooling or job fixtures ensure several dimensions. The workpiece undergoes machining; periodic confirmation of the fixture-ensured dimensions ensures a quality control programme. While planning for production, similar workmanship standards must be made available to the shop or technicians and inspection to help decide on the acceptability without any ambiguity or doubt. This could be a one-time general instruction or display on the shop floor or inspection areas so that every process sheet must not repeatedly explain the same workmanship criteria.

j. Production planning operations include keeping appropriate accountability for the number of parts to be manufactured and the number actually manufactured. Finally, they are accepted for dispatch after quality control checks/verification/inspection to meet the requirements. Suppose there is a discrepancy in the total number at the end of the production line. In that case, it is essential to know what caused the difference rather than who will be responsible for losing material/man/machine hours. Accountability is keeping track of production volumes – as per original work orders and split orders for every item – happening in production shops. Accountability ensures what was initially planned is achieved at the end of the production line. It is necessary to find out the reasons for discrepancy or nonconformity, if any.

k. While planning for production, certain other product-related matters must also be considered. They are mainly related to critical items for which appropriate processes are to be implemented. Where identified, the essential characteristics must be produced with care and appropriate controls. The design and use of tooling should be such that it must be possible to measure variable data. In-process inspection and verification points should be decided for processes where subsequent verification cannot be performed, as in the case of an assembly and closing the panels until the next servicing. Unique processes are to be planned and carried out so that the desired final properties, features and characteristics are achieved consistently.

There may be an instruction or brief description of how the part will be manufactured or fabricated. Such instructions are better known as work instructions. Work instructions may be issued if there is a need to follow the same precisely, meticulously, in the given sequence and with specific identified skills and steps. When the work is known, and one can do it correctly all the time without error, that is, consistently, there is no need to have a work instruction for the work being done. Work instructions can include process flow charts, production documents such as manufacturing plans (master process sheets), travellers, routers (or route cards), job cards, work orders, process cards, quality plan and inspection documents. When issued along with the work instructions or separately, these documents are the controlled conditions or descriptions of the production process.

It is always advantageous to have a master process sheet for every product, describing the process through sketches and descriptions and showing how to proceed with every step with comprehensive information to prevent mistakes. The sequence of steps/operations is laid down in the master process sheets or route cards. Which type of machine is to be used, which machine, who will do it, how long it may take, what are the cares to be taken, what are the tools required, what are the layouts, setups, what software programmes are to be used, in what sequence each step is to be taken and similar other information are given in the master process sheet for the technician to understand clearly and follow it. Specific tools and equipment may be necessary to facilitate production during production work. Identifying specific devices, jigs, fixtures, templates, moulds, software/CNC programs, etc., wherever required in the production documentation, describes the product and controlled conditions for the production process.

Look at the shop floor where hardware production work is in progress, the machine shop where components are manufactured, and the assembly shop where components are assembled. Look at the special process shops where special processes are carried out on the machined components to enhance the engineering properties of the parts. The shops all use various monitoring and measuring devices, such as digital or analogue instruments, gauges, temperature indicators, pressure gauges, etc. Multiple measurements can be taken with various devices and equipment. The shops must have implemented monitoring and measuring of the process parameters, process steps, finished parts, and final products. The production shops or departments responsible for verifying and accepting the final product will release the parts/assembly to the next stage. It may be assembly, testing, inspection or delivery. Look at a design engineering service provider. By using the computer systems, tools/programs, CAD data, and other details as inputs given by the customer or established by the organisation, service delivery evolves the results of analyses and stresses and identifies potential areas that require improvements. There is production work going on. It means that the management has planned and provided necessary resources in terms of human resources with adequate skills, suitable machinery, work environment, essential facilities, process equipment, tools and software programs, utilities, monitoring and measuring devices and documented the method of production through production documents, master process sheets and route cards. Implementation of release, delivery and post-delivery activities are also production tasks. The organisation has planned how to complete the tasks satisfactorily.

Production planning should be able to keep track of and provide evidence through production documentation that all manufacturing operations and inspection operations have been completed as planned, the quantity of parts/products manufactured in the given sequence, given software job is completed, inspected and cleared/authorised for release; and the same is evidenced through records of production.

Anything that does not fit or things that are not suitable to include in an assembly can be called foreign objects. Foreign objects like moisture, dust, sand, metallic wires, swarf, washers, pins, nuts, bolts, rivets, tools and so on, flakes of sealants, waste cotton, cloth and stockinet, polythene sheets, plastic caps, and even food items, physical matters of any size and shape, are enemies or source of nuisance in an engineering hardware product/assembly. The tiny bits of hardware used on our big aircraft, the small tools used to repair them and the small acts of carelessness or neglect have caused large amounts of danger/damage and expensive repairs on aircraft. A glass tumbler with tea stains was recovered from inside an indigenous undercarriage assembly. Bits of cigarettes and sandwiches from integral fuel tanks of a direct supply aircraft are all foreign objects recovered from aircraft in the past! Controlled conditions for the production process include ensuring that the components, parts, assemblies and environment are free from all sorts of foreign objects; methods and actions are taken to prevent them, detect them if there are any encroachments, to remove them before they settle down and cause damage.

While assembling hardware components into a sub-assembly or assembly, like various gears, shafts and bearings are correctly assembled into a gearbox in a production shop, production planning should put necessary precautions, checks and confirmations to prevent, detect and remove foreign objects in place. In critical assemblies/areas, duplicate checks by different persons are typically done to ensure freedom from foreign objects before finally closing the panels/openings. This has to be recorded in production documentation. Foreign objects in software programs could be virus infections and corruption of data! Well, the production process should be planned in such a way as to keep foreign objects at bay permanently.

Several utilities and supplies are used to facilitate production tasks. Some include water, coolant oils, compressed air, electricity and chemical products. It is necessary to ensure that the utilities have no adverse effect on the products when they come in contact with them, for the utilities' quality parameters are to be verified, monitored and controlled. For example, suppose the coolant's pH value (potential for hydrogen ion concentration) is less than 7.0 (pure water is said to be neutral, with a pH close to 7.0 at 25 °C). In that case, the coolant is likely acidic and may cause discolouration and oxidation on metallic parts.

Since electric power failures have caused significant nonconformities like tool digging on aero-engine discs, production planning has to ensure an uninterrupted power supply when such discs are machined.

The equipment, tools and materials shall be those recommended by the designer of the aircraft (OEM) or must be at least equivalent to those recommended by the designer and acceptable to the customer and regulatory authority. The regulatory authority has a scheme for approving substitute materials, tools and equipment that meet the specified quality requirements.

The recall procedure is indicated separately in Annexure D.

Even though organisations take maximum care of all the product lines, several deficiencies, gaps and mistakes were found in various organisations' core production processes during certification audits. Some of these gaps/aberrations are indicated below for the benefit of organisations to verify and improve their Aerospace Quality Management Systems if needed:

a. Controlled conditions not maintained. No evidence of:

    i. use of suitable equipment (jigs, fixtures, moulds);
    ii. accountability for all products during manufacture (e.g., parts quantities, nonconforming products);
    iii. that all manufacturing and inspection operations have been completed as planned or as otherwise documented and authorised;
    iv. provision for the prevention, detection and removal of foreign objects;
    v. control of quality of utilities like coolant oil, DM water, compressed air used in machining parts;
    vi. special process validations;
    vii. use of prescribed work instructions, process management plan, process specification, drawings.

b. The PCBs are stored openly and have accumulated dust; ESD packing is not used during storage. All personnel were not bonded or electrically connected to the grounding/equipotential bonding system when handling ESDS items. ESD wrist straps were not worn.

c. Many critical items and key characteristics (e.g., height, weight, iron loss, etc.) (CTQs) impacting different components have not been identified, and the process to manage them has not been established.

'The auditor was pleased to see the best practice of indicating about 18 process steps in sequence from the raw material stage to the finished product (which included special processes) indicated in the component drawing itself, issued by the Customer in Karuizawa.

Drawing is the final configuration of the part when all the processes are completed and finished as required in the drawing, and it is indicated by the part number/drawing number with revision status. The customer issued traceable raw materials, and the South East Asian manufacturing company (SEA) had all their production documentation, including the quality plan and traveller. The parts have been dispatched to the customer with documentation indicating the part number, revision status, and traceability. The auditor noticed that the traveller told completion of about 12 operations, and the finishing operations were not indicated as done, and the manager agreed. Thus, the part that did not meet the complete configuration was inspected, accepted, and sent to the customer. It was clarified and confirmed that the dispatch was in accordance with the purchase order and Work Order, which called for process steps/operations up to Sl no. 12. The auditor maintained that the configuration of the part dispatched was not complete and a Semi-finished part indicating final part number has been delivered. As the document "Controlling the interface between KZ and SEA factory" confirmed, KZ and the factory agreed that a gap existed. They welcomed a nonconformity (NCR) and later took corrective action to amend the production and dispatch documentation. The corrective actions were verified for effectiveness subsequently and found effective. The system drives the organisation towards excellence'.

Responsibilities are defined at the production process level. Regular meetings are organised with operators at the shop floor level to share results and achievement of objectives of individual manufacturing units, assembly lines, etc. For control of utilities like mineral-oil-based metal working fluids, material safety data sheets given by the manufacturers may be referred to for information regarding physical and chemical properties, toxicological information, stability/reactivity and handling, storage and personal protection requirements. Compressed air checks are carried out per ISO 8573-1:2010 for oil content, moisture/dew point and particle size by the airborne particle counter. Traceability and identification of products/batches to meet requirements are in place. Performance metrics are available for production processes, such as lead time, nonconformities, yield, machine utilisation, etc. The results of measurements are used to drive the activities.

8.5.1.1 Control of equipment, tools and software programs

'*In a quality month celebration, the management displayed a colour poster encouraging workmen: "Don't be a fool; Use the right tool."*

The next day, the poster was corrected as "Don't be a fool; Provide the right tool"'.

The management is responsible for providing controlled conditions, including the right tools for production. Clause 8.5.1.1 of the Standard requires controlling the production equipment, tools and software programs used for production operation. Non-deliverable software used for project management, i.e., product/service realisation, is also a tool like any other production equipment tool. Here, the word tool means jigs, fixtures, templates and work-holding devices, and not expendable tools like cutting tools, drill bits and milling cutters.

Automated processes also need to be monitored. Production equipment, tools and software programs must be inspected for serviceability and satisfactory working conditions before they are released for production work. Validation of the tools and programs is also necessary based on the acceptable first use. The first use may be as part of the first production run of the part to prove the production processes, leading to the first article inspection (production process verification) and clearance of the first part or as one part of the first batch of the production run. When the first part in a batch is found satisfactory and cleared by quality control as an acceptable product, the tools and programs are validated for production work. How long the validation holds good is a matter that can be decided based on what type of tool it is (soft or hard), the likely damages and deterioration it can have during continued use and storage, etc. The frequency of re-validation may be decided on a calendar/periodic basis or the maximum number of parts that could be manufactured after the tool is validated. When hardware-assembled fixtures/tooling are used for production work, disassembly and reassembly for subsequent manufacture make it necessary to re-validate the tooling.

Storage requirements and condition checks during storage and handling will also be decided and implemented for the production equipment, tools and software programs. Organisations may show compliance with the requirements by maintaining documented information – a list of all the equipment, tools (jigs, fixtures, templates), and software programs available in the production shop and records of verification and validation of the production equipment, tools and software programs retained in a tool history card.

SMEs may have many production tools, jigs, fixtures, and software programs for CNC machining components. Numerical control is the operation of a machine by a series of coded instructions composed of numbers, letters, alphabet and other symbols. Computerised numerical control (CNC) is a machine control unit (MCU) having a computer memory with an executive program loaded into this memory by the MCU builder. This executive program enables the machine to perform specified function(s) or motion(s) needed to machine a part. The control also has internal memory for program storage. Reading and storing the information from part program storage (disk) is possible.

Verification/inspection of all hardware tools/equipment has to be established as a practice before every tool/equipment is issued for production tasks. Verification need not be an extensive inspection, provided the tool was verified/inspected earlier after fabrication and proved by using the tool for production work already. There could be a checklist for verification of all the tools/equipment while the ultimate aim is to see that the tool is serviceable and, in all probability, the product will come out good if the tool/program used is suitable. Software digital programs are verified with respect to the CADD/design data/drawing by using simulation techniques, which are again part of the digital programs. CNC machine programs like CATIA V5 and Unigraphics have simulation software. 'Vericut' software simulates tool paths and CNC machining to detect errors, potential collisions or inefficiency. 'Vericut' enables NC programmers to correct the mistakes before loading the program onto the machine, eliminating manual prove-outs. Vericut also optimises the NC program for cutting speeds for more efficient machining – even for high-speed devices.

This verification must be recorded before releasing the program to the shop/machine for the first production run. When used similarly, the program and tools are validated once the first part/product is produced, inspected and accepted. This fact also needs to be recorded as the validation of the software programme/tool. Arrangements must also be made for proper identification, storage, preservation, protection from damage/deterioration and easy retrieval of the hardware and software tools and equipment when not used.

Storage of equipment, tools or software program should be defined for the decided frequency. During proper storage, the equipment and instruments should be protected from damage, corrosion, etc. Re-validation frequency, which the organisation may choose based on the life of the tool, criticality and frequency of use, begins from the initial validation date after the development of the equipment, device or software program. The validation records may be used for process failure modes and effects analysis, gauge RandR study, control charts, capability assessments, etc. These tool validation requirements apply equally to customer-owned equipment and tooling under the organisation's custody.

8.5.1.2 Validation and control of special processes

All processes used for production are required to be verified. Unique processes are processes where the resulting output/product cannot be verified by subsequent monitoring or measurement. Unique processes are needed to be validated and re-validated periodically. If there are any deficiencies during the process, they are revealed only after the product has been used. Metals are subjected to various heat treatment processes like hardening and tempering, annealing and case hardening (carburising) to obtain certain engineering advantages and physical properties beneficial for the parts/products. Case hardening increases the wear resistance of metallic surfaces like gear teeth. Some protective processes are coating, plating or painting to prevent corrosion. Some examples will elucidate the point.

> *'The Quill shaft is used to transmit drive from the aero-engine to its accessory gearbox and subsequently to the accessories like the fuel pump of an aero-engine. The high-carbon steel quill shaft requires copper plating as the final surface finish, a process that prevents surface corrosion. All electroplating processes like acid Zinc and bright Cadmium/Nickel/Chrome/Copper introduce Hydrogen ions underneath the layer of coatings, which may cause embrittlement - a type of deterioration which can be linked to corrosion of the parent metal over some time in service. This ingress of hydrogen ions into the shaft can seriously reduce the ductility and load-bearing capacity, causing cracks and brittle failure at stresses below the yield stress of susceptible materials. The copper-plated quill shafts are therefore subjected to furnace heating up to 150 degrees Celsius for an hour or so to drive out the Hydrogen ions from underneath the layer of copper, a process called "obrabatha" (in the Russian language, which means de-embrittlement - a heat treatment operation). Unfortunately, this final operation was missed because the term "obrabatha" was omitted or inadvertently missed in the English translation of the master process sheet used for manufacturing the quill shaft. As a result, the de-embrittlement operation was not carried out during the production of the quill shaft. Being a special process, the process deficiency of not carrying out the de-embrittlement was not verifiable at the time of final inspection and verification of the part when accepted. The process deficiency on the quill shaft was revealed as a catastrophic failure. At the same time, it was used in a MiG aircraft about 150 flight hours after the initial fitment. Process deficiency was the reason for one of the aircraft accidents!'*

Special processes that require validation are listed at the end of this paragraph. NADCAP (National Aerospace and Defence Contractors Accreditation Program) is conformity assessment and approval of 'special processes' suppliers using a standardised approach. Unlike traditional third-party programs, NADCAP approval is granted based on industry consensus. Some service processes such as training, inspection, testing and record-keeping are also called unique processes.

There are several special processes whose applications find a place in aeronautical and space applications. During aero-engine manufacture and application of overhaul technology processes like SermaLoy J coating (a unique slurry applied aluminide diffusion coating specifically designed for protection of gas turbine hot section components) 20b (pronounced as 20 bay) coating, alitising (aluminising), siliconising, chroming, 'Low Alpha Inorganic White Thermal Coating', etc., are used. The physical properties achieved after the processes and the effect of the process on the physical and metallurgical characteristics of the parent metal cannot be directly monitored or measured on the parts during the process or after the process. The parts will be subjected to destructive tests or micro examinations to ascertain the physical properties achieved afterwards.

It is a common engineering practice to use a mechanically and geometrically similar or representative specimen along with the batch quantity of the product to undergo the process. The sample will be subjected to all the destructive tests to verify whether the desired properties have been achieved on the specimen. Suppose the specimen has been found to meet the requirements. In that case, it is presumed that similar properties and advantages have been achieved on the products, in the batch quantity, as a result of the process. The presumption is followed for routine acceptance of the process quality, where the serviceability of the process equipment (furnace or bath), technician's skill level and the process parameters (temperature, cycle time, takt time, bath concentration/specific gravity, duration of the process, process capability, quenching media, voltage, current, carbon potential, etc., for different processes, as prescribed in the specifications) are known, monitored/measured and controlled.

However, all the processes are required to give consistent results throughout the year, despite the variations in environment, ageing, variations/changes in inputs/controls, resources and any other acceptable variations in physical/mechanical properties of the product. It includes anything that is beyond the control of human capabilities. Proving that the process continues to give consistent results every time is the validation of the process. The validation may include qualification/approval of the process equipment, qualification of personnel doing the process (may not be necessary in case of automated/fully mechanised processes), associated records and periodic re-validation. It is sometimes necessary to conduct all the required tests on quantity one of the batch quantity of products to establish process validation. High-value products can be divided into geometrically similar and smaller samples for validation tests. For example, a bevel gear can be cut into four or eight segments to carry out a carburising special process (for surface hardness) since the material and geometry of the gear undergoing the process are similar in all segments. The organisation can decide on validation frequency depending on the criticality of the process, process stability and the quantity of items processed regularly. Periodicity of re-validations should be based on the process stability achieved annually or for every 25th component. The periodicity may be reviewed to achieve long-term process stability.

During process control of special processes, measuring process parameters and product characteristics is necessary. Measurement is to allow for the adjustments of the process and maintain them in a state of control. These limits/levels should be evaluated, established and documented during process validation to determine the process's robustness and stability. Considerations should include process control limits such as time, temperature, pressure, feed, speed, duration (soaking time) setup conditions, software parameters, raw material specifications, process operating procedures, material handling requirements, process change control, training to personnel, potential failure modes, worst-case conditions (failure mode and effects analysis, fault tree analysis etc.) and use of statistically valid techniques to optimise the process are other considerations for process controls.

Process and product data should be analysed to determine the typical ranges of variation for the process output. Determine whether a process is operating in a state of control and whether the existing controls can consistently produce the specified result. Process and product data should also be analysed to identify any variation due to controllable causes. Appropriate measures should be taken to eliminate controllable causes of variation. Removing the causes of variation will reduce variation in the process output and result in more assurance that the output/product will consistently meet specifications.

The special process validation is safety critical and critical to quality since any deficiency in the process will be revealed only during use. If it endangers safety, the losses will be horrendous. Aerospace engineers have learned several important lessons from their losses because of process errors on critical components like quill shafts and driving gears of aero-engines.

Removing the hardened layer from the surface of case-carburised gears during final gear grinding and not carrying out hydrogen de-embrittlement operation on the copper-plated quill shaft are examples of process inconsistencies resulting in significant aircraft accidents. Operations and conformity checks to product requirements before and after unique processes are also essential in achieving final quality.

All unique processes are essential, even when not carried out at the equipment/component manufacturers' facilities. Organisations and OEMs can often trace the cause of a flawed component back to the special process done at an external provider/supplier's facility by having stringent control of the outsourced special processes. Organisations and OEMs constantly monitor the special process quality to achieve their efficiency and product reliability targets. Organisations that outsource special processes should have a strict procedure for controlling the outsourced special processes. Special process sources shall be bound by contract to ensure the quality and consistency of the special processes outsourced, and make sure the communication is effective.

Outsourced processes cannot be declared as 'not applicable', and the validation of special procedures must be ensured for production and service provision.

Organisations outsourcing special processes have to take full responsibility for the quality and consistency of the special procedure. (Refer to 8.4.1. The organisation shall conform to all externally provided processes, products and services, including from customer-defined sources.) How can organisations fulfil this responsibility? While every organisation is capable of establishing excellent alternate methods, here is a straightforward answer:

a. For parts/equipment manufactured for ASD customers, the special process shall be carried out in-house at NADCAP-approved or customer-designated sources.

b. A copy of the current and valid process specifications shall be held, and a document shall be issued/referred along with the PO to the special process sources.

c. As a routine, get a conformity certificate from the special process sources, which can be used to accept the quality of the process.

d. Obtain a copy of validation records periodically from the special process sources for every process outsourced. The validation report shall indicate the information as in (e) below.

e. The organisation shall carry out a process audit for each of the special processes at the site periodically and note down the following as evidence of meeting the quality and consistency of the special procedure:

    i. Qualification/serviceability of the process equipment, including valid calibration for all the monitoring and measurement equipment (e.g., plotter, time, duration, temperature graph, bath concentration, ammeter, voltmeter, etc.).
    ii. Qualification, experience, skill and training, including any training updates for the technician(s) doing the process.
    iii. Witness/observe one complete process cycle – note down the process parameters being maintained, verify from the process log and process specifications/work instructions.
    iv. Test results were obtained on the representative sample/specimen for the prescribed quality checks in the specification – the tests/results were verified in an approved test laboratory. Compare results from destructive testing with the product specification and ensure conformity.

The ability of the special processes to achieve planned results can be established, and that too effectively and efficiently by using specific methods and procedures, if the organisation/customer approves the special processes based on process capability studies, bringing out Cpk index for the process. Cpk index (Cpk > 1.33) measures how close a process is running to its specification limits relative to the natural variability of the process.

Process failure mode effect analysis (PFMEA), when carried out for special processes, not only identifies the key characteristics but also illustrates potential failure modes and their effects, risk criticality/severity and likely causes of the potential failures. Once the likely causes of the potential failures are identified, appropriate risk mitigation plans can be identified for the process's effectiveness.

Generally, validation of processes for production and service provision is not applicable for software organisations (clause 8.5.1.2, Validation and control of special processes), and they can declare so with justification. Justification for 'not applicable' should be from the point of view of the customer and the requirements of the engineering service or product provided by the organisation. This should not affect the organisation's ability or responsibility to provide products/services that meet customer and applicable statutory and regulatory requirements.

Organisations establish validation of production and service provision processes by identifying and implementing the special processes, verifying the process's conformity and establishing validation and periodic re-validation of the facilities/processes as per specifications. After installation, the facilities/equipment are validated with product testing to ensure the achievement of product characteristics and accordingly manage the process parameters. Qualification and training for the operators are to be ensured to achieve adequate control of the processes. Routine acceptance after the product has undergone the process may be with the testing of coupons and samples. Periodic validation/re-validation includes verification of the characteristics achieved by the product by testing the geometrically similar product samples or one product itself from a batch quantity. Records of all the above activities related to serviceability and calibration of facilities/equipment, qualification and refresher personnel training, routine parameters for the actual process cycle and test results of the product sample are maintained for periodic validation/re-validation of all the special processes.

Some of the specific customer requirements are that the suppliers who provide special processes (not part manufacturing suppliers) must be NADCAP accredited for the following special processes (with checklist) indicated:

- Chemical processing: SAE AC7108 for Anodizing, Chemical Cleaning, Chemical Milling, Conversion/Phosphate Coatings, Paint/Dry Film Coatings, Plating, Stripping, Surface Treatment/Passivation, Etching.
- Coatings: AC7109, for Thermal Spray, Diffusion Coatings, Vapour Deposition, Coating Evaluation Laboratory, Stripping of Coatings, Heat Treating of Coated Parts, Dry Film Lubrication of Coated Parts, Plating of Coated Parts.
- Composites: AC7118, Compression Moulding (CMP), Core Processing (CP), Liquid Resin Processing (LRP), Metal Bonding (MB), Prepreg/Adhesive Bonding/Resin Film Infusion (PAR).
- Conventional machining as a special process: AC7126 for Hole Making, Broaching, Milling, Turning, Grinding, including grinding of coatings or after heat treatment, Edge Treatment.
- Elastomer Seals: AC7115.
- Electronics: AC7119 – Printed Circuit Boards, AC7119/1 – Rigid Printed Boards.
- AC7119/2 – Flexible and Rigid-Flexible Printed Boards, AC7119/3 – High Density Interconnect Printed Boards, AC7120 – Printed Circuit Assemblies, AC7121 – Cable and Harness.

- Heat treating: AC7102, Metal Systems (Carbon and Alloy Steel/Tool Steel/Stainless Steel/pH Steel Cast Iron/Aluminium Alloys/Titanium Alloys/Heat Resisting Alloys/Other Nonferrous Metals), Heat Treating Processes (Normalizing/Annealing/Hardening and Tempering/Solution Heat Treating/Aging/Carburizing/Nitriding/Stress Relieving).
- Heat-treating equipment (Furnace/Pyrometry/Instrumentation/Atmospheric Control).
- Brazing (Vacuum Brazing/Induction Brazing), Hot Forming.
- Non-destructive testing: AC7114 Liquid Penetrant Testing, Magnetic Particle Testing, Ultrasonic Testing, Radiographic Testing, Eddy Current Testing, Digital Radiographic Testing (DDA).
- Nonconventional machining: AC7116 Electrochemical Machining (ECM), Electrochemical Grinding (ECG), Electrical Discharge Machining (EDM) (Fast Hole/Sinker/Wire), Laser Beam Machining (LBM) (Cutting/Drilling/Marking).
- Surface enhancement: AC7117 for Shot Peening.
- Non-metallic materials manufacturing: AC7124 for Manufacturing Resin, Manufacturing Prepreg, Manufacturing Adhesive Films, Manufacturing Core, Manufacturing Fibres.
- Sealants Manufacturing: SAE AS 7202.
- Welding: AC7110 for Torch/Induction Brazing, Flash Welding, Electron Beam Welding, Resistance Welding, Fusion Welding, Laser Welding, Friction/Inertia Welding, Diffusion Welding, Percussion Stud Welding.

8.5.1.3 Production process verification

Production process verification, when implemented, ensures the ability of the process under controlled conditions to produce products that meet requirements. It is integral to production and ensures that production processes consistently produce products that meet requirements. This 'production process verification' activity can be called first article inspection (FAI). It is the verification of the first part manufactured.

Production process verification (FAI) documentation is essential because it reveals your strengths and weaknesses and gives you opportunities and threats. Please do not go by only the name FAI. It is not only an inspection process but also a process to verify the product realisation process, planning, operation and verification process, which gives confidence to the customer and the organisation that the process can consistently produce desirable output/product under controlled conditions.

To prove that the first production run of a new part or assembly successfully passes through the planned activities of production processes, the production documentation and tooling (including software programs) used are satisfactory, and together, the operation can consistently produce parts and assemblies that meet the requirements. The first production run means the first part is produced using production routing and tooling. Parts made before production, such as prototype parts, cannot be used to satisfy the FAI requirement, but using the FAI process for pre-production parts is allowed. FAI is required to be carried out when a new part is manufactured for the first time, when a new supplier does the job, at a new location of manufacture, or when there was no production of the part/component for more than two years at the location or when specifically required by the customer.

'"OEMs and multinationals do not come often. But if they come, they offer you a test by fire, I mean 'FAIR', they ask you for a 'First Article Inspection Report'. If your products pass, you pass and get on with a sustained relationship! Implementation of the AQMS is important, and it should be effective. If not, you must make all-out efforts to make it effective."

See what happened to an organisation which made aggressive marketing to capture a good aerospace business, some Rs. 2000 Million. First, they were told to implement AQMS and get it certified. It took some time. When they got the certificate, they got the order for FAI of some structural assemblies. They struggled day and night and produced it. In the enthusiasm of getting a big business and in the wee hours, they forgot to follow some of the process effectiveness requirements of the AQMS. They were ready to deliver the first lot. The customer said they would come down, have an audit, and verify the FAIR. At the end of the three-day customer audit, there were twenty-three non-conformities, some major! The customer gave them ample opportunities and requested the organisation to carry out all the improvements through corrective actions. Until then, the order was suspended! What an embarrassment! It must be added that they implemented corrective measures. All disciplinary actions were accepted in the re-audit, and the suspension was revoked. It is also true that we may not be able to do things "right the first time. We learn things in a hard way." Plan, do, check and act. Enjoy doing it right the second time and then every time!'

One item from the first batch of production run is used as a representative item. Related documentation includes:

a. All the production documentation, including customer drawings/ specification.

b. Part number accountability for each part.

c. Product accountability – raw material, specifications and special processes, functional testing.

d. Characteristic accountability, verification and compatibility evaluation.

After completing all the activities, this documentation is called the first article inspection report. The report is prepared per the SAE standard AS 9102 Rev C – Aerospace First Article Inspection Requirement guidelines. Once the process is proven approved by the customer, the process documentation and tooling can be used for continuous production until any change in the process is introduced. Process changes could be engineering changes, manufacturing process/methodology changes, tooling changes, and changes in the production location. Whenever any such change occurs, the first article inspection/production process verification must be repeated, fully or partially, to satisfy the requirements. For an aeronautical part, airworthiness clearance is issued by the regulatory authorities based on proven engineering documentation, including configuration and inspection reports, production processes, tooling, methodology and location of manufacture. If, at any time, a change occurs to any of the above, the airworthiness certificate gets invalidated until the change is again proven with a successful first article production and approval. When planning/engineering provides alternate material, at least a partial FAI would be required. This is to account for compliance with the engineering methods used. More details would depend on assessing the potential for affecting form, fit or function. Similarly, a change of manufacturing locations may also require a partial FAI.

For aerospace industries in India, the 'First Article Inspection Report' for a specific configuration is synonymous with the Qualification Test for Type Approval/Type Certification/Provisional Clearance for Flight Testing issued by the civil/military airworthiness and certification authority for declaring the part as airworthy.

The manufacturer may carry out individual production process verifications for standard hardware. Final products and assemblies that include standard hardware will affect airworthiness if the manufacturer's identified hardware was not used in the manufacturing process. Therefore, standard hardware is treated equally as manufactured products when used in an assembly.

While the production planning/methods department releases the production documentation to the shop for the first time, it is desirable to specially notify/call for the production of the first part or one part from the first production batch that it is required to undergo 'First Article Inspection Requirement' or the part is meant for production process verification. When successful, the entire production process could be benchmarked for future reference.

As per AS 9110 C, 'Maintenance Process Verification' is required instead of production process verification. The new maintenance processes must be documented, qualified and approved by the customer and/or regulatory authority. Usually, all maintenance processes documented by OEM are approved by the regulatory authority whenever such tasks are carried out by the MRO organisation by witnessing or scrutinising the activities on aircraft. If satisfactory, the new maintenance procedures are authorised and included in MRO organisations' scope of work. While doing so, the primary objective is to verify that the new processes (including the personnel, documentation and tooling) can perform the maintenance in compliance with established requirements.

The production process verification system significantly improves quality and first-time yields in products manufactured at a facility. While developing and implementing the process, individual work teams can be established to train personnel to build quality into the product and take responsibility for proactive defect prevention. Process technicians may be trained to ensure they have the skill and knowledge necessary to assume individual responsibility for the quality of products they produce. Specific operators can be certified for the processes of manufacturing particular hardware.

The product inspectors are members of the work team and assume the responsibility of quality assurance and monitoring the process, not the technicians' work. The quality assurance personnel become key in helping the process technicians maintain the production process and build quality into the product.

It is necessary to review all applicable documentation for the manufacturing process and the associated quality plan for correctness and to indicate the correct specification references. The quality plan should describe the methods used for production, inspection of individual parts and assemblies, identification of any special manufacturing processes involved, the techniques used to control such processes and the final test procedure for a completed assembly/equipment. Process sheets are to be identified and maintained. FAI forms are to be completed correctly as per guidelines given in Standard AS

9102 B. Documentation may include material certifications, approval of special processes, inspection data, test data and acceptance test results per procedures the customer approves. Inspection documentation must identify the tools and inspection equipment used.

It is necessary to include the identified key characteristics and their measured results. Each attribute per drawing/specification must be placed in the FAI documentation or referred to in a separate inspection record sheet. When FAI of a model-based definition (MBD) derived part is to be carried out, it will have its characteristic values obtained from a fully dimensioned 2D engineering drawing of the model or from the engineering parts list that contains default tolerances/configuration control sheets, or the engineering bubbled characteristics. A ballooned drawing is prepared by identifying 100% of the characteristics and attached to the FAI Form 3. It is necessary to balloon and number all dimensions, surface finish callouts, exclusion and inclusion area callouts, material and hardness callouts and requirements identified in the title boxes at the bottom of the drawings.

No FAI is required for standard parts if the supplier can provide evidence (C of C) that the part has been qualified to procurement specifications such as MS, NAS, etc. Standard catalogue hardware is to be listed on AS 9102 B Form 2. No FAI is required for metallic raw materials like plate/bar/rod, non-metallic paints/sealants/adhesives, etc. Similarly, electronic parts supplied by a supplier listed on the qualified products list (QPL) for that part and commercial off-the-shelf (COTS) items do not require a separate FAIR.

The customer needs assurance that the production and inspection tools and processes used for the process verification (FAI) will also be used in series production until further process improvements are planned. Any nonconformity noticed during FAI should be documented and reviewed for completeness. Inspection should be carried out by inspectors designated and sometimes approved by customers.

It is a desirable practice to specifically verify conformity to all other elements of the quality system being followed in the organisation while clearing the first part since production process verification or FAI is the most critical first step towards the achievement of a satisfactory product for the first time and aiming towards a continuous and sustained business relationship with a customer. The quality elements verified for having met the customer requirements are the purchase process, contract review records, verification of received material, configuration identification while processing, compliance to the required specifications, special processes completed (like heat treatment, NDT), calibration of measuring devices/equipment, inspection records, achievement of interchangeability (ICY), surface finish, performance test results, etc., and final configuration. If the quality system elements conform to the requirements and are effective, FAIR acceptance by customers would be a smooth and rewarding activity. Details of filling an FAI report form as per AS 9102 B are available in that standard. Any format containing similar information is adequate to maintain a FAIR. While completing the FAI report, it must be noted that a 'Certificate of Conformity' must accompany it. Include FAI or test reports for sub-assemblies and/or fabricated detail parts in Form 2. Variable data requirements from the engineering drawing will be provided with numerical results. Do not give wrong information or keep any format box blank. AS 9102 B forms can be downloaded free from www.sae.org/aaqg/publications/as9102b-faq.htm.

Software engineering service providers and such organisations may declare that clause 8.5.1.3, production process verification, does not apply to them since the requirement does not apply to deliverable software.

Some of the customer-specific requirements are that a full or partial FAI shall be performed for affected characteristics when any of the following occurs:

a. a change in design;
b. a change in manufacturing source(s), process(s), inspection method(s), locations of manufacture, tooling or materials;
c. a change in numerical control program or translation to another media;
d. a natural or man-made event that may adversely affect a manufacturing process;
e. a lapse in production for two years or as specified by the customer;
f. the FAI report must include a replication of product part marking (via photograph or sample) that represents production marking.

8.5.2 Identification and traceability

'OEM Customers are so good, and they get to know the ability of the organisation to fulfil their requirements just by doing a cursory audit and asking the organisation to produce and submit the PPV Reports (FAIR). Take, for example, the 'Make in India' reasonable efforts made by a Hyderabad-based organisation that implemented the aerospace quality system and the FAIR components to a West Asian OEM and received a substantial order for supply for the next ten years. What

they supplied was assembled onto defence equipment, and the Indian Defence is purchasing the same!'

Where product realisation happens within the organisation, software or hardware, in whatever stage of production, the products/services must be identified by suitable means. Identification and traceability are necessary to ensure the conformity of products and services. The identification should be available on the product or nearby throughout the product realisation process. Identification of the part/assembly has to be precise as per the configuration identification of the part/assembly. Some organisations do this with the help of unique identification, bar codes, tags, labels or travellers. When the configuration is identified on the part/assembly/product, it must be ensured that the configuration is exactly what was agreed to by the customer when the order was accepted. (Of course, if there is an amendment to the order indicating a change in the configuration, what configuration was finally agreed to by the customer must be produced, and the exact needs to be identified on the product under realisation processes.) The identification of the part/ assembly/ product should be available at any production stage. The identification (required configuration) does not change with reference to the status of the part/assembly/product. The status before and after any monitoring or measurement of the part/assembly/product should also be identified (as parts in process, waiting for measurement/measurement completed/inspected and accepted/waiting for assembly/waiting for dispatch, etc.). Monitoring and measurement are requirements at various stages or before completing the production. After measuring the requirements, the status of the part/assembly/product is decided based on the acceptability. The product must be identified relative to its conformance status throughout the product realisation process (the actual configuration).

In larger organisations where several personnel conduct inspections at different stages of product realisation, involving various areas of work, process and specialisation, inspection personnel are identified by the signature and/or stamp they use or by using passwords to operate their computer system. Production planning department personnel may also be given the authority and identification stamps to release master process sheets/production processes and authorise/control changes, if any. The facsimile of the stamps identifies different departments, areas or grades of inspectors. The stamps with numbers, alphabets and signatures identify the persons responsible for recording the measurements and declare their responsibility and authority for accepting the stages/inspections/tests/document changes and products. Issue of stamps to identified personnel should be a controlled activity since the responsibility for acceptance of products/services/processes through the identification of stamps/signatures cannot be shared and must be retrievable as long as the records are maintained. The identity of the person responsible is traceable. Where records are kept in soft copies, the organisation establishes appropriate controls for the electronic signature of personnel or password control for read-only access/permission to edit, etc. It is desirable to have a procedure for control of acceptance authority media, including stamp, signature, electronic signature and password for personnel who have been given such responsibility in the organisation. The procedure must identify actions to be taken when someone recognised by the stamp/signature/password leaves the organisation or loses the media. The issue of duplicate stamps/passwords and misuse or abuse of the acceptance authority media should be prohibited through personnel discipline. Stamps that are not legible, partially identifiable due to wear out, using other's stamps for whatever reason, putting signatures that are not identifiable, etc., are all misuses and abuses seen in practice. One should be proud of using the acceptance authority media, which declares to the world that the person responsible for the acceptance of the record has done thorough, professional, sincere and praiseworthy work for the cause of safety and quality of the aviation, space and defence products! Facsimiles, areas, types of stamps, etc., must be indicated, apart from records of issue, periodic checks for legibility and withdrawals. An electronic signature is acceptable as long as permitted within the Quality Management System. The Quality Management System must define electronic signature usage and control.

Traceability is the ability to identify using recorded identification, the ingots and furnace or crucible where the original raw material was melted/formed, just as the unique identification of an individual is generated through a person's fingerprints. As it is possible to trace the parents/family, it must be possible to identify the first ingot or crucible where the original raw material was melted/formed. Molten metal from a crucible/furnace is formed into ingots. The ingots are converted into raw materials of various sizes and shapes (such as bars, rounds, flats, rods and sheets) before they reach a machine shop for machining into parts. The nuggets have heat/melt numbers from the furnace; the raw materials formed from the ingots are traceable to the ingots' heat/melt number, and it is the sequential numbering of ingots formed from the furnace/melt.

While the part number identifies the finished/machined part with issue/rev status, the traceability of the metallic nugget, which is transformed into finished/machined components, is also identified by the heat/melt number of the raw material. Different metallic raw materials are tested for their chemical composition and mechanical properties. A heat/melt number is usually marked on the raw material test report, identifying the ingot it came from.

It is mentioned in the Standard that 'when traceability is a requirement', the organisation shall control the unique identification of the product and retain the documented information (record). Perhaps this requirement allows the organisation to manage the traceability of materials without reference to the original traceability from the mill/distributor. When the organisation purchases the material from the mill-authorised distributors or sub-distributers, the material's heat number or lot number is available through the material's test certificate or conformity certificate. When the organisation purchases a small quantity of materials from the local market, this test or conformity certificate may not be available. When they directly give the materials to the organisation for machining the component or part, some customers may not indicate the original traceability of the material or provide a copy of the test certificate/conformity certificate to the organisation. In that case, the customer purchase order number for which the material was given and the internal or customer work order number may be used as the basis of traceability of the material. This is the method of identification and traceability of the material followed by some organisations when they machine parts from the material supplied by the customer (who does not give the original traceability number) or when they machine parts from a small quantity of material bought from the local market. Where traceability of the material is not a contract requirement with the customers, vide customer purchase order, this method of maintaining the traceability of material is acceptable. What best can be done is that the customer may be requested to provide the traceability number of the material whenever they supply any raw material to the organisation for processing. Knowledgeable customers appreciate the efforts made by the organisation to maintain traceability and respond appropriately. Some others may use their authority to ignore the plea from the organisation and may not indicate identifying the traceability as a contract requirement. Due to ignorance, part traceability must be maintained throughout the product's life. It is desirable because the traceability requirement is mentioned as a note to the requirements of the Standard. By the way, all the notes indicated in the Standard are not mandatory requirements but are desirable and can be implemented as the best practices by the organisation.

One raw material (heat/melt number/batch) may be used for machining different parts. Different raw materials (different heat/melt numbers/batches) may be used for machining large quantities of one component. Traceability requirements may include the ability to trace all products manufactured from the same batch of raw material or the same manufacturing batch to the destination where the components/parts finally go. If the parts go into different assemblies, it must be known which parts have gone into which of the assemblies and which have gone into the next higher assembly and the final equipment/aircraft. Aviation, space and defence requirements require that traceability be maintained throughout the life of the products; the life of an aircraft may be for 30–40 years or 10,000–120,000 hours of use.

Material traceability is essential to the aviation, space and defence industries because of the safety requirements and the requirements of high-strength materials that look identical to commercial low-strength versions. Where traceability is a requirement, the organisation should establish controls for the unique identification and traceability of the product and maintain the records. The production documentation of a significant equipment/aircraft should be able to retrieve the information regarding the sequential record of its production (sequence of processes such as components manufacture, sub-assemblies, assemblies, inspection and testing and delivery, etc. An indication of the sequence of manufacture of an aircraft is as follows: components, equipment, fuselages (front, centre, rear), wings, control surfaces (ailerons, flaps, airbrakes, vertical fin, rudder, horizontal tail plane/slab tail, elevators, leading edge slats), undercarriages (main and nose), equipping, assembly of fuselages, assembly (marry-up) of aero-engine(s) to fuselage/wings, various system integrations, system checks, ground run, flight test, final delivery to the customer. Additional equipment/systems for military aircraft have been indicated.

In some industry sectors, like in software, configuration management is a means by which identification and traceability of the deliverables are maintained.

Before they are dispatched to customers, before their integration into an aircraft/aero-engine/electronic/electrical equipment or issue to another organisation/department, all products bear a quality control stamp as evidence of having been inspected to the required standards. The design, control and use of these stamps concerning the acceptance authority media, described in the quality system documentation of the organisation, shall be to the satisfaction of the regulatory authority. The organisation maintains a procedure to ensure effective control and administration of inspection stamps. Authorised stamps on the product itself or tags, labels, packaging, manufacturing plans, inspection records or other suitable means identify the inspection status of the product. Inspection stamps on products or shipping documents indicate the organisation's final product acceptance. Inspection stamps are designed to identify the organisation and the authorised personnel to whom the stamps are issued. The stamps used by the supplier/organisation responsible for inspection shall differ in design (facsimile/size/shape) from stamps used by any manufacturing or other personnel delegated to the inspection authority. Both should differ from stamps used by the Government Quality Assurance authority. Inspection stamps shall identify the person individually responsible for verifying the quality of the product. The inspection stamps allotted to one person shall not be transferable to another. Quality documentation shall include a facsimile of stamp impressions and describe each stamp's configuration, usage limitations and control. A signature shall not be used in the configuration of the inspection stamps.

Periodic checks are to be carried out to ensure the availability and legibility of stamps. Stamps that create illegible impressions shall be removed from service and destroyed. The serial number of lost stamps shall be permanently removed from service with effect from the date the stamp is lost, and the duplicate stamps are destroyed. Stamps returned by superannuating personnel or personnel who cease to be regular employees shall be removed permanently from service.

Comprehensive records are to be maintained for the information such as facsimile of the stamps, stamps allocated to individual personnel, issue date, withdrawal date, reasons for withdrawal, details of stamps lost, inventory record of duplicate stamps and records of periodic checks for legibility of stamps, etc.

All deliveries and releases of aeronautical equipment and stores must be accompanied by 'release notes'. Release notes are to be issued serially and should contain a certificate, the form and wording, which shall be controlled and approved by the regulatory authority/organisation. The release note should include all such particulars to identify the goods easily. Release note signatory, whose names shall be duly authorised and approved by regulatory authority/organisation, shall sign release notes in ink. Release note signatories shall generally be senior personnel in the organisation's quality control department. A panel of three names can be included as release note signatories for the organisation. Signature impressions by stamp shall not be used in release notes.

Again, it may be understood that for MRO organisations, production means maintenance, repair and overhaul activities. Therefore, it can be said that the requirements for identification and traceability are the same for products during manufacturing and MRO activities.

8.5.3 Property belonging to customers or external providers

Clause 8.5.3 of the Standard requires that organisations identify all the properties received from their customers and take care of the customer's property. At the same time, it remains under the organisation's control. All employees must exercise care with these properties and ensure that it is not lost or damaged.

The properties might have been received in the organisation for various reasons, such as a sample of the product or constituent part of the product/equipment, on loan, for fitment or incorporation on the final product. It may be a facility or device to help the production or test equipment for the organisation during production. The property received could be intellectual property or a document/data for use by the organisation. Care for customer property includes identification, verification, protection, safe storage, careful handling and use, accounting for fitment on deliverable products, etc. What should be done with the property after its use at the organisation is a matter decided between the organisation and the customer. Properties owned by customers and held by the organisation must be identified, verified, protected/safeguarded, and status accounted. Usually, the parties will reach an agreement for proper use and non-disclosure of the intellectual property to others at the beginning of the contract. All customer properties need to be protected and safeguarded while they are within the premises or custody of the organisation. Suppose any property, including intellectual property, is lost, damaged or found unsuitable for use. In that case, the organisation intimates the facts to the customer and retains documented information (records) accordingly. If the customer's property was received in the organisation with a letter or given a receipt, a similar letter or receipt/document must be there for the return of the property back to the customer.

Sometimes, certain documents (intellectual property) received from customers may be treated as external documents by the organisation, and there is no harm in treating the documented information as such. Suppose proper control is exercised for the external documents. In that case, the received documentation is treated as master, controlled, and (if the non-disclosure agreement permits) issued as owned copies to the users within the organisation, as per the procedure for control of documents, which will suffice for the exercise of care for the customer properties.

Customer property may include personal data received by the organisation. When organisations receive personal data from organisations or domain experts/stakeholders who are not their employees and the data is being used as a requirement not stated by the customer but needed for the satisfactory performance of the product, the data needs to be treated either as an external document or as a customer property and controlled accordingly.

Manager marketing/customer services generally retain records for all property, tooling and instruments received from customers/suppliers in a format like 'Customer/Supplier Tooling and Instruments - ABCE/MKT/CS/xx'.

For example, customer-supplied raw materials:

For customer-supplied raw materials, stores receive and identify the material per the store procedure. Any discrepancies related to supply are brought to the customer's notice by the marketing manager. At the time of dispatch, stores and marketing ensure accountability for the quantity of the customer-supplied material.

Customer-supplied tooling and instruments:

For customer-supplied tooling, marketing manager receives and identifies the tooling concerning the parts and process as applicable – custodian for customer-supplied tooling and instruments is stores. For the tools that require calibration, say, the QA manager coordinates for getting them calibrated as per the frequency defined by the customer. Further issues and receipt of customer tooling and instruments are maintained per se – 'ABCE/STR/xx – Issue and Receipt Register for Customer Gauges/Tools'.

8.5.4 Preservation

Product preservation includes identification, handling, packaging, storage, protection and conservation of all identified constituent parts/assemblies and the final product during product realisation stages within the organisation. Constituent parts include raw materials, components, life items (with specific shelf life) and items requiring preservation to keep the conformity of the product – each of the constituent products – during the stages of production of the final product/aircraft, during storage and protection, transmission or transportation, and to maintain conformity to the requirements of the final product/aircraft all the time.

'For example, polysulphide sealant is used to manufacture integral aircraft fuel tanks. The storage life of two-part PR-1422 Class B sealant is about nine months when stored at temperatures between 5°C and 27°C in original unopened containers. The two parts of the sealant are thoroughly mixed and applied on the sheet metal riveted joints within the sealant's prescribed application life/pot life. If the storage of sealant is not under prescribed conditions, or if the application does not take place within the pot life of the sealant, the sealant's property deteriorates. The mix will not adhere and cure properly when applied on the surface. It may flake off or form cracks, causing fuel leakage from the aircraft's metallic fuel tanks. Conformity of product (sealants) when not maintained by storing/preserving it at the recommended temperature,

the fuel tank of aircraft leaks (conformity to final product requirements-aircraft is not achieved!)'

Preservation mainly applies to the constituent parts of the final product, as the final product, like aircraft in storage, is usually preserved. Protection is done for aircraft tyres, canopy, the air intake of jet engines, propellers of small aircraft, etc., kept in a hangar or pen, moored, camouflaged or deployed on the field, and always kept ready to take up the mission. Aero-engines are inhibited while being sent for maintenance/repair/overhaul or unused for more than 15 days (preservation).

Preservation of products should be as per recommendations given in the product specifications. Where applicable, the requirements should meet the statutory and regulatory requirements, the requirements like storage of petroleum products (the Petroleum Rules, 1976, Govt. of India), inflammable gases and pressure vessels (Govt. of India Explosives Act: 1884, Explosives Rules: 2008, Gas Cylinder Rules: 2004, as examples).

Product preservation includes provisions for cleaning all parts internally and externally, freeing them from extraneous matters. Rust and corrosion are the naturally occurring enemies of machined components. Moisture, oxygen, dirt, film from other metal-working fluids and other materials can cause severe damage, causing rejection of metallic parts. For machined parts, it may include removal of burr and wiping it clean from dust, moisture and any other undesirable appearance or substance. Then, it must be protected from dust and moisture by polythene-sealed packing or foil wraps.

Hydraulic fluids in sealed containers stored for a long time must be rolled to maintain consistency. Fluid/oil in unsealed containers may gather dirt/dust particles of micron sizes, i.e., contamination. The oil must be checked for contamination levels before being put in hydraulic systems. NAS 1638 and ISO 4406 cleanliness standards are widely used to check the contamination level of hydraulic fluids in aerospace systems.

Prevention, detection and removal of foreign objects are essential during the entire product realisation process, especially during assembly of parts systems (see also description under clause 8.5.1). As any extraneous matter can cause malfunction anywhere, utmost care is needed to prevent its entry. It is necessary to check and detect any foreign object that has entered the assembly and remove it if found. It is essential to ensure that the surroundings are also free from foreign objects. This care to prevent, detect and remove foreign objects is to be established during all types of storage. After the front/centre fuselages are built on jigs, before the fuselages are mated, the aircraft structure is placed on a rotating jig and rotated on its horizontal axis to remove any moving metallic debris, loose rivets and washers and any other foreign objects. This is an example of efforts to remove foreign objects from the aircraft structural assemblies. The hardware/small metallic or non-metallic pieces, which are not readily retrievable, are arrested from movements by trapping them in injected sealants. At the same time, the structure is being rotated, even though it is not a desirable practice. It is a risk management activity, as the aircrew can feel and get irritated by the rumbling any left-out hardware within the structure may cause during flight!

Small things can be sources of significant incidents and sometimes even accidents in aviation, aerospace and defence products. The importance of preventing, detecting and removing foreign objects and the care one should take need no overemphasising. A lot of small things that have caused flight problems and safety hazards and ABCs of reducing small origins of significant issues are indicated in 'Safety Follows Where Quality Leads' (see Ref.1). The tiny bits of hardware used on our big aircraft, the small tools used to build and service them, the small acts of carelessness or neglect, all can cause large amounts of danger and expensive repairs. An incident of rags blocking the fuel pump inside the fuel tank on an aircraft was detected after 600 person-hours of troubleshooting. In another incident, on the inaugural day of flight of a newly developed jet trainer military aircraft, one 8 BA bolt (0.086-inch diameter of screw over thread) was recovered from the engine throttle box, by the regulatory body officer, before signing for the safety of flight F 1090! (form ten ninety is a format of the Indian Air Force). Embarrassment on the face of the managers, a sigh of relief and a smile on the VIPs present on the sidelines of the celebration!

Foreign object damage (FOD): Any damage attributed to foreign objects. The damage can be expressed in physical or economic terms, which may or may not downgrade the equipment's safety and performance characteristics. Foreign object damage is the result of foreign object debris.

Foreign object debris (FOD): Any material or article alien to an equipment or system that could cause damage that may or may not downgrade the equipment's safety and performance characteristics.

It is a fact that foreign objects and debris have contributed to jammed flight controls, engine damage, electrical shorts, fluid contamination, control valve failures, fires and other prominent failures. Some incidents have resulted in the loss of human lives and accidents to aircraft apart from costly material loss and damage. FOD hurts. FOD costs and FOD are preventable. Attitude towards safety is critical in preventing, detecting and removing foreign objects in the workplace.

Finding a loose item in an area already inspected or after final clearance for flight is to be treated as a severe matter necessitating full investigation to prevent recurrence. The importance of aircraft hardware is often overlooked because of its small size. Safe and efficient operation of any aircraft is greatly dependent upon the correct selection and use of aircraft-quality hardware (bolts, nuts, washers, hexagonal screws, cotter pins, safety locking wire, rivets, turn lock fasteners, O-rings, control cable hardware, fluid lines, flare joints, connectors, elbow, lock nut, screws, unions and such end fittings, electrical wiring and connectors). Improper use of these small things or its non-use can endanger safety. FOD comes in all shapes and sizes – a hand tool, dust, oil, metal shavings, machining chips, loose nuts, bolts, spanners, cotter pins, lock wire remnants, drill bits, pencil, pen, packing material, metal or wire clippings, wire scraps, connectors, components, rags, loose coins, parts, gloves, caps, cups, eye-glasses, tape, staples, paperclips, etc. Potential FOD may include personal items such as food and beverages, jewellery, buttons, keys or cell phones. Remember those butts of cigarettes and pieces of sandwiches recovered from the integral fuel tanks of a direct supply aircraft from the UK; foreign objects that travelled 9000 kilometres! Forgive them, for they had no intentions, and it did not cause any harm, but it benefited from lessons learned the hard way.

Never consider a missing item as too small to cause an accident. If an item cannot be found/traced/retrieved, immediately report the fact to the supervisor/shop in-charge. A lost tool procedure must exist in a safety programme, including maintenance/production control flagging, investigation and 're-certification for freedom from foreign objects'. A lapse in concentration or disregard for the rules makes people overlook a missing tool or fastener in an aircraft or aerospace equipment. When the pressure is on to clear an aircraft for flight, close out a panel, finish a job before lunch or clean before going home, people must focus on FOD prevention without complacency.

Training is suitable for bringing awareness but is not always given the priority and forethought it deserves. All employees should be alert while working at the facility after attending an orientation programme on FOD prevention. Periodic refresher training is also required for every aerospace business operation dealing with hardware. The checklist in (Ref. 2) helps the FOD Prevention Programme. It establishes guidelines for the industry to prevent foreign object damage (FOD) to any equipment or its constituent parts. The aim is to prevent the occurrence and risk of FOD and improve the quality of products supplied by the industry by sharing best practices. The checklist may be appropriately modified based on the work scope for specific use in an organisation. The degree of risk for a supplier to introduce FOD prevention in the workplace depends on the nature of the processes.

NAS 412, the National Aviation, Defence, and Space Standard (NAS), which supports the aerospace standard AS 9100, has also laid down a FOD prevention policy and associated procedures of the Aerospace Industries Association (AIA).

Special handling for sensitive products in the organisation is to be established. Sharp products are fragile, explosive, sensitive to touch and affect the skin, or products that give out an odour, noxious fume or electrostatic discharge.

All aviation, space and defence equipment and components need appropriate marking and labelling, including safety warnings. Aircraft markings are generally done per the drawings/scheme for all external areas after painting. Components, pipelines and connections are marked with appropriate part numbers, symbols and labels. Safety warnings on equipment/aircraft are for emergency opening/operations, fire extinguishers, ejection seat operation in fighter/trainer aircraft, etc. Every aircraft is identified by its tail number. Without such identification/marking, the flying object must be considered an unidentified flying object (UFO)!

There are several items, mainly adhesives, sealants, paints, chemicals or rubber products and other perishables, for which shelf-life expiry is marked. Shelf life is the recommended duration that products can be stored from the date of manufacture, during which the defined quality remains acceptable under specified storage conditions. After the dates, the quality may deteriorate or is not guaranteed. Shelf-life expiry dates are guidelines for the maximum product storage period before use. All items with recommended shelf life need to be monitored, and their usage is controlled to be used before the shelf life expires or the shelf life does not expire in storage. When used in leakproof assemblies, shelf-life expired seals do not give satisfactory service life.

Stock rotation is using the first received items first and later received items next, otherwise known as 'first in first out' (FIFO). The purpose is to have newer items always in stock, more so in case of items with shelf life marked on them and use the older items for production (so that it does not get their life expired on storage). There must be a system to correctly identify and monitor stock rotation, FIFO, so that the minimum stock of items for a continuous production run is maintained and no items get their shelf life expired on storage. The standard requires the organisation to provide stockrooms/storage areas that prevent damage or deterioration. Methods for controlling the contents of those areas should be defined. Periodic assessment of stored materials and products should also be conducted.

Special handling procedures for hazardous materials used in storage and production should be carefully and well-thought-out. Some chemicals that are generally considered to be friendly have the potential to be dangerous under specific circumstances. Therefore, chemicals and other hazardous materials must be stored properly to prevent spills and minimise exposures and uncontrolled reactions.

The material safety data sheet (MSDS), which provides information on health, physical and environmental hazards, handling precautions and first aid recommendations, should be retained and consulted often when using hazardous chemicals. Flammable liquids like aircraft fuels, corrosive materials, sealants, explosives (cartridges for emergency door/canopy separation) and all types of compressed gases, in gas or liquid form, contained in a vessel under pressure, cryogenics, such as liquid nitrogen, liquid helium and dry ice, are all hazardous materials, which require careful handling.

Packaging usually applies to products that are being readied for shipment. It can also apply to raw materials, components, sub-assemblies and partial assemblies. Packaging broadly includes the container, packing materials, markings, labels, packing slips, etc. Methods for all packaging need to be such that there is no deviation caused by product conformity. This emphasises the need for documentation, appropriate use of packing/packaging materials and shipping/transportation tests where appropriate. In international shipment, it is necessary to comply with the regulatory guidelines for regulating wood packaging material for fumigation.

The organisation demonstrates compliance with the preservation requirements of the Standard by the method of handling of constituent accepted parts, storage conditions, configuration status identification (work in progress, waiting for inspection, received, quarantined, nonconforming part/assembly, cleared for dispatch, removed for assembly), ESD protection in electronic assembly shops, FIFO (consumable stores), maintenance of environmental conditions in shops and particular areas (temperature, humidity, cleanliness, restricted access, cold storage), monitoring of limited life/shelf-life products, packaging, preservation (of products before delivery), etc.

8.5.5 Post-delivery activities

Suppose the customer requirements include support for the operation and maintenance of the equipment/product at the user's premises after the product has been delivered. In that case, the organisation has to plan and carry out post-delivery support activities as applicable for the equipment/product. The post-delivery activities may also be called 'after-sale service', which may include routine operation/use of the equipment, warranty service, periodic maintenance, repairs, overhaul and disposal after the useful life of the equipment. So that the organisation may undertake the post-delivery activities, the scope should not have made it 'not applicable' with justifications. The organisation must establish adequate resources and service provision for post-delivery activities.

While determining the extent of post-delivery activities, the organisation needs to consider the following:

a. Compliance with statutory and regulatory requirements: For example, MRO organisations have to get approval for their operations both from the OEM as well as the local regulatory authority

b. The potential undesired consequences associated with its products and services: For example, arrangements to prevent incidents and accidents, injury or harm to personnel. Maintenance of ground support equipment and supporting tools, etc.

c. Customer requirements and feedback

d. The organisation should collect information regarding field operations of the equipment, including serviceability, un-serviceability, resources necessary/available at field locations for maintenance, repair and overhaul, need and availability of essential spares for routine and periodic maintenance, etc. The data should be used for analysis, which may give the results as the mean time between failures (MTBF), 'time to repair', 'time between overhauls (TBO), complete repair manual, and field reliability data' of parts/equipment, which are all valuable information for continual improvement and the product lifecycle management (PLM).

e. The organisation should have a system of periodically updating the technical documentation, updating and controlling them and issuing them for use by the customers. Illustrated parts catalogue/parts list, user's manual, maintenance/repair manual, information on modifications/repairs/replacement of spares/any specific issues regarding operation and maintenance, service bulletins, servicing technical instructions, servicing instructions, etc., are all technical documentation issued to customers for the use of the equipment. It is desirable to have a master copy of all such approved documents available for reference and use at designated department/personnel in the organisation, say the customer services department, so that correct and updated information is always available.

f. When repair/servicing parties are sent to the customer's premises/facilities by the manufacturing organisation for repair/servicing/-maintenance of the delivered equipment, the details of the work undertaken by the organisation must be defined. The details may include the scope of work, resources, referred documents and descriptions of work, tools and equipment to be carried including particular tools/devices, verification/inspection/test, reporting of work completed and logistics. This is how the organisation's work at the customer's facilities (off-site work) is controlled.

g. Customer support for the product, such as activities concerning warranties, maintenance, replacement parts, resources, obsolescence, etc., for customers/users answering the queries, training operators/users, etc.

Whether post-delivery support activities are undertaken or not, the organisation should have a system of undertaking investigation and reporting actions to resolve the issues when the customer reports failures or nonconformities on the delivered equipment. Later, nonconformities and customer complaints (related to the equipment/product) are dealt with in detail in the procedure for control of nonconformities and customer complaints under clause 10.2 of the Standard.

A necessity to evolve repair schemes may arise, specifically to address a deviation/deficiency that may not have been envisaged earlier. Therefore, the Standard repair schemes may not be available in the repair/maintenance manuals. In such cases, the original equipment manufacturer (OEM) usually evolves special repair schemes in coordination with the designers. The repair schemes must be approved by OEM, customer and regulatory authorities before they can be complied with on the equipment. Similarly, any modifications to the equipment to enhance its performance or features are documented and complied with on the equipment in service after necessary approval by customers/regulatory authorities (refer to clause 8.3.6 of the Standard).

Organisations would do well to establish and demonstrate arrangements for post-delivery support of products through technical documentation such as operation manuals, maintenance manuals, structural repair manuals, illustrated parts catalogues (IPC), service bulletins and maintenance/inspection instructions. All such documents are issued after approval by regulatory authorities or by approved design organisations. Organisations should have ongoing continual improvement activities such as aircraft and defence equipment modifications throughout the lifecycle. Reliability or maintainability improvement programmes are initiated based on the in-service data received from the operators. In-service data to be obtained from the field service team or customer support process may be mean time between failures (MTBF), field reliability data, unscheduled removal of line replaceable units (LRUs), and mean time to repair (MTTR). Incident/accident information from customers, regulatory/safety authorities or organisations and the lessons learned should also be used to continually improve the products/equipment.

They are more procedural requirements indicated by regulatory authorities during MRO activities of aviation, space and defence equipment.

The production planning and control department will identify and establish controlled conditions with regard to production and service provision during MRO activities. The controlled conditions include satisfactory compliance with reference standards/codes, quality plans, manufacturers' recommendations and/or documented procedures called explicitly for the MRO work. While carrying out work as recommended by the manufacturer or per the referred standard or established guidelines of the organisation, satisfactory compliance shall be certified. This is when the regulatory authorities or OEMs issue urgent servicing notes or airworthiness directives.

Regulatory approval of MRO organisation for the scope of work is usually rated with respect to the capability of the organisation to carry out the required task efficiently and by following an effective Quality Management System. The MRO organisations have to maintain a list of capabilities of approved maintenance and repair processes and maintain the current ratings with respect to the above capabilities/facilities as stipulated by the regulatory authorities. The regulatory authorities may even insist on permanently having the required personnel positioned with facilities, equipment and materials to meet urgent requirements. The organisations also have to certify or assure that the MRO activities, when planned and carried out within the approved scope/capabilities, do not adversely affect the areas outside the scope. It simply means that when servicing the aero-engine fire detection system is carried out on the engine bay, no disturbance should be made on the engine or there should be no adverse effects on the aero-engine installation and controls!

Regulatory authorities may prescribe certain forms and formats to be completed when submitting the documentation after completing the MRO functions on the aircraft or equipment. This documentation submitted by the MRO organisation includes the certificate of conformity, airworthiness approvals/clearances, issue of release certificate of aircraft by authorised personnel, approval for return to service after maintenance, export documentation, etc. The organisation's production planning department has to implement procedures to satisfy customers and regulatory authorities accordingly.

8.5.6 Control of changes

Design changes: Once the design and development activity is completed, and the airworthiness clearance is obtained for the aerospace part or assembly and/or manufacture or use, there will be several occasions where changes may be necessary for the design, manufacture or maintenance of the aviation, space and defence equipment. The changes may be initiated by the customer, user or manufacturing agency based on the experience of use or manufacture, or a proactive continual improvement process initiated by the designer for upgrading technology, manufacturing process or material or an improved design. All these improvements are termed changes, including corrective actions to correct any design deficiency or deviation in design documentation and preventive actions for potential deficiencies determined later. Whenever a change is required in the design output/documents, the needed change must be identified, and appropriate documentation must be created to introduce the change. Even before creating the documentation of change, it is necessary to study the impact of the change on the resulting product, including products already delivered to the customer and the manufacturing process. Examine if the change is required to be reviewed for the total population of the product in use. The proposed changes are to be verified and validated. Once the change is accepted and approved for implementation, the change is documented. The change primarily affects the configuration of the part/assembly/performance. The design office practice must indicate what type of change must be registered. In the aircraft industry, the drawing/part/assembly number change may be indicated by an increment in the issue/revision status. A more detailed change is usually introduced by advanced modification information. After the aircraft has been given an airworthiness certificate or production clearance, several modifications may be introduced based on the field use/maintenance experiences or the need for improvements as the aircraft use continues. These improvements are all identified as modifications, servicing technical instructions, servicing bulletins, inspection instructions, etc. Each of these is finalised in consultation with the customers or regulatory authorities after careful study for the impact of the changes/improvements on the configuration of the part or assembly or its performance, without any adverse effect on safety, quality, reliability and maintainability. Any change should be for the benefit of the product and its use. All records of reviews and changes, accepted and implemented, are to be maintained as documented information. All design and development changes are to be controlled in accordance with the configuration management process, as explained earlier. Retained documented information (records) should indicate where the change has been made in the product's lifecycle and at what stage of the contract with the customer. All design changes, whether made during the development, production, or maintenance stages, should be controlled in line with the configuration management process throughout the aircraft's lifecycle or any other aviation, space or defence equipment.

In military aviation, minor changes that do not affect the configuration of the equipment are introduced as servicing technical instructions, servicing bulletins, inspection instructions, etc., with the customer's and regulatory authorities' approval. All modifications are proposed as 'advance modification information' (AMI) and discussed among the organisation, customers and regulatory officers. Approved changes are categorised based on the criticality and convenience of embodiment on production aircraft and the aircraft already delivered to the customer.

When a change is proposed and implemented, the organisation demonstrates and documents the impact study report, carries out additional testing where necessary and records the list of parts and performance changes on delivered products. Regulatory/customer approval is also recorded before implementing the changes.

Process changes: Planning issues in the production process and connected information to the production process/department with proper authorisation to use the documentation for the product realisation process. If the production planning or engineering department realises that a change in the production process is necessary for improvement or adding appropriate information or a controlled condition in the production documentation, then the planning department may authorise a change in the production process documentation. The documents affected may be the master process sheet, route card, job card, work order or any other similar document issued to production shops. Such changes should be authorised by identified personnel, preferably from the planning/process engineering. The changes affecting the production processes, equipment, tools and software programs should be identified and documented accordingly in the production documentation. Apart from documentation of the changes, the effect of the change on the production processes should be assessed by evaluating the results of the changes to confirm that the desired outcomes have been achieved. No adverse effect was seen on the product quality/conformity. Sometimes, it may also be necessary for the organisation to obtain acceptance of the changes from the customer and/or regulatory authorities in accordance with contract or regulatory requirements.

Auditors, having an eye for details, reach out to the honest 'process owners' and ask why these corrections, overwriting and changes were made. They ask, pointing to the production documentation (route card, quality plan), where they have casually done an honest job. The job was honest, but they did not know the implications of making such unauthorised changes in the sequence of operations, correcting some existing mistakes, writing some operations as not applicable, adding notes and comments, deleting some information, updating the quantity, quantity not tally with the accountability, correcting the part number and revision status, etc. Not getting such remarks/

Figure 8.6 Example authorisation with stamp for corrections made.

corrections/changes in the production documentation duly authorised by designated personnel from the planning/process engineering department is a mistake. The corrections made by the shop or inspection personnel might have been genuine and honest. However, making these changes is a mistake since they are not authorised by the designated personnel from the planning/process engineering department, with their acceptance authority media of signature and/or stamp. People who made these changes did not allow the authorities to verify the impact of these changes on quality and the process, which is incorrect. The auditor writes a nonconformity statement with the production documentation as evidence, and the process owner is convinced and encouraged to take a permanent solution.

Suppose an integrated computer-based enterprise resource planning (ERP) system is used to facilitate the flow of information among the design, production planning/process engineering/methods, production and quality control departments. In that case, only one authority should control the production processes (master processes) and changes to the production processes. That authority is the planning process/process engineering department.

It is appropriate to mention here the recommended way of correcting a genuine mistake in the hard copy of a document or record and authorising the correction. Many aviation, space and defence OEMs (tier 1/prime organisations) recommend the following practice: 'Delete the wrong word or words/data with a single straight stroke, from left to right in the middle. Reading the wrong word or words/data deleted must be possible to meet the legibility requirement of documents and records. Do not use eraser/whitener/multiple strokes to hide the wrong word or words/data. Write the correct word or words/data above the stroke. Authorise it with the acceptance authority media of signature and/or stamp' (Refer to Figure 8.6).

The above-recommended practice can be included in the existing procedure to control documents/records.

The organisation demonstrates compliance with the requirements by preparing a list of authorised personnel to approve changes to production processes. Authority for change may be documented in various ways, such as job descriptions/approved signatories list, document approval authority matrix, etc. Change is indicated by writing details of the change, together with the signature(s) of the authorised individual(s) and date. Usual changes to production processes are manual updating of a production document at the point of use, re-sequencing of production operations, use of an alternative source, amendment to software programs (CNC machine, inspection/testing), replaced equipment/tooling/gauges, update to material/process specifications, revised methods, visual/dimensional inspection, non-destructive testing, destructive testing, functional testing and risk assessment. Process change is formally documented and approved in the 'Production Process Change Request' (or equivalent) format. The requests are dealt with promptly by the planning/process engineer, who checks that the change does not adversely affect the quality of the product. The organisation retains documented information (record) indicating the results of the review and changes, the person(s) authorising the change and any necessary actions arising from the review.

Software organisations may not indicate acceptance authority media (signature or stamp) in their records. It is acceptable so long as the PCs are operated with individual password access and protection. E-mail communication is sufficient.

8.6 Release of Products and Services

Monitoring and measuring the product/service, whether the organisation's product or service, are continuous activities from the planning stage, throughout the production operation stages, to the final stage, where the final product or service is delivered to the customer. All quality parameters are monitored and measured during the stages of production with reference to the process parameters or product features and characteristics against the requirements identified and documented (in the drawing/specification/customer requirements) during the planning stages. All monitoring and measurements, as planned, are carried out in sequence at the pre-determined stages, and the final stage is consolidated at the final stage. The purpose is to see that all the features and characteristics are measured. The record of actual measurements shows the requirements have been met, the evidence of conformity being the record of measurements.

The release of products and services to the customer happens only after completing all the verification activities unless the customer desires otherwise. The customer's or relevant authority's prerogative is to ask for the product without the organisation finishing all the processes or measurements. The customer or the relevant authority may have adequate reason to do so.

Organisations should allow that with appropriate annotations in the release documents and communications.

Occasionally, aircraft or aircraft parts are under development, or certified parts or assemblies are further modified. The same may be released for tests or use before completing all the required measurements/monitoring activities. In such cases, the products released must be identified, and appropriate records are to be maintained. Aircraft parts under development or indigenisation must satisfy various tests on the ground and during flight tests before the parts are certified as fit for flying. Hence, after essential tests on the ground, when released for flight testing for the first time, the identification of the part and clearance for flight test of the part are described as provisional. Airworthiness authorities call it a 'Provisional clearance for flight trials'. Certificate of airworthiness or declaration as fit for flying and production release is done only when all the required ground test results and flight test results, which cannot be obtained from ground tests, are satisfied. Identifying the parts and releasing them for flight trials as provisional allows recall so that replacement could be done if necessary when such (special) specified/trial parts are not found to meet the requirements.

The organisation retains the documented information on the release of products and services. The documented information (records) includes:

a. Evidence of conformity with the acceptance criteria: All measurements must be recorded. Records should indicate the characteristics being measured, how much is the acceptable measurement and how much is the measurement result obtained for each attribute. When the required information is recorded, anyone can decide on acceptance/conformity or rejection/nonconformity by looking at the record. Please refer to the example wherein the dimension as per drawing 25.00 mm dia (with a tolerance of plus or minus 0.01 mm) is measured and recorded as 25.008 mm, using a digital Vernier/micrometre (instrument id: A/01/004). This practical example indicates the criteria for acceptance. They have found an acceptable result after measurement. Putting a tick mark against the measured result is an unhealthy practice located earlier among some inspection staff. The inspector may plead the excuse that the measurement was correctly done; recording the measurement involved additional time/paperwork and should not question the integrity of the inspector for not recording the measurement, etc., which are not acceptable since it does not indicate the acceptance criteria. When called for, measuring the diameter using an attribute gauge (go/no-go gauge) and recording that the diameter conforming is acceptable.

However, if the calculated results are called for, measuring and recording them is essential.

Where measurements are called for in sequence, actual measurements should be done in the prescribed sequence and not be jumbled. Similarly, checks/tests, named after the defined stages, should be done only after completing the prescribed steps.

There were attempts to violate this requirement of checks/tests in the assembly stages of aircraft build due to production pressures and without considering the consequences. For example, a cabin pressure-leak check must be done on a fighter aircraft after installing an ejection seat in the cockpit to maintain the correct air volume and pressure in the pressurised cockpit during flying. The test aims to monitor the air leakage rate out of the cockpit and balance it with the supply of conditioned air into the cockpit. A cabin pressure-leak test before installing the ejection seat was not allowed; as such, a leak check cannot give realistic test results.

All quality plans or inspection/test schedules should call for the measurement results after indicating the estimated/expected results or acceptable range (minimum and maximum) of results. Ensure no absolute measurement results are called for, as a final result is seldom achievable. The technician/engineer who measures the feature/characteristic of the product or observes or evaluates the test result has the right to know the acceptable result. Then, there will be adequate preparation for the technician to anticipate the result while measuring. Confidence is created when the correct result is calculated, and satisfaction is derived. When it is found that it does not meet the requirements, there will be no doubt in the tester's mind to report that the product cannot be accepted in that condition. It encourages the technician to verify the measurements and check for the possible variables in the measurement system.

Tests or measurements are always required or desirable to be done following the test schedules or documented instructions, the absence of which may again create possibilities of error due to changes in persons, environment measuring devices or method of measurement, etc. Suppose there is a need to use any specific measuring device/equipment to measure any product's characteristics. In that case, the test schedule or instructions must indicate corresponding instructions and details of their use. The error of judgement does not happen with experienced persons. However, due to complacency or any other reason, if such an error is likely to occur, it can be prevented with suitable documented instructions.

As seen earlier, identifying critical items and key characteristics is a requirement of the design output of any aviation, space or defence product, and they are generally noted in the drawings, specifications and similar documents. When these design output documents are used for manufacturing or servicing the products, the production planning output document should specifically call for control and monitoring of these critical characteristics during manufacturing/production and maintenance services. Special attention must be given to the critical items and key characteristics while measuring and monitoring the detailed parts or assembly of the product. The quality plan and servicing instructions should indicate the requirement. Inspection record sheets of complex parts and assemblies should indicate the inspection, monitoring and measurement completed for the critical items and essential characteristics of the parts and assemblies as per the design document. The design documents do not indicate the vital items and key characteristics. In that case, the production planning function is responsible for identifying and indicating them in the planning output documents/instructions for measurement.

Measuring product features and characteristics of components or parts in a simple machining process may be as simple as measuring dimensions (linear or angular), surface roughness, etc. Stage inspections of the product at various stages of the manufacturing process are undertaken as per master process sheets, where machining aims to achieve the final dimensions on completion of all phases of operations. When all the operations are completed, a final component inspection is done concerning the relevant drawings.

In software organisations, the product/service delivered to the customer is usually in an electronic media where physical measurements using measuring devices/equipment are irrelevant. Monitoring and measurement of products, which are sometimes in the form of 3D models, drawings, etc., are done in the computer system itself. Results of finite element analysis of structures using software such as PATRAN, NASTRAN, and ANSYS are presented in graphical or analytical form. Studies with regard to aerodynamics, stress analysis, dynamics analysis, aero-elasticity, etc., give results in visual or analytical form. In process verification and analysing, the results with reference to the input data against the acceptance criteria prescribed by the customer satisfy the measurement requirement of products in software processes.

b. Traceability to the person(s) authorising the release: When all the customer requirements are satisfied on the product, and the required verifications/inspections and tests are completed and found satisfactory, the final product may be released to the customer. The product's release will be understood for delivery to the customer, not for the next production stage. The organisation should identify from within the organisation the authority or person authorised to release the final product to the customer, along with the necessary documentation. The final release certificate (or conformity certificate or contract record, which indicates the product transfer to the customer or customer representative) should bear the signature of the authority or person releasing the product. When aviation, space or defence products are initially developed, tested and finally certified for airborne use, type approval or type certification or declaration as fit for use is issued by the regulatory authority based on product qualification. The type certification records ensure that the product meets the defined requirements, with quality conformance by the quality assurance authority during bulk manufacturing being one of such conditions. Any change in configuration of the part/equipment/assembly, location of manufacture or change in the production agency/source, change in production processes, expiry of the certification validity period, etc., invalidates the type certification. The airborne equipment developed in one country may be acceptable for use in another country if approval for the same is accorded by certifying authorities of the country of origin since mutual agreements exist among regulatory authorities of different countries. The records should provide evidence of such qualification of the parts/equipment before release to the customers. Production process verification (refer to 8.5.1.3) carried out on parts/sub-assemblies/equipment is to satisfy one of the qualification requirements of the product.

When any of the requirements are not satisfied on the product (for requirements from among those specified by the customer, requirements identified by the organisation in addition to those determined by the customer, statutory and regulatory requirements and special requirements, refer to clause 8.2.2), the final delivery or release of the product should not be done. The exception is when the regulatory authority or customer approves the product's release, knowing the implications of using the equipment when any requirements are unmet. Documentary evidence must be available with the organisation for approvals of the product's release issued by the regulatory authority or customer. When any of the product requirements are not satisfied at the time of the product's release, and the customer wants to use the equipment for any reason, the organisation can seek a concession from the customer or regulatory authority to enable delivery/release of the product. The acceptance of the request for concession has to be recorded.

All related documents and records must be delivered to the customer, and the product must be available while the product is delivered. This has to be ensured by the authority or person from the organisation who releases the product. This may include compliance to the standard of preparation, work done report, concessions and production permits obtained from customer/regulatory authority, certificate of inspection/quality assurance, ground and flight test acceptance for aircraft and claim for contract price/invoice for payment, etc.

AS 9110 Rev C additionally requires that the maintenance organisation provide objective evidence that all maintenance operations have been completed as planned. Task completion records, when constructed and maintained, would serve this purpose. Measurement requirements for product acceptance should be documented. The documentation should include:

a. Criteria for acceptance and/or rejection.
b. Where in the sequence, measurement and testing operations will be performed, including required customer and/or regulatory authority inspections.
c. Identify which inspection and testing operations are to be verified and/or witnessed by the regulatory authority, such as intermediate and/or final stages.

The MRO organisation should identify defects discovered during maintenance that are outside the scope of the maintenance contract and defects noticed during pre-survey, inspections and while carrying out pre-determined tasks. The defects must be processed per customer and regulatory authority requirements. Non-routine task cards are raised for rectification, and the works are completed as required.

Procedures must be implemented to prepare and complete regulatory authority documentation such as conformity determinations, airworthiness approvals, release certificates, approval for return to service after maintenance, export documentation, etc.

Some customers insist that the customer must approve the use of an operator certification programme or other special manufacturing methodologies (e.g. manufacturing controlling features, die/mould control, method of manufacturing, etc.) before implementation. It is also insisted that the organisation generally select measuring and test equipment with an accuracy ratio of 10 to 1 (product tolerance to measuring and test equipment (MandTE) tolerance) where specified. However, accuracy ratios as low as 4 to 1 are acceptable unless otherwise specified.

8.7 Control of Nonconforming Outputs

8.7.1 Non-conformity

As the results of monitoring and measurement of processes, products, incoming inspections, stage inspections, final inspections, etc., carried out throughout the production processes, snags/deviations/defects or nonconformities may be revealed on the in-house processes or the product/service or on those bought out from external providers/suppliers/customers. In addition to the production operation process, nonconformities may be noticed anywhere in the quality system or documentation, organisation context, leadership, planning, support, performance evaluation or improvement processes. Deviations may be noticed in compliance with statutory regulatory and contract requirements. However, clause 8.7 deals with only nonconformities on outputs/products that do not conform to product/service requirements during the production operation processes within the organisation or after the product/service delivery to the customer. Nonconformities may be identified internally, received from an external provider or identified by a customer. The organisation needs to take appropriate action based on the nature of the nonconformity and its effect on the conformity of products and services.

The term 'nonconforming product' includes nonconforming product returned from a customer. There are various ways for the customer to return the defective/nonconforming product to the manufacturer or organisation/distributor from which the customer has purchased the product. It may be in the form of a complaint along with the part/product or without; it may be a request to investigate the reasons for the failure/defect seen on the part or associated part or a request to replace the defective part/product. When the customer notices a defect/deviation/premature failure on a part/product and withdraws the part/product from the leading equipment and does not continue its use or send it to the organisation, it is again a product returned by the customer. The nonconforming product returned by a customer must be dealt with in a similar manner and procedure as for a nonconformity noticed on the part/product before delivery to the customer.

The standard requires that actions taken by the organisation to control the nonconformities be maintained as documented information (document). The document should indicate the following data:

a. Identify (designate a person or team of persons) and define the related responsibilities and authorities for dealing with the nonconforming product, its review and final disposition. The documented information should also specify who authorises or approves the designated personnel who reviews and decides to dispose of the nonconforming product. The responsibility and authority of the assigned personnel and the requirements, such as competency of the personnel, are also to be defined in the procedure. If there are a variety of products, the procedure shall identify each product type and the person dealing with nonconformities on that type of product.

b. Taking actions necessary to contain the effect of the nonconformity on other processes, products or services. Once nonconformity is identified on a process/product/service, the organisation must determine if a similar nonconformity exists on any other process/product/service within the organisation. Action to contain the effect of the nonconformity needs to be extended to all such processes/products/services.

c. Timely reporting of nonconformities affecting delivered products and services to the customer and relevant interested parties: The organisation's procedure for dealing with nonconforming products should also include timely reporting. After the product has been delivered to the customer, if it is detected or noticed that nonconformity has escaped from the notice of the organisation, and if the product is likely to be affected by the nonconformity while the product is in use by the customer, the organisation must take necessary action to recall the product or take appropriate action to correct the nonconformity of the product at the earliest. This is known as escape notification of nonconforming products to the customer. The organisation has to issue alert notices to customers that the particular nonconformity has escaped detection, which may cause nonconformity on the part/product that has already been delivered to the customer. It is necessary to indicate the details of nonconformity or impact and precautions to be taken by the user to prevent further damage or deterioration. Parties requiring such escape notification can include all interested parties, suppliers, internal organisations, distributors and regulatory authorities, and customers.

d. They defined corrective actions for nonconforming products and services detected after delivery as appropriate to their impacts (see 10.2). Here is an example of how the organisation issued an escape notification to the aircraft users on which failure of a product/part was noticed after the component was delivered to the customer. Recall the 'failure of a quill shaft on an aero-engine gear box' described under clause 8.5.1.2 (page 289). The failure was due to process deficiency of not carrying out the de-embrittlement operation (last process step) on the quill shaft after copper plating during manufacture. Production of the quill shaft was for a batch quantity of 10 numbers, 7 of which were installed on different serial numbers of the aero-engines. As soon as the investigation revealed the reason for the failure of the quill shaft on the ill-fated engine, an urgent escape notification was issued to the customers/users to ground the engines/aircraft on which the remaining six quill shafts were installed. Two of the six quill shafts removed, which have crossed 100 hours of use, had developed micro-cracks, indicating an imminent failure sooner or later. Withdrawing the defective quill shafts prevented their failures and, more importantly, prevented accidents to the aircraft/aero-engines on which they were installed. Grounding the engines/aircraft enabled detailed checks and replacements.

The timeliness of alert notification is critical for items that may affect the safety or reliability of the product/equipment. Safety is vital when the nonconformity may cause malfunction or result in a loss of life or equipment. It is critical to reliability when the nonconformity may cause malfunction sooner or later. In the event of a deviation from the approved engineering process/data, preventing or minimising the potential impact of the nonconformity to the customer with an alert notification is important. The notification of escape of nonconformity to the customer should contain at least a method for identification, recovery and a proposal for the replacement of parts affected, apart from the population of the nonconforming products/equipment delivered, dispatch details, cause of deviation, immediate containment action and actions taken by the organisation to prevent recurrence. It may be recollected that some automobile manufacturers and computer battery manufacturers have done this in the recent past. Through such actions, the organisation suffers the cost of replacements. Still, it improves its reputation, credibility, product quality, customer care and customer satisfaction.

There should be four parts to an organisation's response to any nonconformity identified. They are

i. the correction or containment;
ii. analysis of the root cause;
iii. corrective action to eliminate the root cause;
iv. verification of the effectiveness of implemented corrective actions.

In some cases, immediate correction or containment could be possible. If the cause is analysed in detail, an effective corrective action could be identified to eliminate the cause of the nonconformity.

The organisation deals with nonconforming products/outputs in one or more of the following ways:

a. Correction or containment action: 'Action to eliminate a detected nonconformity'. Action to eliminate a detected nonconformity is defined as correction, the same as rectification/repair/rework/replacement. Correction may be a rework or downgrading of the parts affected. Downgrading is to find an alternate use and not the intended usage. For delivery to the customer, it has to be agreed by the customer. Repair is the action to make it acceptable for the intended use. Rework or replacement is required to be re-confirmed to the requirements by verification. Correction can also be made in conjunction with corrective action.

    Containment action: Action to control and avoid or reduce nonconformity's impact and protect the customer's interests. That is, making sure that the problem does not get worse. OEMs say, 'Stop the bleeding'. Containment includes correction, immediate corrective action, immediate communication to the affected personnel and verification that the nonconformity does not further deteriorate. When nonconformity is identified during external audits, the organisation has to determine and report the specific containment actions and appropriate corrections within seven calendar days after the audit. The time frame could be shorter in an efficient organisation. It can be adapted for acting on the findings during an internal audit or otherwise when necessary. Not all nonconformities need containment. Typically, if the product is involved, containment is mandatory.

b. Segregation, containment, return or suspension of provision of products and services: The organisation needs to ensure that a product that does not conform to product requirements is first identified and segregated.

The nonconforming product is to be prevented from continuing in the production line, knowingly or unknowingly (even when there is no intention to use it in the production line). Neither the product should be delivered to the customer. Control exercised for the nonconforming product is deliberate and conscious to ensure that a product with deviation is not used or allowed to continue in the production line without the authorities' authorisation. That is, the nonconforming product should be segregated, identified and quarantined. Correction or containment action should be done immediately after that. It has to be returned to the sender if it was wrongly received or received without appropriate documents/PO reference. If the nonconformity has resulted in production hold-ups, or the production itself was affected by the nonconforming product, the containment action is to stop the production and evaluate the situation.

c. Informing the customer: If the nonconformity found internally is likely to affect 'On Time Delivery' or the product's final quality. Suppose the promises are not expected to be fulfilled, or the planned results are not likely to be achieved. In that case, informing the customer to avoid more significant embarrassment on the delivery date is always desirable.

d. Obtaining authorisation for acceptance under concession by a relevant authority and, when applicable, by the customer: In some cases, correction/rework may not be possible, or rework may not restore the original acceptable condition for the part/product. In the opinion of the concerned personnel, if the nonconformity is not severe enough to affect function/performance, life, interchangeability, maintenance, strength, and safety (FLIMSS) of the part/final product, the nonconformity may be referred to the customer and regulatory authority for acceptance by authorising its use as is or for permitting release to the customer. The organisation supplying the product should have the data/authority required to support any permission granted by the customer/designer or regulatory authority for the dispositions of use-as-is or repair.

Any superficial corrections would lead to wasted time and resources. For example, the car is moving on rough terrain, and then you realise that the tyre pressure is not holding on one of the tyres. Correction is to stop the vehicle on the wayside, with all the precautions, and recharge the affected tyre with adequate nitrogen pressure. However, the correction may not be practical if there is already a leakage in the tyre tube or at the charging port. Unless the tube is rectified, any nitrogen pressure filling will not hold for long. So, replacing the suspected tyre with the spare tyre (stepney) you carried along is a better correction.

The organisation, which looks at nonconforming products seriously and aims to reduce or eliminate the nonconformities, should first analyse why the nonconformities have happened. Once the reasons are known, eliminating the detected nonconformity is the right action, which is known as the corrective action.

The organisation can take action to segregate the affected part and divert it for any other use or application other than the original intended use. It must be ensured that the nonconforming part/product so segregated must be permanently marked with clear visibility that it is a scrap/not intended for the original use. All nonconforming parts are to be held in an identified and controlled area. Control also includes barcoding used on parts to remember, which are confirmed before assembly. Scraps will have a different barcoding, which prevents it from being used as a good part. Parts typically held/stored outside, exposed to Sun and rain, should be permanently marked so that the visible markings do not fade away. Suppose the part is adjudged as not feasible/economical for repair/rework and disposed of as rejected. In that case, the part shall be made unusable, that is, unusable for the original intended use, before disposal. This ensures that the rejected parts are clearly differentiated and never used unintentionally. It is a customer expectation that the organisation's quality system provides for scrap disposition and is done according to its internal or customer requirements. Customers may even request a copy of any objective evidence provided by the organisation supplying the product. Examples of the evidence are records of decisions made in the internal material review board and records maintained for the actual destruction or permanent marking carried out on rejected components, which are disposed of as scrap.

When a product is returned from the customer, and a root cause analysis or defect investigation is carried out for the reported/detected nonconformity on the part/product, necessary corrective action must be taken. The action should be decided according to the effect of the nonconformity or the potential effect of the nonconformity that can have on the product for continued use.

Effective corrective action should prevent nonconformity recurrence by eliminating the cause. However, corrective action should not be confused with preventive action. The definition for preventive action is 'action to eliminate the cause of a potential nonconformity or other undesirable situation'. It should be noted that preventive action, by its definition, does not apply to already detected nonconformities. However, analysing the causes of detected nonconformities may identify potential nonconformities on a broader scale in other areas/processes of the organisation and provide inputs for preventive action/risk management. The organisation shall determine actions to eliminate causes of potential nonconformities to prevent their occurrence. This means that a risk assessment is required.

> '*Dispositions of use-as-is or repair shall only be used after approval by an authorized representative of the organisation responsible for the design. When the design organisation has placed an authorized representative in the manufacturing organisation, such as a design liaison engineer (DLE), disposition of use as is or repair may be given by the design representative. However, the design must define in a document the criticality and the type of deviations for which the representative can grant dispositions. A periodic summary of the total dispositions given by the representative needs to be brought to the notice of the customer and/or regulatory authority to review the kind of dispositions granted and controls exercised. Unless expressly authorized by the design organisation or having a delegated power of approval from the design organisation, the design liaison engineer, positioned in manufacturing organisations to liaison manufacturing issues with the design organisation, has no role to play for dispositions of use as is or repair of a non-conforming part/product.*
>
> *Moreover, such dispositions should not be given if the nonconformity results in a departure from the contract requirements, such as specified in the purchased order issued by the customer or an agreement signed between the customer and the organisation. The alternatives left to the organisation are to rectify the defect/nonconformity, replace the affected part, or seek concession from the customer and/or the regulatory authority to retain the nonconformity. DLE writing "acceptable" on the snag sheet referred to him, and taking no further action is unacceptable. Regulatory authority prescribes appropriate actions when minor deviations are noticed during production or maintenance. If it is a requirement of the contract, agreed between the customer and organisation, a deviated part can be used only when authorised by the customer. The organisation has to convince the customer with adequate technical details and the potential risks of using such a deviated part. Regulatory authorities generally prescribe the method of review of deviations and the route for acceptance of concession for the deviations*'.

Any product marked for scrap that is finally rejected as unsuitable for the intended use shall be conspicuously and permanently labelled, segregated and kept separately at a place where the part or product cannot get into the production line. When the component or product is disposed of from the organisation's premises, it should be physically rendered unusable for the intended purpose, damaged beyond repair, defaced or mutilated. The purpose of doing so is to make sure that, deliberately or unknowingly, the defective part should never be used on aviation, space and defence equipment/assemblies, as safety cannot be left to the risk of chances or possibilities.

When a nonconforming part or product is corrected/reworked/rectified, it should be subjected to inspection/re-verification to demonstrate conformity to the requirements. Where everybody in the production line is conscious of the defect indicated by the line/stage inspection, immediate rectification may be carried out, and re-verification or re-inspection of the rectification may sometimes be missed. Re-verification and indication of acceptance by any acceptance authority media by the inspection responsible for the work will only bring out objective evidence of satisfactory completion of the rectification. The standard requires that appropriate records of the nature of nonconformities and any subsequent actions taken, including concessions obtained, are to be maintained.

Control of counterfeit or suspected counterfeit parts with respect to preventing re-entry into the supply chain, refer to para 8.1.4.

Verification/inspection of actions completed: All reworks must be verified before acceptance. Once the corrections and corrective measures are completed, the exact needs to be verified and accepted, with the evidence of the availability of acceptance authority media (signature/stamp) employed in the organisation. Password control for the PC and tracking the review of rework/data would be adequate in software organisations.

It is necessary to understand some of the specific terms used in 'nonconformity management' to control nonconforming products while maintaining the documented information (documents) for regular practice and retaining the records, including concessions.

Nonconformity may be defined as non-fulfilment of a requirement. The requirements can be related to the Quality Management System, product or service. Therefore, when nonconformity is identified in a Quality Management System or its elements, it is the system's nonconformity. Similarly, when nonconformity is determined by a product or service, it is the product nonconformity. It may be mentioned here that all the following definitions apply equally to the Quality Management System elements and products or services appropriately. Therefore, when nonconformities are identified in the quality system through internal or external audits, the following definitions of terms apply equally and appropriately in relation to the Quality Management System.

Major nonconformity: A non-fulfilment of a requirement, which is likely

a. to fail some elements/parts of the Quality Management System or fail to meet a provision of the Standard or

b. reduce its ability to assure controlled processes, or

c. reduce its ability to meet requirements for the products/services, or the nonconformity is likely to affect the conformity to the requirements of the product or product quality (FLIMSS—function, life, interchangeability, maintenance, safety, strength), fit, form, flight characteristics, weight, balance, service life or installation on to the next assembly.

If the nonconformity is major, any of the following situations may arise:

a. Planned arrangements are not being implemented consistently.

b. Nonconformity affects the product or service seriously.

c. Absence of or total breakdown of a system to meet AS 9100/9110/9120 requirement, or absence of an organisational procedure, or not meeting a customer Quality Management System requirement.

d. Nonconformity is likely to result in the possibility of delivering nonconforming products to the customer.

e. A condition that could result in the product or service's failure (or reduce the usability) and its intended purpose.

Minor nonconformity: A non-fulfilment of a requirement that is not likely to result in the failure of the Quality Management System or reduce its ability to ensure controlled processes or compliant products/services. It can be either one or more of the following situations:

a. A single observed quality system element failure or lapse in meeting the AS 9100/9110/9120 standard or customer Quality Management System requirement.

b. A single system failure or lapse in conformance with a procedure is associated with the organisation's Quality Management System.

c. A deviation on the characteristic of the part/product, with respect to the design documents, which may not affect the product quality (function, life, interchangeability, maintenance, safety, strength, etc.) but could not be reworked/improved due to constraints identified and accepted in risk management (examples: workmanship, tool wear out, aesthetics).

Several minor nonconformities against one requirement of the Standard/QMS document (e.g., similar nonconformities associated with different locations or different departments/functions/processes within a single area) can represent a total breakdown of the system. They, therefore, may be considered as a major nonconformity.

The defect is non-fulfilment of a requirement related to specified product use.

Root cause analysis: An analysis carried out to find out why the deviation or defect has happened and reach the root of the nonconformity/problem is generally known as the root cause analysis. There are many methods and tools available to an organisation for determining the cause of nonconformities, from simple brainstorming to more complex, systematic problem-solving techniques such as 'why-why analysis', 'five why analysis', five-part corrective action request, 8D methodology for root cause analysis, use of Pareto analysis, fish-bone diagrams/Ishikawa diagram, etc. Many such methodologies address nonconformities in the product or Quality Management System. Used in a 'reactive' mode, it can prevent problems from recurring. Used in a 'proactive' mode, it can examine current operations and help identify areas and activities that can be improved through risk management.

As the name implies, the root cause is the most fundamental reason a problem 'has' (reactive) or 'could' (proactive) occur. Root cause analysis techniques are often used in the reactive mode to uncover the reason for problems that have already happened. Usually, the quick but incorrect response is a failure or human error. It is only the symptom and not the root cause of the problem. The root cause is often more than the symptoms, such as process or programme failure, system or organisation failure, poorly written work instructions, lack of training, etc. That is why root cause analysis is considered a handy and productive tool. RCA is a method that helps to determine what happened and why it happened and provides an opportunity to learn from past problems, failures and accidents.

Corrective action: 'Action to eliminate the cause of a nonconformity and to prevent recurrence'. Action to eliminate the cause of a detected nonconformity or other undesirable situation is known as corrective action. The action is to see that the nonconformity does not repeat on the affected product or similar products or processes. There can be more than one cause for a detected nonconformity. For a detected nonconformity, the possible causes must be identified to the extent possible. Therefore, any corrective action methodology will have the arrangements to determine the root cause(s) first. Correction and corrective actions are proposed to eliminate the root cause(s) once the causes are known. Documenting the root cause as an isolated case, lack of instruction, operator error, etc., may not be appropriate as there may be system causes or inadequacies in other processes, such as insufficient support (training or logistics) or management processes. When nonconformity is identified during external audits, the organisation has to agree with the external auditor on the appropriate corrective actions and the plan to complete the corrective actions within 30 days after the audit while taking adequate time to evolve appropriate corrective actions. Completion of corrective actions should not exceed 90 days from the last day of the external audit. The time frame to close the findings of an internal audit could be shorter in an efficient organisation. Corrective action has three 'C's to be completed, i.e., containment, correction and elimination of the cause.

On completion of the proposed corrections and corrective actions, the features and characteristics of the product are to be verified to meet the product requirements. It is also necessary to evaluate the effectiveness of completed corrective actions at a later stage. The later phase can be during the subsequent batches of production in a series production line, when the same product is manufactured next time or when similar products are measured. Characteristics ensured the absence of a similar nonconformity.

Preventive action: Action taken to eliminate the cause of a potential nonconformity or undesirable situation is known as preventive action. It is the same as action taken to mitigate or reduce the impact/severity of a risk. The risk is a potential nonconformity. The nonconformity has not happened but will likely happen when favourable conditions/environment and situations occur. Design failure mode and effect analysis/design failure mode, effect and criticality analysis, process failure mode and effect analysis, etc., carried out during design stages of products/processes; identify potential failures or defect/deviation occurrences under the manufacturing or working environment and situations. Then, the designer improves the design to make it fail-safe with a high margin of safety. The design may introduce redundancy or support it with a backup system to ensure safety and reliability and complete the mission during usage. Preventive action or risk management is to prevent occurrence or reduce its impact when it occurs. Corrective action is to prevent recurrence.

Sometimes, an analysis of the causes of a detected nonconformity may identify potential nonconformity in a broader sense in other areas and may provide some input for preventive action. When similar repeated nonconformities on a product are placed, or similar root causes of many detected nonconformities are arrived at, design reviews and improvements are essential and urgent. In consultation with the customer/regulatory authority, production engineering (production planning and methods) and production and quality groups, the designer identifies product modifications or establishes preventive actions.

Verification: Confirm through verification/inspection and provide objective evidence of having done the repairs/rework to restore the acceptable conditions so that the product can be cleared for the intended use. Verification may be done to downgrade the product and alternate usage.

Objective evidence: Objective evidence can be verified by anyone not directly involved. Data supporting the existence or verity of the stated facts is factual evidence.

Therefore, any procedure to control nonconforming products should have the following essential elements to eliminate detected nonconformity.

1. Identify the nonconformity/deviation/snag/defect on the product/part during inspection/verification or in the Quality Management System during internal audit. While identifying the nonconformity, it is essential to indicate the acceptance criteria or reference where the product or quality system requirement is specified. The requirements should have been specified and may be available in related quality system standards/documents, product drawings, specifications, instructions, quality plans, acceptance standards, test schedules, or similar documents. There are three constituent parts required for each statement of nonconformity:
    i. What is wrong, what is missing, what is the gap and what is not established, documented or implemented with reference to the requirements of the Standard?
    ii. Reference to the applicable requirement in the product document or the quality system standard clause number.
    iii. Objective evidence to substantiate the finding of nonconformity: where, in which part or at what stage the nonconformity was found. What feature or characteristic shows that it is not meeting the requirements?
2. Containment, correction and corrective action proposed with a probable date of completion of action: Appropriate rectifications and elimination

of the root causes are necessary within a defined time frame, which should take adequate time to implement the activities and should not be too long to delay the actions.

3. Carry out root cause analysis: Any suitable quality tool could be used to determine why the deviation happened. The reason for nonconformity/deviation/snag/defect could be any one or more of the six 'M' letter words, which control the inputs and steps in any process, which seeks to give a desired output (refer to Process Approach, six 'M' letter words – men, materials, machines, methods, measurement and environment (mother earth). Asking 'why the deviation happened five times or at least three times repeatedly (Why - Why - Analysis)' may help find the root cause of the nonconformity. Arrive at a reasonable cause to eliminate or reduce it to the minimum over time.

4. Verification/inspection of actions completed: Once the corrections and corrective measures are completed, the exact needs to be verified and accepted, with the evidence of the availability of acceptance authority media (signature/stamp) employed in the organisation. Password control for the PC and tracking the review of rework/data would be adequate in software organisations.

5. Evaluation of the effectiveness of corrective actions already implemented: When planned results (requirements and objectives) are consistently achieved, it is said that the corrective action taken in the past is effective, and no recurrence is observed.

The above methodology and sequence are known as the five-part corrective action request. An appropriate format is given in the examples.

When a corrective action is requested by the customer, after completion of corrective actions by the organisation, the customer may like to verify the corrective actions to determine how effective the corrective actions are. The customers generally use a checklist to do the verifications. The verification checklist contains questions related to immediate correction, root cause statement, corrective action plan, verification of corrective actions (when corrective action is required to be carried out by a supplier) and follow-up audit.

Initiating a format for control of nonconformity every time for trivial or minor observations is a cumbersome task. Therefore, a thorough root cause analysis for a repeated nonconformity or a customer-identified defect/complaint would be a more practical control method. Any of the methodologies indicated can be used, which gives a fruitful improvement. The suitability of the method is to be determined by the organisation. Trivial or minor snags detected on a daily routine for every type of product/process/service could be accounted for and analysed on a weekly/monthly basis. Pareto analysis reveals which are the vital few and trivial many. These defects need systematic actions to eliminate the root causes. If the effort is continued for a reasonable period, implementing corrective action becomes a habit, and a quality culture develops in the organisation.

8.7.2 Non-conformity report (NCR) format

The organisation retains documented information (record), which includes the following:

a. Description of the nonconformity: Statement of nonconformity and objective evidence
b. Description of the actions taken
    i. Containment/correction
    ii. Root cause analysis
    iii. Corrective action identified, completed, verified
    iv. Verification of effectiveness of corrective actions
c. Description/reference of any concessions obtained from customer/ regulatory authority
d. Identification of authority deciding the action in respect of the nonconformity

A typical nonconformity report (NCR) format (process/product) and customer complaints resolution (example) is attached in Annexure E.

> *'Some quality personnel say, "Ours is a thankless job. We are always blamed for rejecting items. We don't have sufficient people to make this paperwork of going after non-conformities; we reject it or get it reworked. Too much paperwork is infructuous and does not add value to the products or improve revenue. We can afford a certain percentage of rejections, so we don't have detailed actions for rework and rejections."*
>
> *All the non-conforming products may be found within the factory, viz., the re-works, rejections and customer returns. Generally, two mistakes are being committed, one not identifying the non-conforming product and the other not dealing with the nonconformity that resulted in the re-work or rejection. The latter is more alarming. Suppose a re-work or correction resolves the deviation. In that case, the organisation*

complies with the problem being solved, and no thought is given to the recurrence of the deviation that may happen sooner or later. Every time the deviation happens, they may fall back on the proven method of re-work or correction, which may be possible sometimes but not always. Improvements will seldom come. There is the saying – "Give a man a fish, and you feed him for a day. Teach a man to fish, and you feed him for a lifetime". It is more worthwhile to deal with the nonconformity once the recurrence of the nonconformity is periodically monitored than to make corrections every time'.

9

Performance Evaluation

The standard requirements/clauses are arranged in accordance with the bigger picture 'Plan-Do-Check-Act' cycle (Deming's wheel). Chapter 9 is the 'Check' section (including measurement and analysis). The requirement details are indicated in the respective sub-paras of the chapter.

9.1 Monitoring, Measurement, Analysis and Evaluation

9.1.1 General

Monitoring, measurement, analysis and evaluation deal with the check part of Deming's wheel (PDCA cycle). The organisation that has established and implemented the Quality Management System evaluates its performance and effectiveness. The organisation determines the following with respect to monitoring the measurement of the processes:

a. What processes need to be monitored and measured?

b. Method of monitoring, measurement, analysis and evaluation needed to ensure valid results

c. The frequency at which these monitoring and measurements are to be done

d. The frequency at which these monitoring and measures are to be analysed and evaluated.

Evaluating the performance and the effectiveness of the Quality Management System is mandatory. It is left to the organisation which processes must be evaluated for effectiveness. Products must be assessed to conform to the product requirements, customer, statutory, regulatory and contract requirements.

The organisation must retain documented information (records) with respect to the results of measurement of processes, products, performance and effectiveness of the Quality Management System

9.1.2 Customer satisfaction

Like love is the essence of all religions, customer satisfaction is the essence of all business processes within the framework of a Quality Management System. Peace and tranquillity are achieved when there is love all around. (Some may call it achieving the nirvana, the ultimate peace or reaching heaven!) The organisation's very existence is to satisfy customers, shareholders, authorities, suppliers, employees and other interested parties through mutually beneficial business relationships. One of an organisation's performance excellence measures is the quantum (how many or how much) of the requirements and objectives met.

Customer delight happens when all the requirements are met or sometimes exceeded, and the organisational objectives are met. The organisation, which measures the bottom line in the balance sheets, may not survive for long. For sustained business excellence, the Quality Management System is a good model. The organisation has to monitor customer perception of whether the organisation has met customer's requirements. As they say in legal language, 'Justice must not only be done but must be seen to be done'. It is not only ensuring conformity to the requirements of the product delivered to the customer or exceeding the customer expectations that matters but also manufacturing a product that meets the requirements cost-effectively and efficiently, meeting the organisation's objectives. So, for all organisations, the performance indicators in relation to their customers are the real measurements of the success of their own business. Analyse and review for continual improvement, plan and take appropriate actions, periodically evaluate the effectiveness of actions for continued success and achieve business excellence.

Two perceptions must be determined when the organisation has met the customer requirements. One is the organisation's perception, and the other is the customers' perception. The two perceptions need not be the same. The needs and expectations of the customer are 'the requirements', which are clearly understood by the organisation. It takes all the effort to plan and operate the process to give the results to the customer meeting 'the requirements. The Quality Management System standard requires that the organisation monitor information relating to customers' perception of the degree to which their needs and expectations have been fulfilled. This can be determined. The organisation's perception (real introspection) of the degree to which customer's needs and expectations have been fulfilled can also be determined. Total transparency and full sincerity are required here. 'Justice must not only be done but must be seen to be done'. Business excellence is achieving the two perceptions equally, sustaining and improving it!

When the customer requirements are met, the customer's perception is that 'justice is seen to be done'. Being genuinely transparent and doing excellent business, the organisation's perception is 'justice is done'. That is when the resources are adequate, people are competent, processes are effective, there are no nonconformities within the processes, no rejections, no reworks, no returns, no complaints, there are no issues against smooth planned production operations, there are all-round improvements, the quality system is conforming and the QMS is effective and improving. The organisation itself can determine and monitor both the customers' and the others' perceptions, a self-appraisal of the organisation's business processes.

The organisation must determine the methods for obtaining and using this information about customers' perceptions. Organisations depend on customer responses to a survey sheet sent out to the customers. The customer records his reaction in the survey sheet, sometimes quickly working without much thought, casually after much delay, occasionally depending on how busy the customer is and how serious he is in answering the survey sheet. The one who fills out the survey sheet may not know the history of excellent supplies from the supplier and the supplier's responses to some irrational or obscure inquiries from the customer. So, depending entirely on the customer's response or rating, determining the customer's perception of satisfaction by the organisation is incomplete, insufficient, spontaneous and may not reflect the accurate/realistic perception. Therefore, more practical approaches are necessary to determine the customer's perception of satisfaction.

Customers often pretend to be knowledgeable when they are not. For the job being done, the process owner should be more knowledgeable and proactive. The organisation is there not just to meet the customer's requirements. It is there to educate the customer if required and always give them better choices, alternatives and the best possible solutions, products or services. The organisation, which shares knowledge with the customers, is looked upon respectfully and is eager to cooperate and share benefits. Customers have confidence in the organisation and depend on them to do the business. (You should marry the one who loves you, not the one you love!). If the customer is happy with what he gets, the organisation that supplies it should not be too pleased and complacent. It should not derive the customer's satisfaction and rest on its laurels. The organisation needs to be proactive in gaining more and more confidence/satisfaction from the customers through their business relations. Then, the customers will be pleased to retain the organisation, which supplies precisely what they want and is perhaps more valuable than expected. Because ultimate customer satisfaction should be derived from the results or the benefits the customer gets through the product/service delivered to them by the organisation. Therefore, monitoring information relating to customer's perception of satisfaction has to be a two-way process.

The organisation has to determine how to arrive at a more realistic measurement of customer perception and how to use the results for the organisation's benefit. In light of the above, information to be monitored and used to evaluate customer satisfaction should include product conformity, on-time delivery performance, results of customer satisfaction surveys, scorecards, report cards, customer complaints regarding product quality, corrective action requests received from customers and service deficiencies. Note that the Standard additionally advises including examples of monitoring customer perceptions such as customer feedback on delivered products and services, meetings with customers, market-share analysis, lost business analysis, compliments, warranty claims, dealer reports, etc., as applicable. It can include the in-house detected nonconformities, rejections, reworks and corrective action requests made to its suppliers. These field failures have a bearing on the quality of supplies by the organisation (number of defect occurrences) and delay in responding to customer requests. On the positive side, it can include the number of times the customer has appreciated the efforts made by the organisation, awards/recognitions won by the organisation and other state/social/customer recognitions.

Complaints indicate a low CS indication, but its absence need not imply satisfaction. Even when requirements are agreed upon and fulfilled, they need not necessarily ensure the satisfaction of the customers. For example, some customer satisfaction surveys invite customer responses on the cost of the product delivered. Since the cost of the product was agreed upon between the customer and the organisation before the commitment to supply the product, there is no point in asking for the customer's response on the cost agreed upon after product delivery. Suppose the dissatisfaction with the price lingers after the agreement. In that case, the customer may give a poor rating on that score, even when the quality and OTD performances are excellent. It also reveals the folly of including cost in the customer satisfaction survey!

A better and more realistic way to determine the customer's perception of satisfaction is to have a weighted average score for all the positives and negatives based on the actual figures gathered during a calendar/financial year and evaluate it objectively to improve it further from an achieved goal.

If the planned results are not fulfilled, actions to improve the process will be identified and implemented without delay. Respond to the customer positively if a suggestion or complaint is made, or thank him for being supportive. Not responding on time may be considered as undesirable.

Any deficiencies or gaps found during the analysis should be addressed by developing and implementing improvement plans. Address deficiencies pointed out by the customer or evolve proactive measures to reassure the customer to show improvements. The effectiveness of all the actions taken should be reviewed periodically with the results achieved.

9.1.3 Analysis and evaluation

Measurement, analysis and evaluations are necessary to demonstrate the organisation's performance concerning the conformity of the product, which underpins customer satisfaction conformity of the Quality Management System for efficient and effective functioning of the organisation's business. And efforts to continually improve the effectiveness of the Quality Management System to be at the pinnacle of business excellence. All these processes consider various methods of measurement analyses using quality tools and statistical techniques, and the evaluation provides opportunities for improvement on multiple counts.

Every organisation needs to measure, analyse, review and improve performance using data and information derived from all functional levels within and from all processes.

'Factual approach to decision making' is one of the Quality Management System principles, and data collected through measurements form the basis of making decisions related to quality characteristics. Effective and efficient measurement of various data, collection and analysis using multiple statistical techniques or quality tools help monitor the organisation's performance. Data analysis within the Plan-Do-Check-Act cycle ensures that fact-based decisions are made for the benefit of all organisation stakeholders.

As a result of various measurements and monitoring carried out for the effective implementation of the Quality Management System in the organisations, over time, there will be enough data available to indicate the health and performance of the organisation. Trend analysis can often track the performance of an organisation. It is important not only to collect data but also to use that information to improve the products and processes in the organisation continually. The data quantify process performance. When collected, analysed and evaluated periodically, the quantitative data about various business objectives, KPIs and Quality Management Systems indicate whether the organisation is improving. The data and analysis on the performance of the process, as appropriate, are submitted to top management and other relevant personnel in the organisation by the management representative.

Data analysis is one thing in which raw data is collected and organised so that certain conclusions can be derived. The information must be used for continual improvements. Some of the data that can be collected are the following achievements over a defined period:

The organisation analyses and evaluates data on:

1. Information on problems on products and services reported by external sources (e.g., government/industry alerts, advisories) and
2. Information arising from monitoring and measurement of processes as per para 9.1.1, General

   a. Conformity of products and services; process measurements (to check whether processes are correctly done so that the product (output of the process) is correct and meeting the requirements). This can be the percentage of products that meet the quality requirements versus the total production. It can also be measured, item-wise, in terms of the number of:

      i. Nonconformities/defects/deviations found internally throughout the organisation at any stage versus total production,
      ii. reworks/rectifications carried out to rectify defects/nonconformities internally, throughout the organisation at any stage,
      iii. internal rejections of parts made during in-process or final inspections throughout the organisation,
      iv. defects/customer complaints/deviations reported by customer or product/part returns from customer after the product was delivered,

   b. Percentage of on-time delivery of products according to the delivery schedules prescribed by the customer and accepted by the organisation
   c. Percentage delay in delivery of products against the customer requirements of delivery schedules
   d. Measure customers' perceptions of how their needs and expectations have been fulfilled. Analysis of quantified achievements against targets set, reasons for dissatisfaction, analysis and actions related to that, communication to such customers as to what actions are being

taken to improve the product quality and other contract requirements, observations or suggestions made by customers, the actions decided up on them and follow up, re-assurance and communication of continued commitment to satisfied customers.

e. The performance and effectiveness of the Quality Management System: Internal audits (Monitoring of the health of the Quality Management System implemented in the organisation). Any other data collected to evaluate the organisation's performance can also be analysed. Charts, graphs and brief write-ups of data are all forms of data analysis. Interesting information emerges from the data. The summary of data often supports arguments. Using appropriate quality tools from among the seven quality tools is desirable to analyse the data. Use any statistical methods to study and prepare charts and graphs accordingly.

   The seven quality tools are briefly indicated below:

   i. Cause-and-effect diagram (Ishikawa or fishbone chart): Identifies possible causes for an effect or problem and sorts the ideas into functional categories.

   ii. Check sheet: A structured, prepared format for collecting and analysing data; a generic tool that can be adapted for various purposes.

   iii. Control charts: Graphs used to study how a process changes over time.

   iv. Histogram: The most commonly used graph for showing frequency distributions, or how often each different value in a data set occurs.

   v. Pareto chart: This shows which factors are more significant on a bar graph.

   vi. Scatter diagram: Graphs pairs of numerical data, one variable on each axis, to look for a relationship.

   vii. Stratification: A technique that separates data gathered from a variety of sources so that patterns can be seen (some lists replace 'stratification' with 'flowchart' or 'run chart').

Data from all sources are analysed and issues are resolved using a variety of statistical techniques and tools, including Six Sigma methodologies, brainstorming activities, process mapping, capacity planning, value stream mapping (VSM), characteristic selection matrix (CSM), failure mode and effects analysis (FMEA), statistical process control (SPC) charts, Gauge RandR, Root cause analysis (5 Whys) and trend analysis.

When the results of process performance information are compared with the pre-determined requirements of the process, it either confirms the process's effectiveness and efficiency or identifies any need for corrective action. The result of analysis, in fact, always identifies opportunities for process improvement.

f. If planning has been implemented effectively: Planning also requires monitoring and measuring these actions and gathering, analysing and evaluating appropriate data and information to determine the effectiveness of such actions.

g. The effectiveness of actions taken to address risks and opportunities: Refer to the risk register.

h. The performance of external providers; Performance of the suppliers (to evaluate the performance of suppliers to maintain the supply chain management and to retain the suppliers for continued business) Improvement of available directory of approved suppliers, with the addition of multiple sources and improved scope of supply, retaining the suppliers, who perform well and meet the requirements of the organisation, advising or controlling those suppliers, who consistently fail to meet the requirements of the organisation, communicating the performance rating of all suppliers so that it is an encouragement or advice to improve further the performance, a re-assurance for continued business or caution to erring/under-performing suppliers to take efforts to further enhance and meet the benchmark set by the organisation, etc. are intended by analysing the performance of suppliers.

i. The need for improvements to the Quality Management System: Trends on quality performance, individual process performance, on-time delivery performance constraints, if any, in achieving them consistently, etc., are considered. When corrective or preventive actions are ineffective, or there is a random recurrence of nonconformities, the analysis should lead to permanent preventive actions such as error-proofing, design improvements or modifications.

Measurements will indicate whether the process requirements have been achieved or not. What normally happens is that the process owners do not identify the process requirements to be measured and the objectives of the process, however minor they may be. Suppose it is possible to measure each process's requirements and objectives. In that case, that is the final achievement of the result of the processes. Usually, what happens is this. The process owner knows what results will be achieved and does not care much about the specific objectives and requirements of the result/product achieved. How much was the delay on the way to achieve the results, what were the constraints faced, what compromises were made, how much time and cost additionally involved in attaining the results, etc., are left to be evaluated by someone else. Unless the process owner knows where the hiccups were, what the constraints were, what alternatives were for a better and more efficient way of achieving the results and what improvements the process owner thinks of. That is the importance of laying down the process requirements/objectives and measuring them to determine how effectively and efficiently they are achieved. When a deficiency in the requirements or objectives is identified, an effective process owner grabs the opportunity to analyse the results and try to improve the process. So, the measurement methods adopted should demonstrate the ability of the processes to achieve planned results. Measurement helps the process owner realise that when planned results are not fulfilled, correction and corrective action should be taken as appropriate. That is the way for improvements. When the organisation determines suitable methods, it is advisable to consider the type and extent of monitoring or measurement proper to each of the processes in relation to their impact on the conformity to product requirements and the effectiveness of the Quality Management System.

In the case of any process nonconformity identified through monitoring or measurement, the organisation has to take appropriate action to correct the nonconforming process. Correction is possible only when the reason for the nonconformity is identified. Next, when nonconformity is detected in a process, it needs to be examined whether the process nonconformity has resulted in any product nonconformity, which is a more severe issue.

A more detailed examination is necessary to identify whether the nonconformity identified in a process could have affected other processes or products. Suppose any product is affected by the process nonconformity. In that case, actions have to be taken as per the procedure for control of nonconforming products, per the requirement in clause 10.2 of the Standard.

Assessing the cost of nonconformities in all the processes is a method of identifying the performance negatives in an organisation. It may be called the cost of poor quality, directly or indirectly affecting the organisation. Cost of poor quality (COPQ) is associated with providing poor-quality products or services. There are four categories of costs:

a. internal failure costs (costs associated with defects found before the customer receives the product or service and costs due to inefficiency of the process equipment),
b. external failure costs (costs associated with defects found after the customer gets the product or service),
c. appraisal costs (costs incurred to determine the level of conformities to quality requirements) and
d. prevention costs (risk management costs, costs incurred to keep failure and appraisal costs minimum).

While assessing the cost of nonconformities, it is necessary to add all the costs associated with the prevention, appraisal, external and internal failure costs, including the expense of inefficient equipment.

Prevention costs (activities specifically designed to prevent poor quality of products) are mainly design inadequacies or inherent errors (cost of not doing the risk management activities). Mistakes in design documentation that lead to manufacturing errors are also prevention costs. Prevention costs arise from efforts to ensure potential defects do not occur. When effort is made to strengthen the design and prevent failures or damages revealed during use, it may add to prevention costs but is unavoidable. Design reviews are partly prevention and partly appraisal. Quality planning, establishing appropriate Quality Management Systems, adding quality-related information for reference, training and development of personnel, statistical process control, design verification, etc., are also essential and unavoidable prevention costs. If more prevention costs are incurred initially, that will taper down to nil in the long run.

Appraisal costs are activities designed to find quality problems, such as supervision, inspection and testing. When an effort is made to look for errors in the proposed design and a review is made, it is an appraisal. Finding out and correcting defective products/services through inspections/testing and identifying and correcting gaps/errors in the practice of Quality Management Systems through internal or external audits are also appraisal costs.

External failure costs result from delivered poor quality, such as correcting the problems on the delivered product, unscheduled after-sales service and dealing with customer complaints. Examples of the same are attending to technical support calls from customers, cost of technical support, preparation of support answers or adding to the manuals, cost of replacement with improved product or refund, cost of public relation work to soften harsh opinions of customers, impact of loss of customer goodwill and organisation's reputation, etc. Warranty costs recall of products due to quality deficiencies or failures to meet statutory and regulatory requirements are also external failures, which are part of the quality costs.

Internal failure costs arise before the organisation supplies its product to the customers. These costs go beyond the obvious costs of finding and fixing deviations and defects internally by discarding (scrap) or repairing (reworking) the defective items. Many internal failure costs may sometimes have to be borne by personnel outside of product design and development, apart from the cost of faulty products developed or produced. The cost of incorporating configuration changes on the product is because of failure to identify the exact configuration the customer wants. Internal work environments causing injury or inefficiency are also failure costs. Internal failure costs may also be due to wasted time in realising or delivering the right product, loss of data, embarrassment, lost business and the cost of replacement of frustrated employees who quit service. Failure to demonstrate satisfactory product performance to customers and the cost of those tests or tasks that could only be done once are also internal failure costs.

9.2 Internal Audit

9.2.1 Introduction to audits

It is mandatory that all organisations conduct internal audits of the Quality Management System implemented in the organisation. Internal audits are planned at defined intervals. Very dynamic organisations have frequent audits. The audit aims to evaluate whether the Quality Management System implemented in the organisation conforms to the organisation's requirements for its Quality Management System. It includes customer and applicable statutory and regulatory Quality Management System requirements. The Quality Management System implemented in the organisation must also conform to international standards (AQMS).

Internal audits are first-party audits, usually conducted by the employees of the organisation, who have been trained to perform the internal audits. Sometimes, some organisations prefer to get the internal audits conducted by external providers, consultants or third-party auditors.

A detailed internal and external auditing chapter is included later in Chapter 11. Auditing is 'a systematic, independent and documented process for obtaining audit evidence and evaluating it objectively to determine the extent to which the audit criteria are fulfilled'. (Refer to ISO 19011: 2018 Guidelines for Auditing Management Systems). When the processes are audited and the evidence (in the form of statements and documented information (documents and records) is evaluated objectively, it is verified whether the organisation's Quality Management System conforms to the Standard's requirements and, therefore, is effectively implemented and maintained. Suppose the verification reveals gaps/aberrations/mistakes in practice against the requirements of the Standard. In that case, the auditor brings out a finding in the form of nonconformity or observations. Corrections and corrective actions are to be taken by the organisation without undue delay to eliminate the detected nonconformities and their causes.

While conducting the internal audits, performance indicators of the processes and QMS in the organisation can be evaluated to determine whether the Quality Management System is effectively implemented and maintained. Performance indicators measure the performance achievements of the processes over some time against a target. If the targets are achieved, it is said that the methods/QMS is effective. The organisation can plan and implement improvement actions if the targets are not fulfilled.

9.2.2 Internal audits using criteria

The organisation plans and prepares an audit programme (a set of one or more audits intended for a specific time frame and directed towards a particular purpose). An audit programme includes all activities necessary for planning, organising and conducting the audits, covering all areas and processes to be audited for the first time. Next time onwards, important areas (importance based on the status of the production processes and the other regions) where more nonconformities were found during the previous audits, where the process was not effective in the earlier audits, where the process performance KPIs have not identified or achieved earlier but no actions have been identified or implemented, audits of such processes can be planned for more number of audits per year, which means more number of times in a year with lesser time intervals in between.

An audit is always done with reference to audit criteria. Audit criteria are the requirements in the international standard and the organisation's requirements, customer and applicable statutory and regulatory Quality Management System requirements. The scope of the audits can be the scope of the QMS in the organisation, which means all the required and established processes in the organisation.

Internal auditors are selected by the organisation and given adequate trainings for performing the audits. The audit process should ensure objectivity – provide an unbiased result, be impartial and have no prejudice (pre-conceived opinion not based on reason or experience).

The results of the audit are to be reported to the relevant management. Here, the management is the process owner and the one who has assigned the audit process. The audit reports are to be submitted to the management.

The management (process owner) takes appropriate corrective actions without delay. The process owner is responsible for making corrections and corrective actions in the process audited. However, the internal auditor can verify the measures and close the nonconformities if reported during the audits.

Retaining documented information (records) for the audit programme and audit results is mandatory.

Defining the responsibilities for planning and conducting internal audits, establishing audit records and reporting audit results is desirable. The audit criteria, scope, periodicity and methods of audits can be defined. The organisation may determine the competency required for the internal auditors. The Selection of internal auditors can be based on internal auditor training given to the organisation's employees. When they perform audits, they should be able to do the internal audits professionally, identify nonconformities against the criteria/standard requirements with objectivity and maintain impartiality of the audit process.

There is no specified or approved training programme for internal auditors prescribed by the Standard or IAQG. So, the organisation can get the necessary training for internal auditors from any professional. ISO 19011:2018 guidelines are included for the internal auditor's training module and correctly interpreting the AS 9100 series standards requirements. If the internal auditors fail to identify nonconformities, which third-party auditors may find during their audits, the inefficiency of the internal audit process may be due to a lack of:

a. adequate training for the internal auditors,
b. knowledge of the Quality Management System standard
c. audit experience (Experience gained by being an observer with experienced auditors) and
d. objectivity and auditor attributes (refer to ISO 19011:2018).

A list of internal auditors duly approved by the top management must be maintained. The internal auditor being an employee of the organisation does not violate the impartiality. However, when they do audits, they should perform them impartially and have no bias or preconceived notions about the effectiveness of Quality Management System practices within their organisation. Internal auditors are not permitted to audit their work. This aspect should be taken care of when auditors are selected to audit specific processes/areas, as some of the auditors may be process owners or auditees. Suppose the auditor is a process owner in an area. In that case, he should be given another area/process for audit, not his own. In many cases, particularly in smaller organisations, independence can be demonstrated by the freedom from responsibility for the audited activity.

To conduct a good, complete, thorough and professional audit, auditors should use a checklist describing the procedures, resources and controls for the processes; a flow diagram or similar tools to verify the input to the process, steps/stages and output, controls to the inputs and steps, verification of requirements for the output and measurement of objectives for the procedure. AS 9101E Quality Management Systems Audit Requirements for Aviation, Space, and Defence Organisations, which the certification body auditors use, can be used as checklists for internal auditing after customising it with reference to the documentation used by the organisation.

Any audit, if done thoroughly and professionally, will likely reveal gaps, aberrations and mistakes in practice; Audit may indicate them as nonconformities or observations. All findings are to be agreed upon with the auditee and conveyed to the auditee to help him plan and carry out corrections and corrective actions. The management responsible for the audited area should ensure that necessary corrections and corrective activities are taken immediately. The root cause of the detected nonconformity has to be identified, and corrective action to eliminate the root cause has to be completed. The person responsible for management should take follow-up activities. Follow-up actions include verifying the steps taken to correct the nonconformity and eliminate the root cause of the detected nonconformity. The results of the verifications are to be recorded. The person responsible for conducting the organisation's internal audit is usually the 'Management Representative'. He should present the summary of internal audit findings to the management for review. Management review makes decisions to improve the effectiveness of the Quality Management System.

Regular internal audits in an organisation are a health monitoring process that monitors the health of the organisation's quality management system. Therefore, periodic health check-ups must be done effectively to identify any unidentified or inherent system deficiencies or product deficiencies, acquired illnesses of harmful practices, aberrations or contaminations, or infections. Hence, the internal auditors in an organisation have a role nothing less than that of hospital resident doctors. As the doctors are trained to diagnose the deficiencies or gaps in the Quality Management System, the internal auditors can manage the health of the Quality Management System in the organisation. When appropriate corrective actions are taken, the QMS health conditions improve and full health is restored. Quality Management System becomes effective for continued excellence.

> *'The common practice gap of some organisations is that "Processes such as Top management, statutory and regulatory requirements, customer QMS requirements and MR processes are not indicated in the internal audit plan, and no evidence is recorded in the checklists." It appears that the MR in such an organisation is apprehensive about conducting an audit on the management processes, or the top management is not very concerned about the effectiveness of the Quality Management System in the organisation. Sometimes, the audit questions boomerang on the MR, saying, "You should know it"!'*

9.3 Management Review

9.3.1 General

A formal review of the Quality Management System for its continuing suitability, adequacy and effectiveness is integral to the organisational processes. It is indeed an integral part of the organisation's business management process. Formality brings added impetus to the process. The overall management review is a complex process at various organisational levels. It is always a two-way process generated by top management with inputs from all organisational levels and functions. Periodically check that the organisation is moving towards achieving its objectives in accordance with the corporate policy, and help the organisation evaluate the strengths, weaknesses, opportunities, threats and alignment with its strategic direction for achieving the objectives. These reviews and activities may be happening informally daily, weekly or monthly. Several organisational matters are covered in the discussions or reports. Formal reviews on all the agenda points may be happening at periodic intervals. The whole process is for the management to know the business, progress and future needs and accordingly give directions to the organisation based on facts and figures. A factual approach to decision-making, one of the principles of Quality Management Systems, is employed here. The organisation decides the periodicity of the formal reviews as quarterly, half-yearly or annually.

9.3.2 Management review inputs

Top management decides the list of participants in the regular reviews. MR is the member secretary for the meetings if meetings are held for the reviews. Agenda points can be circulated to all members of the management review, and the views can be recorded. Meetings are held as it is the most convenient method. MR circulates the agenda points, participates in the discussions and records the decisions made at the review meetings. Advance information given to all the members allows them to prepare for their thoughts and presentations, evaluate changes and help them participate effectively in the review meetings. The review meeting/discussion among the participants occurs at an appointed date and time and at defined periodic intervals. The planned intervals for the review must be determined and documented, and the reviews must be conducted accordingly. Conducting the reviews every three or six months is desirable, preferably in line with the internal audit frequency.

The reviews by management are part of a continuing process and are not to be considered as a one-time event. A meeting occasionally cannot be regarded as an effective way of meeting the requirements of communication and review. Such a meeting should be part of the ongoing review. Many organisations have specific periods for formal reviews, and records of the reviews, discussions and decisions are always made.

The management reviews are intended to assess opportunities for improvement and decide, if at all necessary, how the changes could be made to the Quality Management System (without affecting the integrity when changes to the Quality Management System are planned and implemented; see 5.4.2), whether any changes or improvements are required in the existing quality policy and quality objectives. Records of management reviews are required to be maintained. This depends on whether management review is conducted as part of the other top-level organisational reviews like pre-scheduled 'Committee of Management' meetings or the divisional Quality Management System reviews are performed separately. The head of the organisation/CEO must chair such meetings to make the top management commitment to the Quality Management System visible. In the absence of the director or CEO, the next one in the hierarchy can preside over the meetings. Consistent absence or deliberate delegation could be viewed as a lack of commitment to the quality system and organisational improvements.

Incidentally, the chairperson and most participating committee members may remain the same. Records maintained in either meeting are acceptable, provided all required agenda points are covered through the sessions held at defined frequencies. Records can be maintained electronically and circulated to all concerned.

> *'Some CEOs say, "What meeting? What periodicity? We meet every day in the office, factory. We discuss daily issues, so why keep a record of them? Everyone is busy. How do you say that a management review did not take place? We manage it."'*

Is a meeting necessary for the management reviews? A meeting is neither required nor always the most effective way of accomplishing management review. Dynamic organisation management reviews activities at periodic intervals or on a day-to-day basis. All matters are considered; some are discussed among a few seniors and some are not. Management drives the organisation due to their perception of its day-to-day functioning. Managing by walking around, managing by interactions, ordering through telecommunications, pushing, pulling and keeping an eye on the total production tasks or balance sheets do not reflect a true managerial approach or commitment to quality. Considering the customer as the most critical entity whose requirements are supreme and every effort will be made to meet the customer's requirements also does not satisfy the criteria of good management. The point is seldom received by such management that proactive measures or analysing the day-to-day issues to find the root cause and taking appropriate corrective action to prevent recurrence (remove the root cause) will only bring improvements. Management reviews prescribed by Quality Management Systems are recorded periodically and become somewhat rhetorical rather than constructive.

In most cases, the reviews and decisions have been taken informally after discussion with available people. There is no opportunity for review. Consider the root cause's pros and cons to permanently remove the problems from the system or organisation. Such reviews are seldom effective and cannot give the desired results of following a proper QMS and bringing in improvements. Whatever the periodicity prescribed, management review sometimes does not occur in time or the top management is away on tour or delegates the responsibility to MR since everyone is 'busy' otherwise. Also, whenever it is held, Management reviews often miss out on the continued suitability of quality policy and objectives for a long time. Management review is, ideally, a preventive process before issues mature into problems.

Holding management reviews once or twice a year does not allow frequent data review, leading to timely resolution and preventive actions. A two-tier recorded management review is desirable. Tier one review is on a weekly or monthly basis, including the shop floor meetings and tier two review is on a quarterly or half-yearly basis. Tier one review would have addressed all the required inputs and outputs as and when they occurred. The advantages of this approach are:

a. The information is contemporary, and the resulting actions are likely to be preventive and

b. It happens smoothly and effectively, without formalities.

The agenda for the management review meeting indicates the inputs, the details of which are presented by the participating members and process owners. Quality Management System standard makes it compulsory that the agenda points for discussion in the management review include, first of all, the status of actions from previous management reviews. Some decisions are recorded in the last management review with responsibility and the target date for completion of actions. The current review will bring out all follow-up actions and the status of the steps on those issues. Next, in the agenda points, the management review considers all the changes in external and internal issues relevant to the organisation's Quality Management System. External issues could be the market response for the products, some issues the regulatory authority is investigating, trading in a stock exchange, etc. Internal issues may be related to hold-ups in production operations, security issues, some unexpected breakdown of machinery, fire alarms and some behaviour issues of employees or common loss.

Major reviews happen on the information on the performance and effectiveness of the Quality Management System, where the following trends are considered and discussed:

a. Customer satisfaction and feedback from relevant interested parties: Statistics of customer satisfaction score, complaints and feedback (customer-wise/product-wise) to be presented by the marketing process

b. The extent to which quality objectives have been met: All measurements against a target for a period for all functions/processes to be presented by MR. Actions to be decided where planned results (effectiveness) are not achieved.

c. Process performance and conformity of products and services: All measurements for conformity to process/product (quality) requirements for every process/process output are to be presented by the process owners – actions to be decided where planned activities are not realised. If actions have already been decided, share them in the management review for information.

d. Nonconformities and corrective actions: Discuss all nonconformities in all the processes and products (internal/customer) and decide on the corrections and corrective actions after root cause analysis. Discuss a plan for implementing the corrective actions (internal/customer) for every nonconformity (matters could be decided and recorded by the respective authority/disposal authority beforehand). The action taken is presented by the same authority or the process owner for quality control in the management review for information. This planning must be periodically reviewed and updated during corrective actions or management reviews. These actions must be proportional to the potential impact on the conformity of products and services.
e. Monitoring and measurement results: The process owners could present a brief of all monitoring and measurement results for every process/product for information on the management.
f. Audit results: Results of all types of audits (internal/customer/third party (CB), nonconformities and opportunities for improvement could be discussed. If any actions are pending, the same could be decided. Evaluate the effectiveness level of all production operational processes and decide on the actions for improvement. MR presents the results of audits.
g. The performance of external providers: Results of periodic evaluation of the performance of the suppliers/external providers with whom the organisation had business during the intervening period could be assessed, and actions decided for improvement of the externally provided processes (purchase of materials, labour, special processes and services). This information can be presented through the purchase or planning process.
h. On-time delivery performance: The on-time delivery performance of all the products to each of the customers may be tabulated and presented for an overall performance review with respect to OTD, one of the very important customer requirements. The data collection and presentation are simple periodic activities the marketing process enjoys.

Management reviews are planned and conducted considering the following points:

a. The adequacy of resources: Review the adequacy of the available resources with respect to personnel, infrastructure and environment.

Appropriate decisions may be taken with respect to any special requirements from the customers and process needs. The HR and maintenance teams are to provide the inputs and take action as necessary.

b. The effectiveness of actions taken to address risks and opportunities (see 6.1): Review the risk register for which inputs and assessments must come from various process owners. Suppose mitigations are to be decided and approved. In that case, the same can be done during the management review, where all the members of the risk review committee are also present. The effectiveness of all mitigation actions completed in the past for high-risk items has to be verified with respect to any recurrence of the risks identified.

c. Opportunities for improvement: All opportunities for improvement in all the areas/processes and the objectives are discussed in the management reviews, and decisions are made for continual improvements.

Any other points as appropriate could be discussed depending on the nature of the product or service being realised, as an MRO organisation or a stockist distributor. Input and outputs of the management review process are related to the organisation's size and complexity and how the organisation's management uses the review to identify improvements to the quality system and the business. Reviews always include assessing improvement opportunities and the need for changes in the quality management organisation. One of the agenda points of this review is to analyse changes that could affect the Quality Management System. The review ensures that the changes have no negative impact on the integrity of the quality system or organisational processes.

Other points to be included in the agenda for management review meetings are the achievement, adequacy and effectiveness of the personnel training programme and changes to any statutory and regulatory authority requirements that could impact the organisational activities.

9.3.3 Management review outputs

Minutes of the meetings are the most common type of documented information (records) retained for the decisions and action points recorded during the management review. Other reports of reviews, data/information, statistical charts, including trend charts, etc., given by the process owners, a summary of the presentations, an overview of the decisions and attendance/evidence of participation of all the management review members, etc. are also retained as records. The records created by the MR or member secretary of the management review committee could be retained electronically. The overall status of the business process is reflected in the review records. The review records reflect improvements and changes and include decisions on other important issues, such as plans to introduce new products, marketing strategies, business expansions and new/additional sets of objectives. The records of the management review should include all decisions and actions related to each of the agenda points reviewed by the management committee, especially:

a. All opportunities for improvement, process-related, product related and those related to customer requirements,

b. Any need for changes to the Quality Management System, the improvement of the effectiveness of the Quality Management System,

c. Additional resource needs and their provision

d. Review of risks managed for the effectiveness of actions taken and any new risks identified for assessment.

Brief discussions on the indicated agenda points and their respective decisions are recorded, highlighting the action points, with identification of responsible persons and probable date of completion of actions.

MRO organisations additionally include a review of safety policy as to whether it is appropriate to the organisation's purpose.

The policy must indicate the method and periodicity to review safety objectives for continued suitability and effectiveness. The safety policy and objectives established must be communicated within the organisation, which must also ensure that the personnel adhere to the safety standards prescribed.

10

Improvement

The standard requirements/clauses are arranged in accordance with the bigger picture 'Plan-Do-Check-Act' cycle (Deming's wheel). Chapter 10 is the 'ACT' section (including corrective actions and continual improvements). The requirement details are indicated in the respective sub-paras of the chapter.

10.1 General

Improvement is the ultimate goal of every human activity. Continual improvement is a mandatory requirement of Quality Management Systems. Continual improvement is a set of recurring actions that are carried out to enhance the organisation's performance to meet customer requirements and thereby improve customer satisfaction. Organisations must determine and select improvement opportunities and implement actions necessary to enhance performance.

During the operations of various processes in the organisation, many opportunities come up for improvements. Data collection with reference to the identified process deficiencies, observations or defects on products, root cause analysis for the identified defects, etc., provides ample opportunities for analysing them and implementing corrections and corrective actions. Corrective actions are real improvements since they eliminate the root cause permanently.

Identification and selection of the opportunities can be with reference to the actions that can be taken:

1. To meet customer requirements and thereby enhance customer satisfaction.

   a. For the process, the products/services: Examples include – an improved methodology, which will complete the work faster, use of an alternate machine tool, which will consume less power, use of automation, use of the latest version of a software tool, which will help to do the finite element analysis more accurately, fixtures, which reduces setup time for machining etc.

b. To address future needs and expectations of the customer and the organisation: Examples are – establishing PPV for every product delivered to the customer. Manufacturing extra quantity of parts with the available material (so that small quantities of raw materials are not left over, contingency and immediate requirements of the customers can be met without delay, when needed), procuring minimum order quantity of raw materials and stock it for repeat orders from the customer, buying raw materials and spare parts together for the total quantity required for different parts and their respective quantities added together, etc.

2. To correct, prevent or reduce the undesired effects: Examples are the use of jigs, fixtures, and templates for accurately machining complex parts while machining, the use of automated controls for switching on and off the furnace at the right temperature, the use of alarms and warning devices to indicate the status of the process and the use of covers for protecting reducing contamination of baths for plating processes.

3. To improve the performance and effectiveness of the Quality Management System. Examples are addressing nonconformities found during internal audits effectively, planning to conduct more frequent audits in processes where there were more nonconformities in the past, verifying the effectiveness of corrective actions taken, carrying out a trend analysis of the effectiveness of the procedures over some time, etc.

Based on the selected opportunities and the actions taken to implement improvements to the process and product, the Quality Management System provides incremental benefits.

Improvements may be brought about in an organisation through breakthrough change. Juran defined breakthrough as a 'change, a dynamic change to new, higher performance levels'. Improvements may be brought about through innovation, that is, implementing changes in the production processes, products or the organisational business strategy. Innovation is a significant positive change that brings successful outcomes for the organisation.

A re-organisation of the existing infrastructure, processes, people and their functions aims to bring about improvements. The merger of separate units into one entity, establishing independent departments for diversified products, etc., can be considered re-organisation.

Adoption of new technologies can bring about improvements. In emerging technologies, the 'Agile' methodology for the software sector is prominent, and some software engineering services use it to manage their projects. Agile is a set of methods and practices used to improve software development. Agile methodology was formulated based on other management methods such as lean, Kanban and mentoring.

(Lean: The core idea of lean is to provide perfect value to the customer while minimising waste throughout the value streams or using fewer resources. Lean creates processes requiring less human effort, less space, less investment and less time to realise products or services at lower costs and with fewer defects than traditional business systems. With lean, information management becomes much simpler and more accurate.

Kanban: The part of a just-in-time production system is a label attached to each box of components as they go to the assembly line (A) from the previous process (P). The label acknowledges the delivery of a given quantity of specific parts to assembly (A). It is returned to (P) as a record of work done and an order for new components. Thus, Kanban coordinates the inflow of parts to the assembly line, minimising several steps in the process, control and documentation.

Mentoring: This is a process that enables learning and professional development. Mentor has the knowledge and understanding of process as well as the variety of styles, skills and techniques that are relevant to the *context*.)

'Agile aims to reduce waste and promote and prioritise activities that add value to the customer. Agile supports project management, time management, quality improvement and team performance. Agile methodology provides a more flexible approach to business and improves communication, workflow, and efficiency. Agile provides high priority to satisfy customers, welcomes changes in requirements later in the development stages after the initial identification of customer requirements, and delivers working solutions frequently and in terms of minor incremental improvements. Instead of the PDCA cycle, Agile has the "Learn -Act- Reflect" cycle, where -

Learn: Agree on the goals and objectives, metrics and measures are established to measure success, map the existing practice, estimate and prioritise activities, identify the options for the next cycle of activity, reflect on the previous actions to establish key learning points, determine the future needs and decide on the next activities.

Act: A period of action and development of a set amount of work, work prioritised by reflection and learning, activity is tracked and measured to provide defined metrics, limited change to scheduled work

Reflect: Test, release, and obtain feedback; review and reflect on the last activity cycle. Reflect on what went well and what can be improved, review progress and output through metrics and measures, reflect on overall direction and progress, and identify options for the following work cycle.

(From "Being Agile in Business", Pearson Education Limited, UK)'

10.2 Nonconformity and Corrective Action

10.2.1 Nonconformity

Nonconformity is non-fulfilment of requirements. The requirements are stated in the Quality Management System standard: documented information maintained or retained in the organisation, statutory/regulatory/customer QMS requirements to be followed by the organisation. Nonconformity occurs in processes, products or services within the organisation. It can happen in the external providers' process, product or service. It can occur when a product or service is delivered to customers. If the nonconformity applies to the organisation, the organisation needs to react to the nonconformity.

The organisation needs to react in two ways – one immediate containment action or correction to arrest the nonconformity from aggravating, to control it and to prevent it from getting worse or more serious. Secondly, the organisation has to deal with the consequences that nonconformity may lead to. The likely consequences are how the nonconformity will affect the product or Quality Management System. It may affect the integrity of the product or service or result in the breakdown of a system to meet the Standard requirement. It can also result in the probable shipment of a nonconforming product or result in a condition that could fail or reduce the usability of the product or service and its intended purpose. Other consequences are that it may result in a single system failure or lapse in conformance with a standard requirement or a lapse in conformance with a procedure in the organisation.

The organisation has to take action to ensure that:

a. the nonconformity does not recur wherever it has happened and

b. the nonconformity does not occur anywhere in the organisation or at the supplier's/customer's premises.

To accomplish that, it is necessary for the organisation to:

a. evaluate the cause of the nonconformity and

b. identify if there is a need to act on the cause or not. And if so,

c. identify the action to eliminate the cause of the nonconformity.

The organisation does this by:

a. Reviewing and analysing the nonconformity: The nonconformity is reviewed and analysed by appropriate tools (refer to 7 quality tools vide para 9.1.3).

b. Determining the causes of the nonconformity, including those related to human factors: The cause of the nonconformity is generally identified by why–why analysis or cause and effect diagram. It may be some aberration, mistake or uncontrolled condition concerning men's materials, machines, methods, measurement, environment and exposure to risk. Human errors, a concern of discipline, need to be analysed further to ascertain what caused the human to err. Controlled conditions established by the organisation include implementing actions to prevent human error. Those conditions would have changed, or mistakes could have crept in. Any investigation needs to stretch beyond the person to analyse the root cause of human error, such as stress, inadequate training, bad habits, behavioural deficiencies, etc.

c. Determining if similar nonconformities exist or could occur: Exploring if any similar nonconformities live elsewhere in the product or any other products is a proactive action the organisation takes to control the nonconformity. Any potential nonconformity that could occur is also explored.

The organisation must maintain documented information (procedure) defining the actions and management of detected nonconformities. Corrective action is to prevent recurrence by eliminating the cause of the nonconformities. The key word here is to eliminate the cause of the nonconformity. Corrective actions decided and taken by the organisation to eliminate the causes of the nonconformity should be appropriate to the consequences of the nonconformities, i.e., relevant to the effects of the nonconformities detected on products or QMS.

The procedure must first include identifying and reviewing the nonconformity concerning the criteria for acceptance, be it product-related or process-related, whether the nonconformity has been detected in-house or reported by the customer after product delivery.

The organisation requires the following action to identify and implement any action needed to eliminate the identified cause of the nonconformity. That is implementing the corrective actions.

Once the corrective actions are completed, the cause of the detected nonconformity has been eliminated. Hence, similar nonconformity on the product/service, within the organisation or at the supplier/customer premises, should not recur. The production processes/products have improved from the previous status of having nonconformity. While continuing with the production processes after implementing the corrective actions, it is necessary to review if there is any recurrence of the nonconformity. Implementing corrective actions expects a planned result of no recurrence of the nonconformity. Suppose the desired outcome is achieved due to a review of the corrective actions. In that case, it is said that the corrective actions implemented are found effective.

Suppose any potential nonconformities have been identified during the review and causal analysis (root cause analysis) for the nonconformity. In that case, the ordinary course of action is to include the risk in the risk register, assess the risk using the same criteria, identify mitigation/corrective action to reduce the risk if the risk level is high and verify the effectiveness of actions taken. This control of potential nonconformity becomes part of the planning activity.

If the corrective action is intended to eliminate the cause of a nonconformity related to the Quality Management System, necessary improvements and changes have to be implemented in the system and documented information maintained by the organisation.

While arriving at the root cause of the nonconformity, it is also necessary to determine the factor responsible for the nonconformity. Suppose it is determined that the control exercised by the external provider is responsible for the detected nonconformity on the product or service from a supplier. In that case, it is necessary to flow down the corrective action requirements to the external provider/supplier as a requirement for root cause analysis and corrective action. The flow of information/requests is based on the responsibility for the nonconformity with the supplier.

The nonconformity detected might have been caused by the organisation's indicated requirements to the external provider/supplier, or an essential requirement was not indicated to the supplier by the organisation. Then, the organisation should own the 'responsibility for the nonconformity'. Corrective action is to be taken by the organisation itself.

It is necessary that the corrective actions, including alerts and positive recall wherever necessary, be taken in a timely manner so that the improvement action is primarily beneficial to the customer. If the containment/ corrections/corrective actions are not timely, the deficiency may impact customer satisfaction and sometimes the business itself. If the actions are safety critical, it is all the more important that the improvements are implemented in a timely manner. The organisation has to define specific actions whenever any lapse is noticed with respect to timely and effective corrective actions. Verification of the effectiveness of corrective actions is to be planned and completed in every corrective action wherever it is implemented.

It can be summarised that while defining the actions and management of detected nonconformities, the following must be clearly defined in documented information maintained by the organisation.

A root cause analysis, such as why the deviation/nonconformity has happened (why–why analysis), may identify the cause. Once the root cause is identified, actions to eliminate the cause of the nonconformity can be considered and evaluated so that the defect/nonconformity does not recur. Complete the corrective step so decided, and the results of actions taken can be recorded for reference. It is necessary to watch for consistent results as desired during repetition or the production process. The effectiveness of the corrective action taken is verified during the subsequent production batch or where continuous production occurs after three months.

When a more severe nonconformity is required to be investigated, with the involvement of a team of process owners, organisations generally adopt a team effort to analyse the root cause and identify appropriate corrective actions, the methodology commonly known as following the 8D steps, which is indicated as a corrective action procedure.

Doing corrections and implementing corrective actions in-house because the organisation urgently requires the supplied product/service to proceed with the production/assembly processes deprives the supplier of an opportunity to learn and establish improvements in the supplier's controls. Worse, the supplier may continue to supply nonconforming products/services! The organisation adds cost of poor quality into its accounts!

It is also necessary to identify and document actions where timely and/or effective corrective actions are not achieved due to the corrective actions taken. Corrective action is taken based on the nonconformity noticed on one or more products or services. It is prudent to explore or take action to determine if additional nonconforming products exist based on the causes of the nonconformities. The requirements include checking all the remaining products and taking further action when required. The common pitfall of practice in some inefficient organisations is that the corrective actions end with the correction/rectification of the product. Improvements cannot come when the corrective action process is not practical.

Nonconformity Management and Corrective Action

source: Internal/Customer/Supplier

(a) Identification of nonconformity:
(b) Objective evidence/criteria for acceptance for the non-conformity:
(c) (i) Affected quantity: (ii) Rest of the quantity verified and found:
(iii) Quantity of other products affected/likely to be affected:
(iv) Element of QMS affected:
(d) Containment/correction - immediate fix-: quantity/date
Containment/correction completed on : quantity/date
(e) Root cause analysis (identify/show the quality tool used:)
(i) Detected cause of Nonconformity:
(ii) Potential cause of Nonconformity (if any):
(f) Is there is a need to act on the cause or not : Yes/No, reason:
(g) If yes, identify the corrective action to eliminate the cause of the nonconformity: ....
(h) Action completed : on Process: on product:
Flow down to External provider:
Information to Customer :
Alert to Customer:
Positive recall to customer:
concession request to customer/regulatory authority:
concession granted by customer/regulatory authority:
(i) Were the actions timely? : Planned dates: Completed dates:
(j) verification of effectiveness of corrective actions:
Was there any recurrence of the Nonconformity : Indicate details if Yes/ No
Reviewed on:
Was there any recurrence of the Nonconformity : Indicate details if Yes/No.
Reviewed on (after one year) :
Approved by:
(Authority deciding the action in respect of the non-conformity)

Figure 10.1 Documented information on nonconformity management and corrective action.

10.2.2 Nonconformity report (NCR)

A typical documented information on nonconformity management and corrective action is indicated in Figure 10.1

Examples of investigation findings, and technology/process/product audit findings are given in Figure 10.2

It is mandatory that the organisation retains documented information as evidence of actions taken:

a. The nature of the nonconformities and any subsequent actions taken, such as containment, correction, root cause analysis, corrective step identified and implemented.

b. Review the effectiveness of the corrective actions completed, i.e., to check whether the planned results are achieved.

Effectiveness and efficiency are two words often used in discussions and confused with respect to achieving the results of actions. Effectiveness is achieving the results as planned or meeting the requirements (quality, delivery schedule, customer satisfaction, etc.) as defined. In a product/service realisation process, meeting requirements and achieving results are always when the product/service delivered is to the entire satisfaction of the customers and the product/service delivered is exactly as per the customer requirements.

Efficiency is achieving the results as planned or meeting the requirements as defined with reference to the resources used by the organisation. The objectives achieved when a product/service is delivered to the customer are compared to the resources used in similar units of measure, indicated as a percentage of the efficiency. The actual cost of production and delivery versus the selling price × 100, the cost of service provided versus price realised from customer × 100, or effort used in person-days versus effort planned in man-days × 100, etc., are examples of efficiency. Organisations aim at higher efficiency as an objective, and the customers want higher and higher effectiveness of product/service for satisfaction by raising the planned targets.

10.3 Continual Improvement

Continual improvements result from a proactive approach, and enhancements by managing nonconformities and corrective actions are the reactive approach.

As long as the organisation is in business, excellence is the ultimate aim, and all the efforts are made to achieve the objectives by meeting the targets set. Once the targets are completed, the processes are said to be effective. Further steps are necessary to achieve excellence; therefore, the targets are raised. It is a yearly activity to set new targets, a 'SMART' way of goal setting (S = Specific, M = Measurable, A = Achievable, R = Relevant, T = Time-bound) and working towards continual improvements. Further efforts may be necessary to ensure the planned activities are fully realised and conform to the requirements of the Quality Management System. It is essential to review periodically whether the planned actions are fully realised, and the organisation's internal audits are carried out to confirm this aspect. Internal audits further confirm the suitability and adequacy of the Quality Management System. Addressing nonconformities found during internal audits improves the suitability and adequacy of the Quality Management System. The processes are fully effective if the set targets for the goals, objectives and key performance indicators (KPIs) are achieved without any nonconformity (deficiency in achieving conformity) to the planned activities. The organisation thus continually improves its Quality Management System.

Analysis and evaluation results described under para 9.1.3 (Analysis and Evaluation) can be considered to determine the needs or opportunities to address continual improvement. By walking around, the management cannot fully view the organisation's performance. Periodic management reviews, including meetings, allow the organisation to take comprehensive stock of the situation's performance/health parameters and review them for desired results, continual improvements and effectiveness.

Proactive product improvements in the form of modifications on aircraft and equipment, technological improvements in the systems, controls and practices, etc., are all improvements. Managing risks is a proactive improvement activity.

Implementation of continual improvement activities in the organisation and evaluation of the effectiveness of the Quality Management System are monitored through the following:

a. Quality policy: Periodic review for continued suitability and adequacy of the policy concerning the changing dynamics of the business operations provides an opportunity for continual improvements in the organisation.

b. Quality objectives: Periodic measurement and monitoring of the quality objectives and key performance indicators (KPIs) and reviewing the achievements against pre-set targets indicate opportunities for continual improvements.

c. Audit results: All types of audits effectively conducted in the organisation directly indicate the health of the Quality Management System in the organisation; i.e., the number and criticality of the nonconformities and other observations are inversely proportional to the effectiveness and health of the Quality Management System implemented and practised in the organisation. Internal audits are a direct measurement by the organisation. Customer and regulatory authority audits are second-party and third-party health assessments, respectively, deemed as super-specialty assessments, in which the criticality/severity of nonconformities identified may put the organisation and the stakeholders under intensive care for improvements.

d. Corrective actions: Business processes, even if they are effectively planned and executed, may occasionally indicate deficiencies or gaps because human beings own every process. The progress and improvement of the organisation depend on how quickly and effectively the deficiencies or gaps are identified. These deficiencies or gaps are reviewed as opportunities for improvement, and adequate actions are taken rapidly and effectively to overcome the deficiencies and gaps. That is continual improvement, which the organisation can demonstrate to all the stakeholders. Hiding the deficiencies or gaps or not taking appropriate actions to overcome the deficiencies and close the gaps in the quality management system may cause the organisation to fail indefinitely.

It is said that if anything is certain, it is that change is certain. (Change is the only constant in life.) Asking 'what if...' is an excellent start to pursuing excellence through increased flexibility and openness to change.

People: What if... a member of staff leaves the organisation? Having an interim plan to handle their workload and a succession plan to deal with the situation more permanently may help the organisation react to the change without disrupting productivity.

Financial: What if... sales drop? There needs to be robust financial contingencies in place to ensure that the organisation can continue to operate efficiently and effectively in the short term to allow time to consider the next steps for the ongoing stability of the business.

Customer: What if... customer complaints go up? In an ever-changing world, customer expectations of excellence also move with the times. Keeping in regular contact with your customers to ensure you understand and anticipate their needs may help the organisation adapt quickly and pursue excellence for its customers. There are many more such areas to consider.

Suggestion schemes and getting the employees' views for process/product/resource improvements are proactive measures for continual improvements. It encourages continual improvement of the Quality Management System and is a key tool in achieving business excellence. A business that encourages staff input at all levels stands a better chance of survival in the current economic climate.

The situation may lead to undesirable results if improvements are not visualised, quantified/measured and reviewed at appropriate intervals. The reviews provide organisations with excellent opportunities to take appropriate corrective actions so that the planned results will be achieved with additional efforts if not fully achieved. The organisation must monitor the implementation of improvement activities and evaluate the effectiveness of the outcomes. Retaining appropriate records is a desirable way to demonstrate the efforts and evaluations of the organisation for continual improvements.

The continual improvement opportunities can result from the lessons learned in the past, resolutions of problems encountered and the benchmarking of identified best practices prevalent in similar organisations worldwide. Benchmarking compares the organisation's processes and products to the best practices followed by reputed organisations. Perhaps spreading the improvement culture in the aviation, space and defence sectors results from the lessons learned from accident/incident/defect investigations and the solutions to specific problems found in various space, aircraft and defence equipment. Adopting the best practices is a natural progression of the industrial world. It has become faster due to the revolutionary improvements in the innovation and proliferation of communication methods and media.

The regulatory involvement varies from natural value addition with the knowledge to flow along using the authority. In the high technology domain, for the regulatory authority, there are always challenges to problem-solving, contributing to better quality and safety. Working to resolve the issues/problems with the responsible organisation and not just identifying the difficulties and moving away.

> *'A regulatory inspector inspected a front-line fighter aircraft's ejection seat in the seat servicing bay. The pull-off test for the Personal Equipment Connector (P.E.C.) on the ejection seat was easily passing on the application of much less force than the 15 pounds specified. Per the test schedule, any force under 15 pounds was acceptable, and the minimum required force to pull out the PEC assembly was not specified. It is a safety-critical requirement. The purpose of this PEC assembly on the ejection seat is that when the seat ejects, the supplies of anti-g, oxygen and air ventilation, as well as the tel/mic, are automatically disconnected and sealed, whilst the emergency oxygen is turned on and admitted via the primary oxygen supply line to the pilot's mask/breathing line.*
>
> *To the inspector, the situation created a potential risk that if the PEC assembly comes out from its location due to any inadvertent touch/force by the pilot's hand while moving his hand in the cockpit, the emergency oxygen is turned on and after ejection, the pilot may not get sufficient oxygen supply.*
>
> *An observation was raised stating the risk and requesting to specify the minimum load necessary to pull out the PEC assembly in addition to the maximum load defined. Inspectors can suggest solutions*

to the problem, whereas auditors, even if they know the solution, are not to suggest any solutions. The inspector suggested using a tensile load-tested, single strand Copper or Aluminium wire of appropriate American Wire Gauge (AWG) and wire-lock the PEC assembly to the location on the ejection seat if the manufacturer of the ejection seat and regulatory authority permitted to do so. The suggestion of the regulatory authority inspector was accepted by the American ejection seat manufacturer and Indian military aircraft design regulatory authority, and the organisation implemented the same on the ejection seat as a Service Bulletin (SB). This is among the several examples of value addition by the regulatory authority while conducting surveillance inspection/quality assurance. As a government-appointed authority, the regulatory authority carries out its functions on behalf of the interested party, i.e., mainly on behalf of the customers and the citizens of the country at large. Therefore, through product and process audits, regulatory authorities can contribute a lot more to help the organisation fulfil its responsibility for meeting the requirements of the product to the satisfaction of customers, stakeholders, and interested parties.

It is like the regulatory involvement in production. When the regulatory authority uses its control or does not allow certain stages, they say - "Putting spokes in production's wheel". Knowingly or unknowingly, in the heat of production pressure, the production agencies hurry up on certain things.

Clearance of stages of manufacture by inspection is the central part of the quality assurance functions carried out by the regulatory authority for quality assurance (military aircraft). Most of the clearances are made without any observations. Ego and other undesirable personal traits have rarely created a situation of fear instead of friendly cooperation between the agencies. At extremely poor use of authority, "Is everything okay? I am busy now. I will come and see the stage next time", was also heard, signing off on the bureaucratic chair, flowing along without effort'.

Experience gained from the investigations of accidents, incidents or defects of fighter aircraft and associated systems, equipment and accessories reveals that the sources of defects, errors, mistakes and aberrations were due to design deficiency, material imperfections, heat treatment and special process deviations, assembly/test errors, quality control lapses, operation and maintenance aberrations. There are hundreds of examples of failures, defects and mistakes revealed during investigations carried out in the past. Every finding points to one or more instances of non-compliance with a requirement specified in the aerospace Quality Management System standard.

On a review of all the results of the technical investigations carried out in the past, where the cause of the event was identified, it could be seen that the root cause identified was due to the gap, aberration or mistake or technical error made in compliance of the quality system requirements. To reiterate the facts, out of many examples, a few are indicated in the table, the details of which are available in the history of Indian Military Aviation and known to the interested parties.

Every accident/defect investigation finding is an input for system/product improvement. The interested parties can update their professional knowledge based on the feedback to prevent the recurrence of the defects. Evidence leads to findings. If there are findings by the investigators, there are recommendations for improvement. If there is no finding, no recommendation is to be made. Recommendations and exhortations such as 'to be more careful', 'technicians have been advised', etc., may be avoided as nobody commits an error with an intention.

Lessons learned from defect investigations and failure analyses are valuable tools for product improvement and higher reliability. Strict compliance with the verified design or manufacturing/maintenance requirements is vital for ensuring product quality.

Performance history of aircraft, findings in various defect investigations and accident investigations completed in the past reveal that all the processes followed for design, development, manufacture, maintenance, repair or overhaul were not fully effective; it means that the planned activities were not fully realised or/and planned results were not fully achieved. The absolving statement that the auditors make at the end of the audit – that the findings are from a selected sample and not of the whole organisational processes – cannot be applied to the inspection or quality assurance process by the organisation. Every product requirement has to be satisfactorily met before clearing the stage or final assessment of the ASD equipment for delivery to the customer. Organisations cannot share the responsibility for quality with the regulatory authorities while explaining the lapses or errors to the customers. The design/manufacturing/maintenance organisation alone is responsible 'to verify that the product and service requirements have been met' (Clause 8.6) before releasing the product to customers after their respective works. 'The organisation is also responsible for the conformity of all externally provided (outsourced) processes, products and services, including from sources defined by the customer' (Clause 8.4.1).

Examples of Investigation Findings, Technology/Process/Product Audit Findings						
Sl No.	Technology /Process / Product	Number of failures noticed	Obcerved Defect/Devietion	Requiremen ts AS 9100 D Standard Clause / Reference	Objective Evidence/ Technological Requirements	Failures noticed after implementing corrective ections (verification of effectiveness) (1996-2000)
2	Cast iron seeling Rings - Aero-engines	56 Nos	copper molybdenum coating thiclmess 6-15 microns	8.5.1.2	copper molybdenum coeting thicimess 80-120 microns	3
2	Cest iron seeling Rings - Aero-engines	56 Nos	Updated drarring not used	7.5.3	Process spectficetion /drewing obsolcte	1
3	Chromium plating: electrolyte temperature cariotion permitted was 2 deg C & the Thermometer used had a least count of 10deg C	Process errors	Measuring equipment used had e least count of 10 deg C, whereas the required minimum mearurement wes 2 deg C	7.1.3.1	The organization shall ensure that the resources provided are cuitible for the specific type of monitoring and measurement activities being undertaken and are mainteined to ensure their continning fitness for their purpose	Nil
4	Hydroulic pipe line installetion on aircraft Rubbing was noticed between e hydroulic pipeline & a wire braided hose.	Rubbing was noticed between a hydraulic pipeline & a wire braided hose. (Rupture happened in the past after similar rubbing)	Adequate clearence not maintained between hydraclic pipeline & a wire braided hove.	7.5.1	The availability of evidence that all preduction and inspection/verification operations have been completed as planned, or as otherwise documented and authorised; Installation instructions not rollowed.	Defective hydreolie pipe line was replaced with serviceable one. 'P' clamp was provided on the braided hose assembly. One time check was carried out on all aircraft during mejor servicing. Subject hoses were secured with 'P' clamps as corrective action.
5	Fatigue failure of dowel bolt on MBG engine main bevel gear/spur gear assembly (six bolts) modified to seven bolts arsembly.	9	Seven bolt essembly wes a design modification/ change which wes not validated for the service life	8.3.6	Design changes to ensure that there wes no adverse impact on conformity to requirements	Correct solution was by 80-100% blue contact on flat surfaces.
6	Aircraft boild records	Records not available for verifications, when required	Retention and disposition of documented information was not defined.	7.5.3.2	Records not retained for any specific period	Condurive evidence not found during investigations

Figure 10.2 Examples of investigation findings, technology/process/product audit findings.

11

Auditing

This chapter explains the types of audits and principles, auditing methodologies and certification procedures.

There are three types of audits:

1. Internal audit (first-party audits): The process of internal audits and the role of internal auditors are described here
2. External audit
   a. (Second-party audits): How do the auditors perform the audit?
   b. (Third-party audits): What is the role of certification bodies? What does the organisation have to do to face the audit, and how do external auditors perform audits?

The organisation's primary objective while implementing the 'Quality Management Systems – Requirements for Aviation, Space and Defence Organisations' is business excellence. The practices aim to improve quality, on-time delivery, customer satisfaction and cost-effectiveness by applying the best practices and requirements indicated in the Standard. It motivates adopting a Quality Management System and is considered an organisation's strategic decision. Aviation, space and defence equipment are designed, manufactured and maintained with much interdependency among organisations' supply chains worldwide. IAQG envisages the Quality Management System to establish standard requirements for use at all levels of the supply chain. Therefore, compliance with the requirements of the Standard has also become a customer-driven requirement for all organisations, in keeping with the policy of IAQG.

Compliance with the requirements of the Standard and the process effectiveness are verified during internal audits in an organisation. Further improvements are brought out in the organisation through management reviews, risk management and corrective actions. Internal audits and management reviews are a form of self-verification, review and self-evaluation.

External audits are independent, systematic and third-party verification for conformity/compliance with the criteria and effectiveness. Customers depend on the results of external audits to gain confidence in the quality system implemented in the organisation. It is the confidence that the products/services delivered by the organisation will likely meet the requirements of the customers and the product. External/third-party audit is a means to achieve customer certification/registration and recognition. To earn certification, an organisation has to implement the requirements of the Quality Management System standard and simultaneously deliver products on time that satisfy customer expectations.

External audits happen in a strictly controlled process of certification to AQMS standards (AS/EN/JISQ 91xx series of standards) as defined by AS9104/1A Requirements for Certification of Aviation, Space, and Defence Quality Management Systems, including direct oversight of the AQMS certification process by Industry Members of the IAQG. Certification body-approved auditors carry out assessment and audit of the Quality Management System implemented in the organisation and report two things – the system conforms to the requirements of the Standard and the system is effective. Suppose there are nonconformities, or the system is ineffective. In that case, the organisation can address the nonconformities and implement corrective actions. Once the corrective actions are implemented and accepted by the certification body, certification of the organisation is recommended.

Customers and 'Industry Controlled Other Party' (ICOP) organisations view third-party certification as satisfactory compliance with the Standard in an organisation. OASIS registration through certification to AS/EN/JISQ 91xx series of standards makes the organisation visible to all the existing and prospective customers who desire to do business with the organisation.

The process of certification to AQMS standards (AS/EN/JISQ 91xx series of standards) is further controlled as per AS9104/1A. Requirements for Certification of Aviation, Space, and Defence Quality Management Systems under direct oversight by Industry Members of the IAQG. AS9104/1A has three separate sections:

- AS9104/1A Requirements for Certification of Aviation, Space, and Defence Quality Management Systems

There are four basic objectives for any audit. They are:

a. Conformity to the criteria of audit (Reference standard, Organisation's QMS documentation, requirements),
b. Compliance with statutory, regulatory and customer QMS requirements, contract requirements
c. Ensuring the effectiveness of the Quality Management System and
d. Identifying opportunities for improvements.

Certification is a status symbol, like the crown on the head of the queen on official occasions or a medal won by a winner. It is a visible sign of their possessions or credentials at that moment. For an organisation, AQMS certification is the sign/credential for effectively implementing the Quality Management System. The certification could be described as an outward sign of the inner grace of complying with the Quality Management System requirements. If the compliance is effective, the organisation achieves excellence and continues to grow in business and reputation. Certification is not the end of the implementation or practice of the Quality Management System. It does not guarantee that the products delivered by the organisation will all conform to or meet the requirements. The organisation is responsible for ensuring conformity with the requirements and ensuring effectiveness in meeting customer expectations. The certification body assesses the same, on a sampling basis, for granting the certification. It does not necessarily guarantee that the organisation will consistently achieve one hundred per cent product conformity, even though it should be a permanent objective. Certification also does not imply that the organisation is providing a superior product. The product is not certified as meeting the requirements.

Third-party certification bodies, after assessment, issue certificates to the organisation on satisfactory completion of an audit process. Certification body auditors perform the audits in all the organisational functions/procedures. All the audits are always based on sampling within an organisation's management system. It is not a guarantee for the organisation to claim one hundred per cent conformity with the Quality Management System requirements or a claim that the product has met the quality requirements.

International Accreditation Forum, Inc. governs and controls all audit processes and programmes for the certification bodies that provide conformity assessment services. Such certification facilitates trade/business among organisations and customers as the certificate of compliance provides confidence to the customers that the requirements for the product will be met. The main challenge for all organisations is to maintain the established Quality Management System with a focus on achieving the organisation's objectives, including the quality of the products, and not merely maintaining the certification.

Shortfalls in the organisation's performance and the quality of its products result in customers' loss of confidence. Certification can be viewed as the approval from the customers to an organisation since the certification body conducts an audit for registration/certification and maintenance of the certification as if it is conducted on behalf of a typical customer. It is not on behalf of any existing or potential customer or supplier; therefore, the certification assessment is always a third-party, independent assessment.

The 'International Accreditation Forum, Inc.' (IAF) operates programmes resulting from organisations' certification and maintenance. Accreditation reduces the risk for business by assuring the prospective customers that accredited 'Conformity Assessment Bodies' (CABs)/Certification Bodies (CBs) are competent to carry out the audit within their specified scope of accreditation. 'Accreditation Bodies' (ABs), which are members of IAF and their accredited CABs/CBs, are required to comply with appropriate international standards and IAF mandatory documents for the consistent application of those standards, in addition to meeting the requirements as per AS9104/1A Requirements for Certification of Aviation, Space, and Defence Quality Management Systems under direct oversight by Industry Members of the IAQG.

The International Standard Industrial Classification of All Economic Activities (ISIC) is the international reference classification of productive activities. United Nations Department of Economic and Social Affairs has published the International Standard Industrial Classification of All Economic Activities, Revision 4, 2008. (ISBN: 978-92-1-161518-0). Accreditation bodies have also made such Industrial Classification (scope sector) of All Economic Activities. The certification body should have approval in the following scope sectors/industry groups from the accreditation body to conduct an assessment/audit of the industry organisation.

Refer to the Tables given in 'The International Standard Industrial Classification of All Economic Activities (ISIC) Rev 4 '(UNITED NATIONS PUBLICATION Sales No. E.08.XVII.25 ISBN: 978-92-1-161518-0) (Freely downloadable from the website https://unstats.un.org/unsd/publication/seriesm/seriesm_4rev4e.pdf").

The NACE code is a statistical classification system that organises economic activities into a hierarchical structure. It is used by European Union member states to collect and present statistics on economic activity). The ANSI National Accreditation Board (ANAB) code classifies organizations based on the management systems they use, the products and services they offer, and the personnel they employ.

The ISIC code uses a hierarchical structure. The categories at the highest level are called sections. The first two digits of the code identify the division, the third digit identifies the group and the fourth digit identifies the class."

The International Aerospace Quality Group, Other Party Management Team (OPMT) accepts that for the management of any extraordinary events or circumstances affecting accreditation bodies, certification bodies and certified organisations, information and communication technology (ICT) can be used to support and maintain the integrity of the audit/assessment process. Modern ICT optimises assessment and audit and their effectiveness and efficiency.

IAF MD 4: 2018 is the IAF mandatory document for using ICT for up to 100% remote auditing and assessment purposes. It includes software and hardware such as smartphones, handheld devices, laptop computers, desktop computers, video cameras and others to gather information.

Levels/tiers of an organisation in the supply chain are indicated in Figure 11.1 Supply chain (Tiers 1–6)".

IAQG Structure: The IAQG – International Aerospace Quality Group – is a cooperative global organisation of prime aerospace Original Equipment Manufacturers (OEMs). They provide aviation and space products, services and land and sea-based systems for defence applications. IAQG Sectors are a sub-structure of the IAQG. There are three regional sectors for IAQG. They are Americas Aerospace Quality Group (AAQG), Asia-Pacific Aerospace Quality Group (APAQG) and European Aerospace Quality Group (EAQG). In most cases, the sector representative is part of the IAQG oversight team.

Numbering system for the Aerospace Quality Management System Standards developed by IAQG is as follows:

- International Standards – 91xx, planned for harmonisation across all three aerospace sectors and are recognised globally
- Americas Standards – 90xx, published for use by AAQG
- 'AS' Standards – Americas, Published by Society of Automotive Engineers,

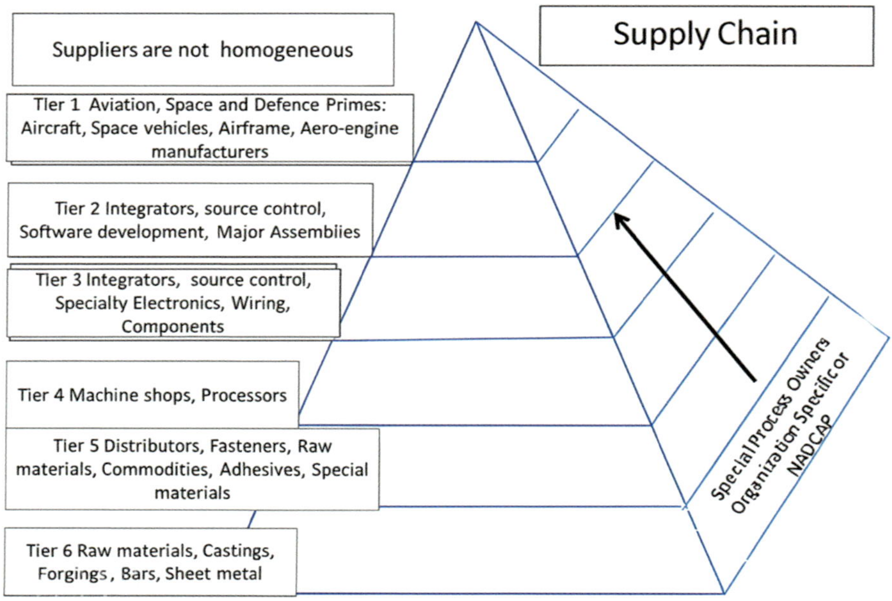

Figure 11.1 Supply chain (Tiers 1–6).

- 'EN' Standards – Europe, published by AECMA, European Association of Aerospace Industries. (AECMA, Association European des Constructeurs de Materiel Aerospatiale)
- 'JIS Q' or 'SJAC' Standards – are the Japanese/Asia-pacific equivalent.

One accreditation body insists that when quoting the relevant aerospace standard in audit reports, the revision is a year, not a letter. (Example 9100:2009 rather than 9100 C)

AS and EN, the primary AQMS standard shall be referenced as EN9100: 2016. However, the AS standard is also required to be referenced by the customer, then it shall be written as:

- AS9100 Rev D (technically equivalent to EN 9100:2016/2018 and JISQ 9100:2016/2018) or
- EN 9100:2009, AS9100C, JISQ 9100:2009 or
- JISQ 9100:2009/EN 9100:2009/AS 9100C.

The form AS/EN/JISQ 9100:2009 or similar shall not be used.

AS9104/1A Requirements for Certification of Aviation, Space, and Defence Quality Management Systems introduces several new things about how the AS9100 series certification programmes are administered. One of them is, for example, how the number of audit days is estimated. The certification bodies will have to follow the given table very carefully. (Refer to Annexure G: AS9104/1A Requirements for Certification of Aviation, Space, and Defence Quality Management Systems Table 2-Audit Duration Requirements.)

The number of person-days required depends on the number of employees and the complexity of processes in the organisation. The certification body has documented procedures for determining audit time. The certification body determines the time needed to plan and accomplish a complete and adequate audit of the Quality Management System by considering various aspects apart from the organisation's strength. The requirements of the relevant management system standard, size and complexity of the organisation, technological complexity, regulatory authority involvement, special processes carried out, outsourcing of activities included in the scope of the management system, the results of any prior audits and the number of sites included in the scope of certification. An approximate minimum number of man-days required for certification/annual surveillance/renewal audits are indicated for guidance. Audit time required for surveillance audits is approximately half the number of man days of the certification or renewal audit for a three-year certification cycle.

IAQG develops standard requirements and guidelines for quality improvement by the aviation, space and defence industry. In keeping with the vision of IAQG and providing greater emphasis on following the process approach, AS 9101 REV G has been published, which supports the engagement and evaluation of an organisation's Quality Management System process approach, as required by the AS 9100 series standards.

American National Accreditation Body (ANAB) is recognised by an IAQG Sector primarily responsible for accreditation certification bodies/registrars issuing certifications/registrations to Quality Management System standards. OASIS – Online Aerospace Supplier Information System is the website's name that would represent the online application of a comprehensive assessment data repository representing the output of business processes undertaken by IAQG. OIN – OASIS Identification Number is a unique tracking number assigned to all OASIS organisations. Once assigned, this number does not change.

An IAQG sector recognises an Auditor Authentication Body (AAB). It is primarily responsible for auditors' approval of aerospace Quality Management System standards. A national auditor authentication panel approves auditors working for one or more approved certification bodies/ registrars.

Aerospace experienced auditor qualifications and approvals: There are two grades of auditors for certification audits of aviation, space and defence sector organisations: Aerospace experienced auditors and aerospace auditors. The length of industry experience an auditor possesses makes the difference. Competency requirements (education, training, work experience and audit experience) of auditors for certification audits are indicated in the following guideline standards:

- ISO 19011:2018
- ISO 17021:2018
- AS9104/1A Requirements for Certification of Aviation, Space, and Defence Quality Management Systems

A summary of the auditor qualifications and competency requirements (competency is the demonstrated ability to apply knowledge and skills) are as follows:

Competency	Aerospace experienced auditor	Aerospace auditor
Education	Graduate	Graduate
Work experience in quality	Four years of aerospace industry experience within the prior ten years	Less than four years of aerospace industry experience
Auditor training	Lead auditor training in ISO 9001 (40 hrs) Lead auditor training in AS 9100 (40 hrs) OR Aerospace auditor's transition training – (AATT) Authorised by IAQG	
Audit experience	Four audits of a minimum of 20 days within the last three years for ISO 9001, witnessed by a lead auditor Two full AS 9100 audits within the last three years, witnessed by an AIEA	

Requirement of Continuing Professional Development: 15 hours of CPD and four audits every three years. Details may be obtained from auditor authentication bodies approved by IAQG.

11.1 Type of Audits and Principles

The audit is a systematic, independent and documented process for obtaining and evaluating evidence objectively to determine the extent to which the audit criteria are fulfilled (Refer to ISO 19011:2018).

Auditing is a basic tool to assess effective implementation and conformity to Quality Management System requirements like AS 9100 standards. In addition to determining conformity, the audit process evaluates the effectiveness of the Quality Management System and its associated procedures. ISO 19011:2018 provides guidelines for the auditing management system. IAQG has prescribed an auditing tool/template to be used by all external auditors and certification bodies via AS9101G Quality Management Systems Audit Requirements for Aviation, Space, and Defence Organisations. This document standardises requirements for conducting and reporting the Quality Management System audits. It provides requirements for an audit and reporting process based on the methodology and continual improvement approach defined in the AS 9100-series standards. The template uses common audit tools and permits uniform, transparent, standardised reporting of audit results. The standards can be used by aviation, space and defence organisations at all levels throughout the global supply chain. The standards can also be used for internal audits in an organisation.

a. Internal audits, sometimes called first-party audits, are conducted by internal auditors on behalf of the organisation itself. The findings in an internal audit are considered for management review and other internal purposes of evaluating the effectiveness of the Quality Management System. The information may form the basis for an organisation's self-evaluation of conformity and effectiveness. When the organisation's staff members carry out the internal audits, they demonstrate independence by the freedom from responsibility for the activity being audited.

b. External audits include those generally termed second- and third-party audits. Second-party audits are conducted by stakeholders interested in the organisation, such as customers or by others on behalf of the customers. An audit on the external providers/suppliers carried out by the organisation is also a second-party audit. Third-party audits are organised by Certification Bodies and performed by competent and approved auditors. The Standards govern external third-party certification audits for aviation, space and defence organisations – ISO 19011:2018 gives the guidelines for Auditing Management systems and ISO 17021:2018 (for ISO 9001:2015 certification), AS9104/1A Requirements for Certification of Aviation, Space, and Defence Quality Management Systems and AS 9100 series standards (AS 9100 Rev C/ AS 9110 C/AS 9120 B). Since all AS 9100 series standards include ISO 9001:2015 requirements, aerospace certification bodies must also follow ISO 19011 and ISO 17021 standards. Suppose there is any conflict or difference in the stated requirements between the AS 9101 REV G and ISO 19011: 2011. In that case, IAQG has resolved that the former shall take precedence.

Audit criteria are a set of policies, procedures or requirements established and implemented in an organisation as documented information (i.e., the quality manual, documented processes, departmental manuals, work instructions and similar internal documents (ISO 19011:2018 gives the guidelines for Auditing Management systems). Criteria should also include applicable statutory and regulatory requirements, Quality Management System requirements from customers, contractual requirements or industry/business sector codes of conduct. Audit criteria are used as a reference against which audit evidence may be indicated or compared during audits. For internal audits of the Quality Management System, the Standard AS 9100/9110/9120 is the audit criteria according to which the policies and procedures are established, documented and implemented. During product/process audits, product/process characteristics described in drawings, specifications and other documents are used as the audit criteria. Aviation, space and defence organisations can use AS 9101 REV G templates for internal audits, apart from the AS 9100 series standards as the criteria. A customised checklist/template would be desirable for internal auditors since the internal audit criteria are specific to the organisation.

Audit evidence is records, statements of fact or other information relevant to the audit criteria and verifiable (ISO 19011:2018 gives the guidelines for auditing management systems). Checklists, questionnaires used for audits, notes, documents and records of objective evidence are also verified by the internal or external auditors during the audits. When required, anyone can verify this evidence to ensure its availability; therefore, the evidence is called objective evidence. The audit evidence that the auditor may verify can be qualitative or quantitative.

Audit findings are the results of evaluating the collected audit evidence against audit criteria. Audit findings can indicate either conformity or nonconformity with audit criteria or opportunities for improvement. (ISO 19011:2018 gives the guidelines for auditing management systems).

A nonconformity report (NCR) is documented information (a record) retained for stating results and providing objective evidence of nonconformity against audit criteria, including the information on containment of the nonconformity/correction, identified root cause, implementation of corrective action and verification for effectiveness of corrective action and final closure.

A process effectiveness assessment report (PEAR) is a documented information (record) stating results and providing evidence of determination of the effectiveness of the process.

Principles of auditing are evidence-based conformity or nonconformity, ethical, fair, independent and professional. These principles make the audit an effective and reliable medium, providing information on which an organisation can improve its performance. The desirable auditor qualities are:

- diplomatic, i.e., tactful in dealing with people;
- observant, i.e., actively aware of physical surroundings and activities;
- perceptive, i.e., instinctively aware of and able to understand situations;
- versatile, i.e., adjusts readily to different situations;
- tenacious, i.e., persistent and focused on achieving objectives;
- decisive, i.e., reaches timely conclusions based on logical reasoning and analysis and
- self-reliant, i.e., acts and functions independently while interacting effectively with others. (ISO 19011:2018 gives the guidelines for auditing management systems para 7.2 Personal attributes)

These are specific personal qualities required for anyone engaged as an internal or external auditor. Internal and external auditors must adhere to the principles as a pre-requisite for the audit activities, including arriving at similar conclusions in similar circumstances. Apart from the competency to conduct an audit, the above qualities, which can be summarised as qualities of a good citizen, are expected from all internal and external Quality Management System auditors. All competency requirements of auditors and training imparted to auditors ensure these principles.

11.2 Audit Methodologies

Basic principles of QMS, vide para 0.2, are important in supply chain management, any trade or business, especially in aviation, space and defence industries. The principles guide the application of Quality Management Systems even after revisions and additions. All quality management standards promote adopting a process approach when developing, implementing and improving the effectiveness of a Quality Management System to enhance customer satisfaction by meeting customer expectations and requirements. The standard requirements aim to achieve long-term, sustainable success for an organisation through two vital aspects of conformity and effectiveness (CandE) of the Quality Management System. When implemented, planned activities and processes meet the requirements of the Quality Management System and the customer requirements. That is conformity. Planned results, when measured and found, achieved as targeted, meet the quality of the product/services and on-time delivery expectations of the customer and achieve the defined objectives of the organisation. That is effectiveness, described as 'the extent to which planned activities are realised and planned results achieved' (ISO 19011:2018 gives the guidelines for Auditing Management systems).

These two vital aspects of conformity and effectiveness of the processes in the Quality Management System are verified through any audit. The process approach methodology of audit helps evaluate the conformity to quality system requirements. The process approach helps find the effectiveness of meeting product quality and on-time delivery performance of products and achieving the objectives and performance goals of the organisation. If the organisation uses a process approach to manage the processes in the organisation, using the process approach is the ideal methodology for evaluating conformity and effectiveness of the Quality Management System in the organisation.

1. Audit customer focus: Customer focus and customer feedback are audited by verifying actions taken by the organisation based on the available performance data related to on-time delivery, after-sale service, corrective action requests, results of customer satisfaction surveys, nonconformity, complaints received regarding the product quality, responsiveness to internal and external customer requests, etc. Organisations should have adequate procedures to deal with and manage these processes. The ability to make meaningful decisions and implement actions to resolve issues effectively shows the overall effectiveness of the Quality Management System and organisational excellence in satisfying the customers.

2. Audit top management (leadership): Top management is interviewed to evaluate the continued suitability of the organisation's quality policy and objectives. Management commitment to the customers, revealed by actions taken on nonconformity to quality of products (internal and external), customer performance expectations and on-time delivery performance are examined. Other points examined are how the organisation has established measurements of their documented objectives and actions taken to improve the Quality Management System's performance and effectiveness.

3. Audit process management: Process performance and effectiveness, which directly impact the customers, are audited concerning the process map/planet diagram for the processes. Specific emphasis is given to measuring process objectives and organisational targets and any actions arising from not meeting the objectives/targets. Audit trails are pursued for customer concerns, corrective action requests, process controls, performance against objectives and continual improvements.
4. Audit process performance and effectiveness: Processes are audited in detail to ascertain whether the processes can achieve the planned results and performance levels. Identification of processes, sequence and interactions, responsibilities, process controls, availability of information to operate and support activities and competency of people are evaluated. Process measurements and analysis of results are evaluated through performance information, percentage of nonconforming parts and on-time delivery. This is to determine effectiveness, continual improvements and efforts to achieve desired outcomes.
5. Audit Quality Management System performance and effectiveness: Procedures for handling customer complaints, including products returned from customers, customer feedback data and other relevant data such as customer survey results, are examined with the management representative. Results of internal and external audits, feedback from regulatory authorities, records of management reviews, preventive actions and their effectiveness, records of actions taken to address process/product nonconformities, their corrective actions, etc. are verified. The organisation's current performance data against set targets, including customer-specific targets and the follow-up actions taken to address any shortfalls, are reviewed during the audits.
6. Audit continual improvement: Continual improvement of the quality system is evaluated during audits by focusing on conformity and effectiveness of processes and issues, which are essential to the organisation, their customers and regulatory authorities. The organisation needs to have a structured approach for continually improving process performance. Audit trails are found from internal audits, customer complaints and management reviews. Improvement actions are evaluated based on audit findings, customer satisfaction data, process performance, problem resolutions, benchmarks and lessons learned. Evaluating the effectiveness of corrective activities and deploying actions where desired results are not achieved are indications of the organisational effectiveness for continual improvement.

The process approach is one of the most effective ways of evaluating an organisation's Quality Management System. The basic inputs to the processes are the identification and documented information (as required) of all the related processes, responsibilities and resources. Steps or activities are the implementation of the processes. Output of the processes is the results achieved, including the products delivered to internal or external customers. Verifying adequate controls for the inputs and steps is necessary to achieve the desired outcomes. Suppose the results, when measured, meet the requirements of the Quality Management System requirements, product/process quality and test requirements, process/product objectives, customer requirements of meeting special requirements and on-time delivery, the functional/departmental and the organisational objectives. In that case, the process is confirmed as effective. This audit methodology is planned and formulated from the process map/planet diagram for all processes in the organisation. The process-based audit pursues trails by following actual products, customer orders and related documentation (e.g., customer contracts, drawings, shop orders, inspection records) through the organisation's product realisation and related management and support processes. The audit also evaluates whether the organisation's processes can achieve planned results and performance levels, including applicable customer-specific requirements and targets. The Quality Management System effectiveness, product quality and customer satisfaction are interrelated, which will be reflected in the evaluation/audit findings. It is emphasised here that an effective internal audit conducted in the organisation following the above methodology will bring out findings and results similar to those of a certification audit, which will benefit the organisation for effective implementation and excellent performance.

11.2.1 Certification audit process

Assessment or audit is an evaluation carried out by an auditor at the (supplier) organisation's facilities in accordance with the IAF/ISO/IAQG requirements for certification. Both expressions of assessment and audit are used interchangeably, carrying similar meanings. To gain certification/registration, an organisation must conform to the Standard's Quality Management System requirements while delivering products that meet and satisfy customer expectations. Both conformity to the requirements and effectiveness in achieving the results are evaluated for all the processes, especially for product realisation processes, during all audits.

Certification audit processes and the organisational preparedness/ readiness required to respond to the needs of each of the stages of the audit processes are as follows:

A. Pre-audit activities: The organisation should prepare for the certification audit. Identify the scope and applications with respect to the quality manual and the quality system implemented for certification separately. Exclusions, if any, claimed, are to be with supporting justification. Collect information regarding the percentage of revenue for the scope of business activities, the total business and the number of employees (i.e., full-time, part-time, contract, temporary) working for the scope among the entire workforce. Identify the number of shifts specific to production and/or maintenance activities.

Identify information regarding the statutory and regulatory requirements and their compliance status for the scope of the Quality Management System and processes. Registration, license, environmental clearances from the government, legal aspects of the business operation, safety measures, etc., are some of the legal requirements. Maintenance organisations need to identify the regulatory authority approvals and recognitions obtained for the scope of activities. Any limited approvals, probation, suspension or withdrawal of already granted approvals, etc, if any, are to be brought out.

Suppose the organisation does not wish to include any customer-restricted areas in the scope of certification. In that case, the audit process does not have the areas. Vice versa, if the audit process is not completed for any product/service realisation area or any required information, which may be proprietary and confidential, is not to be provided to the certification auditors; the same may be excluded from the scope of certification.

The regular presence of customer or regulatory authority representatives at the organisation to provide guidance or quality assurance activities needs to be brought to the notice of the certification body.

B. Stage 1 audit: The certification audit commences with the stage 1 audit. Pre-certification visits conducted by auditors earlier have been replaced with stage 1 audits with the implementation of ISO 17021/AS9104/1A Requirements for Certification of Aviation, Space, and Defence Quality Management Systems AS9101G. It is mandatory for the certification body/ auditor that at least part of the stage 1 audit be carried out at the organisation's premises in order to collect adequate information and commence direct communication by the auditors with the organisation. This helps the certification body understand the organisation's business and processes and prepare an audit programme. The organisation as a whole has to be ready with the following information and compliance status, which, when compiled by the management representative, will serve as a checklist for the readiness of the organisation to face the certification audit:

1. Address the organisation's location(s) and confirm their general conditions, housekeeping, etc. are satisfactory. Ensure that special work environments for production/maintenance areas, if any, such as clean rooms, electrostatic discharge sensitive areas and x-ray radiation areas, are kept according to the requirements.

2. Provide the following Quality Management System documentation for review:

    a. Quality Manual, Documented Information (Procedures), Description of processes showing their sequence and interactions (Example: Sun and planets diagram or business process map) and list of any outsourced processes. Identify the special functions performed within the organisation or outsourced to external parties.

    b. Confirm that the requirements of the applicable AS 9100/9110/9120 standard are addressed in the organisation's documented information.

    c. Ensure that internal audits have been planned and performed. Records of internal audits of all processes, including statutory/regulatory/customer Quality Management System requirements and contract requirements with customers, are available. Results of external audits carried out by existing and prospective customers, regulatory authorities and certification bodies, if any, should be available, in addition to the records of audits of suppliers carried out by the organisation.

    d. Make the records of the latest management review available. Ensure that the findings of the internal audits are discussed in the management review and the management decision substantiates the management system's implementation level.

    e. List the top five or more aviation, space, defence and other customers with whom the organisation operates. Customers requiring AS 9100/9110/9120 standard compliance in the organisation and any other customer-specific Quality Management System requirements may be identified as such, customer-wise. Indicate the volume/amount of their business with the organisation. Customer-specific Quality Management System requirements generally are the 'First

Article Inspection (FAI)' requirements (production process verifications), quality records to be created and maintained by the organisation, coordination of document changes; defined special requirements/critical items/key characteristics; approval of design changes by the customer; flow down of requirements to sub-tier suppliers; customer notification of production process changes; traceability; handling of nonconformities, control for software, control for customer supplied data/documents, etc. Some customers include other IAQG Quality Management System standards as applicable in contracts. Examples are AS 9115: Quality Management Systems – Requirements for Aviation, Space and Defence Organisations – Deliverable Software, AS 9131 Rev C: Quality Management Systems – Nonconformance Documentation, AS9102 Rev C: Aerospace First Article Inspection Requirement and ARP 9107: Direct Delivery Authorisation Guidance for Aerospace Organisations. (See reference)

f. Documented information/documentation (both documents and records) as indicated in paras 7.5.1.1 and 7.5.1.2

g. Identify any customer-delegated verifications and materials review board (MRB)/salvage board authority,

h. Indicate the current organisation chart showing hierarchy and reporting structure.

i. Risk register-identified risks, assessment and control/management with periodic reviews during planning activity and operations, including main risks related to process, people, product and risk while selecting and using external providers/suppliers. Risk management checklists, Failure Mode and Effect Analysis (FMEA), etc. are examples which could be included in the risk register.

3. Maintain records of customer satisfaction determination, customer satisfaction survey scorecards and their trends, special status declarations granted by customers, awards won by the organisation, feedback from regulatory authorities or other interested parties, etc.

4. Maintained customer complaints register, including verifying customer reports, feedback and communication after resolving the complaints and suggestions.

5. Complete the quality objectives, organisational performance measures and trends for the previous 12 months. Include 'On Time Delivery' (schedule variance) and quality performance measures (meeting product requirements, effort variance, and in-process and final rejections/ rework). Maintenance organisations are also to indicate the safety objectives, measurements and trends.

6. The top management should be ready, during all stages of an audit, to explain the following when requested:

   a. The organisation's strategic decision is to follow the Aerospace Quality Management System for organisational excellence, continually meeting customer expectations and making efforts to improve performance goals.

   b. The organisation's commitment to all the stakeholders, mission, vision and strategy for the business.

   c. Establishment and continued relevance of the organisation's quality policy, objectives and periodic reviews.

   d. Establishment of performance measures aligned to the quality objectives; Performance measurements (key performance indicators) against set targets, with respect to the organisational goals, quality system performance and effectiveness and customer expectations (such as delivered product quality, on-time delivery performance, etc.). Explain actions taken to review periodically and initiate improvements if there is any shortfall in the achieved results. Steps taken to address the issues of not meeting the organisation's and customers' performance expectations must be shown.

   e. Identification of customer needs and expectations and efforts taken to translate them precisely into acceptable products and services within the organisation for delivery to the customers to their entire satisfaction.

   f. Method of communication about meeting customer requirements to the entire organisation and the effectiveness of the communication.

   g. Resources provided for realisation processes, including positioning competent personnel, competency mapping and their training, infrastructure and environment necessary for improvements in the operations.

h. Leadership taken to develop and implement the Quality Management System and efforts being taken for its continual improvement based on the current performance and effectiveness.

i. Organisational performance with respect to customer expectations such as quality of products/services, on-time delivery, supplier performance rating, business scorecard, key performance indicators (KPI), audit results and customer perception.

j. Periodic management reviews and reporting the results of the reviews to the concerned stakeholders.

7. Make a general assessment of the status of awareness among personnel within the organisation regarding meeting the requirements of the Standard in general. Awareness, in particular with respect to the quality policy, identification of key performance indicators and measurable objectives, core production/maintenance processes and operations of the management system and support processes.

8. A brief description of the following processes established in the organisation through documents, procedures, work instructions or checklists and their compliance records are to be made available:

a. Project/program management;

b. Configuration management, including planning, identification, change control, status accounting and audit processes. Arrangements established in addition to the identification and traceability of parts/assemblies during manufacture and/or maintenance are to be indicated.

c. Establish identification of special requirements/critical items, including key characteristics during design and development activities, production/maintenance activities and verification of quality requirements.

d. Establish methodology and programmes for prevention, detection and removal of foreign objects for control of foreign object debris/damage (FOD) regularly for all areas where production/maintenance activities and testing are carried out.

e. Indicate continual improvement activities have a structured approach for enhancing customer satisfaction, improvements in the organisational efficiency, improvement of the effectiveness of process/product quality, on-time delivery, supplier performance and organisation's objectives. Corrective and preventive actions completed for the products and services and the resulting trend can indicate improvements. Established procedures for verifying the effectiveness of corrective and preventive actions should also be available. Increases in customer base, product range, business volume and profitability also indicate improvements.

9. Make preparations to discuss the Stage 2 audit plan with the auditor. Confirm the audit programme and the related details. The audit programme may be similar to the one described for the internal audits in the organisation. Determine the interval between Stage 1 and Stage 2 audits. The period depends on the time and resources required to resolve areas of concern identified and communicated by the auditor.

10. Guide the audit team/auditor on a tour to all facilities and locations within the scope of audit/certification for familiarisation and to give a feel of the activities being performed.

11. The organisation should be ready to receive all communications, written or otherwise, from the visiting auditor. The auditor may not issue nonconformity reports (NCR) at this stage but shall indicate all matters of concern, issues and observations, which should be promptly acted upon, completed, verified and conveyed to the certification body before the Stage 2 audit. It is not advisable to ask auditors to give solutions to the concerns they express. Auditors will encourage the auditee to find appropriate solutions on their own by only exchanging understanding and awareness of the Standard requirements since they are not expected to provide consultancy or direct advice in finding solutions to the problems. The organisation's prerogative is to work out long-lasting solutions and substantial improvements by conducting root cause analysis and implementing appropriate corrective actions.

C. Stage 2 Audit: The organisation offers its Quality Management System to evaluate its conformity and effectiveness during the Stage 2 audit process. The audit programme gives the details of the audit process. All the elements of the Quality Management System and the associated organisation's functions will be offered for audit for conformity assessment and determination of effectiveness. All issues or concerns indicated during the Stage 1 audit must be satisfactorily resolved before proceeding with the Stage 2 audit. It is necessary to appraise the audit team about any revision in the plan/

programme due to organisational changes since the Stage 1 audit. The differences may be in the personnel, department/business unit reorganisation, any new developments regarding customer-related processes or scope or any other issues that may impact the audit process. The organisation should be ready with the following information, in addition to continuing with the normal functions:

a. Conformity to all requirements of the applicable Quality Management System standard as documented and practised regularly, the evidence of which is shown in the records maintained.

b. Compliance with all statutory and regulatory requirements within the Quality Management System and its performance.

c. Management responsibility for the organisation's functions shall be visible through the declared policies and links to performance objectives and targets, responsibilities, personnel competency, procedures, performance data, internal audit findings and continual improvement of all Quality Management System processes.

d. Monitor and measure all quality and performance objectives against the set targets, reporting to and reviewing the results at the top management level.

e. Controls are exercised for the organisational processes to meet customer requirements and any applicable legal requirements. In contrast, the procedures are being complied with.

Detailed audit findings, including reference to the audited processes, process documentation and associated records, are available to the organisation after the Stage 2 audit. The audit team will present the findings of the Stage 2 audit at the end of the audit process. In a formal presentation to the organisation, the audit team will indicate the identified 'nonconformity reports' (NCR) in writing, observations and the opportunities for improvement of the Quality Management System.

All nonconformity reports (NCR) are to be planned for actions, which include containment actions, corrections, root cause analysis and corrective action(s). Planned completion date(s) are also to be indicated. The time frame for the planned actions and completion is available in AS9104/1A Requirements for Certification of Aviation, Space, and Defence Quality Management Systems/1A Requirements for Certification of Aviation, Space, and Defence Quality Management Systems-1 or may be indicated by the audit team. It stipulates the requirements for the closure of nonconformities.

All the corrective actions need to be implemented, reviewed, accepted and verified within 90 calendar days of the date the nonconformity was raised. The audit team will verify NCR before closing the same, and then a recommendation for certification may be made by the certification body. It is for the organisation to decide actions on all other issues, observations and identified opportunities for improvement.

D. Surveillance audit: In a three-year cycle of certification validity, the organisation is subjected to periodic surveillance audits. The periodicity may be annual. The date of the first surveillance audit following initial certification must not be more than 12 months from the last day of the stage 2 audit. Surveillance audits ensure continued compliance and effectiveness of the Quality Management System requirements. The management system's scope covers representative areas, and functions are monitored regularly during surveillance audits. It also considers the changes, if any, to the certified Quality Management System and the organisation. Surveillance audits are not necessarily full system audits; individual audits must cover the whole system. During every surveillance audit, the audit programme mandatorily includes the following areas/processes, which confirm the continual management dynamics of the Quality Management System in the organisation:

a. Review of any changes to the management system and documentation,

b. Periodic Internal audits and management reviews, including management involvement and communication processes,

c. Effectiveness of the Quality Management System concerning achieving the organisational objectives. Evaluation of the effectiveness of the Quality Management System is indicated by the health and wellness of the established system revealed through effective internal audits. Measurements carried out by the organisation demonstrate achievements of customer satisfaction and organisation quality objectives.

d. Review of actions taken on nonconformities identified during the previous audit to ensure that the corrective actions taken are practical,

e. Treatment of customer complaints, feedback and determination of customer satisfaction levels,

f. Progress of planned activities aimed at continual improvement, corrective/preventive actions and analysis of data

g. Purchase process

h. At least half the number of the production/engineering service/MRO activities and their operational controls/processes.

E. Maintaining certification: The organisation can maintain certification by demonstrating that it satisfies the management system standard requirements. Certification can be continued till the validity of the certification period ends. The recertification process should start at least three months before the certificate's expiration so that the recertification process can be completed before the expiration date on the existing certificate.

F. Feedback process: Audit findings are available to customers as information. Feedback processes are available to customers on the certification process, and audit findings are brought out by the certification body. Feedback to the suppliers may be raised on the problems related to the quality of the product received by the customers from the certified supplier. Based on the audit findings, suppliers can raise feedback on product nonconformities to the supplier. This feedback intends to request the supplier to perform root cause analysis, take appropriate corrective action and improve the quality system. Customers can give feedback to the certification body when the product's nonconformity is related to possible system lapses, identified or not identified, during the audit process. Customers can also request the certification body to focus on the potential weak areas of the quality system. The certification body can give information on the audit findings to the customers to determine the possible impact on the product quality or any other actions to be taken by the customers in view of the audit findings.

IAQG oversight through AS9104/1A Requirements for Certification of Aviation, Space, and Defence Quality Management Systems permits that 'Accreditation Bodies, Other Party assessors, regulatory agencies, or customer representatives may accompany the audit team as observers of the audit process at any time. When customer or government representatives are accompanying the audit as observers, the audit team leader shall have the option of including any comments/concerns brought forward by these representatives in the audit report. Visitors who accompany the CB audit team shall be coordinated with the client before the start of the audit'.

11.3 Internal Audits Using 9100 Series Standards as the Criteria

Effective internal auditing adds value to the organisation's continual improvement efforts by improving the Quality Management System, product quality and customer satisfaction. An organisation's top management always wants to maximise the value internal auditing brings to organisational improvements.

Every organisation has strengths and weaknesses that enhance or inhibit its effectiveness in providing quality products and services. Indeed, in a proactive organisation, the weaknesses are likely targets of continual improvement efforts. However, organisations know that they need to constantly identify and improve upon the weaknesses that could limit their competitive position. To do this, an organisation has to create an environment of transparency where weaknesses are displayed for open problem-solving without fear of finger-pointing or retribution. It is not the individuals at fault. It is the processes and the system that need correction. Management commitment is required to solve the weaknesses of the processes and system and not identify who is responsible for the failures.

Ensuring that experienced internal auditors are instrumental in bringing effectiveness to an organisation's internal audit programme is necessary. This is where the auditor's information gathering, observations and interpretations of the Standard can add to the organisation's value. The internal auditor is the change manager between the organisational weaknesses and the quest for continual improvement for an effective overall Quality Management System that ensures the quality of products/services and satisfied customers. Internal auditor performance is often found impractical due to various constraints. Internal auditors do not possess the advantage of diverse field experience in conducting process-oriented audits that the external auditors include. Therefore, organisations sometimes have found that outsourcing internal audits to consultants can be productive and cost-effective. However, it is necessary to ensure that such consultants have the requisite exposure/experience in aviation, space and defence organisations as regular employees have the knowledge about the specific standards and their appropriate interpretations and application of auditing techniques using the AS9104/1A Requirements for Certification of Aviation, Space, and Defence Quality Management Systems AS 9101 E and AS 9100 series standards as the criteria. Ordinary consultants with QMS consultancy experience in aviation, space and defence organisations may claim expertise in the domain but are seldom effective in correctly interpreting the AS 9100 series standards while carrying out consultancy or internal audits, at least in the early years of their profession. Some consultants have documented areas like configuration management, risk management, outsourcing, work transfers, etc. quoting the standard requirements, using the verb 'shall' and not specifying anything the organisation has done to meet the requirements! Therefore, the organisation's personnel fail to implement the requirements correctly.

External auditors and qualified/experienced consultants in the highly regulated aerospace industry have learned the techniques that enhance the value of their service. The techniques are very useful and effective for internal auditing, too. Auditors should possess those personal attributes described under the principles of auditing to enable them to act in accordance with bringing benefit to the organisation. Senior experienced internal auditor needs to take up the role of a mentor by interpreting audit observations and findings in a way that adds value to the organisation through collective continual improvement initiatives. Experienced auditors are responsible for shaping an inclusive culture that encourages new auditors to participate effectively in the quality system improvement process to maintain the organisation's standard of excellence.

The auditors provide significant value-adding activity to the audited organisation during audits. Instead of simply looking for formal compliance with the requirements of the Standard, auditors look at the actual effectiveness of the management system. They discover the benefits of adopting the Quality Management System for the organisation and its customers. In other words, they try to determine whether the Quality Management System is applied to maintain and continually improve the organisation's performance. They verify that the organisation can deliver the expected results by meeting customer requirements/expectations and addressing statutory and regulatory requirements. For this purpose, internal auditors should have fundamental knowledge and understanding of the technical and managerial issues concerning the business and the industry sector in which the organisation operates.

Mentors must be able to guide other auditors to put process owners at ease so that the required information can be gathered from them during internal auditing. Role-play during internal auditor's training is helpful in this regard.

Internal auditors must know what information to look for, have the critical thinking skills to find and interpret it and learn how to extract information from their colleagues. Since the internal auditors need to gather information from people they work with in different areas, the rapport should not affect the independence of the audit process.

11.4 Audit Planning

AQMS certification audit is governed by ISO 19011, AS9104/1A Requirements for Certification of Aviation, Space, and Defence Quality

Management Systems AS 9101F and ISO 17021. Relevant ISO 17021 (paras 9.1.2–9.1.14) requirements are indicated below:

a. 9.1.2 – An audit plan shall be prepared based on the processes defined by the organisation and documented in the QMS Process Matrix Report. Relevant guidance is provided in ISO 19011. An eight-hour audit per day is the norm. Extra time must be planned to verify the effectiveness of the corrective actions taken for the nonconformities reported in the last audit and the night shift operations. To have an effective audit, the audit plan should be based on the sequence and interactions of the organisation's processes; the criticality of products and processes, including special processes; the risks associated with the product or process maturity, product-related safety issues, results of internal audits; previous audit findings (e.g., CBs, customers, regulatory authorities), performance measures and trends for quality and OTD (e.g., KPIs, scorecards, dashboards), previous management review results; customer QMS requirements; statutory/regulatory requirements (e.g., ITAR, RoHS), customer satisfaction/performance data; and the proportion of aviation, space and defence business each customer represents. Samples for core business activities are picked up in proportion to each customer's business.

b. 9.1.3 – CB appoints an audit team based on competence and as per relevant guidance provided in ISO 19011. Management representative or the manager in charge of conducting the internal audits in the organisation selects the trained auditors.

c. 9.1.4 – Determine the auditor time for an effective audit by the CB with justification to be recorded.

d. 9.1.5 – CB indicates multi-site sampling and documented rationale for the sampling plan, as applicable.

e. 9.1.6 – The audit team verifies the structure, policies, processes, procedures, records and related documents, determines processes are established, implemented and maintained effectively to provide a basis for confidence in the client's management system and communicates to the client any inconsistencies between client's policies, objectives and targets and the results. Internal auditors do the job in first-party audits.

f. 9.1.7 – CB shall provide the client with the name and, when requested, background information on each audit team member, time for the client to object and time to reconstitute the team.

g. 9.1.8 – Audit plan to be communicated as per the agreed dates/time.

h. 9.1.9 – On-site audit of documented requirements in accordance with the relevant guidance provided in ISO 19011, AS9104/1A Requirements for Certification of Aviation, Space, and Defence Quality Management Systems and AS 9101E. Internal audits can also use similar but customised formats for reporting.

i. 9.1.10 – CB shall provide a written report for each audit based on relevant guidance in ISO 19011 and may identify opportunities for improvement (OFIs) but not recommend solutions. Internal auditors should submit checklists with objective evidence and nonconformities in the relevant format.

j. 9.1.11 – CB shall require the client to analyse the cause and take corrective action to eliminate the identified nonconformity within a defined time frame. Follow-up actions after the internal audits are also similar.

k. 9.1.12 – CB/Auditors shall review the corrections and corrective actions to determine if these are acceptable.

l. 9.1.13 – CB/Auditors shall inform the client what is needed to verify effective correction and corrective action such as (i) additional full audit, (ii) additional limited audit and (iii) documented evidence that will be confirmed on future surveillance audits. The management representative or the manager in charge of conducting the internal audits in the organisation reports the findings/results to the management with trend analysis and ensures follow-up actions with the process owners for nonconformity and opportunities for improvement (OFIs).

m. 9.1.14 – CB shall ensure the decision for certification or recertification is made by persons or committees different from those who carried out the audits.

Internal auditors should practice auditing as per the audit methodologies described above. The process approach for auditing is the proven effective method. Focusing on the actual processes in the organisation, verifying their conformity and effectiveness and checking their ability to meet the quality and organisational performance objectives consistently give the best internal auditing results. Internal auditors should audit risk management within the processes, looking at the identified risk factors. It is necessary to discuss the process initially with process owners to learn about the process. Auditors need to pursue a line of questions, focusing on each function with respect to the process owners who are normally responsible for the risk management.

Interviews should be performed in such a way that the auditor should get an understanding of the process from the process owner. The auditor should not get side-tracked by superfluous details, which may not add value to the auditing process. Getting too much into the process details by the auditor also may not benefit the organisation since auditing is not meant to improve the auditors' knowledge. The auditor must recognise the process's inputs, stages/major steps and outputs and how the process is being realised. Keeping in mind the auditee's process description/process map, the auditor should be able to determine the importance of the process he is auditing. The auditor must keep a trail of the information provided by the auditee in the quality manual or other documentation where the auditee has defined the interaction of processes. Therefore, without getting into the nitty-gritty of the process, the auditor should ask the most important question of how the process is measured. How is the performance? Is there performance monitoring and measurement? What are the objectives and targets? Is it achieved during a specific period? If not, what actions are planned to achieve the objectives/targets? The answers to the questions above give the measured results of the process. If the planned outcomes are achieved, it more or less confirms that the process is effective. Verify whether there are any criteria for satisfactory operation and control and what methods are used for the operation and control of the process. It must be possible to check whether the process is carried out in accordance with the requirements of the Standard, especially the conditions given under planning and operations.

The auditor should take at least three samples to see if the practices properly reflect the descriptions in the document or statements by the process owner. While doing so, conformity to the Quality Management System standard and other requirements (procedures stated, documented or otherwise, work instructions, statutory/regulatory requirements, contract requirements, customer-specified QMS requirements, etc.) must be verified. If any practices are not according to the provisions, they can be identified as nonconformity after further verification.

It is also important that auditors/internal auditors not use jargon or expressions like 'I think', 'probably', 'in my opinion' and 'may be', etc. while doing audits because all such expressions indicate uncertainties and unverified matters. The auditing process has no room for such uncertainties and opinions or sharing the knowledge of the auditors or auditees. The auditing process only evaluates matters, facts and their compliance, established/laid down criteria/documents and their implementation. If the auditor is unsure about his knowledge or of the facts noticed, the checklist or standard carried by the auditor can be used as a reference. The auditor has no role to educate, correct or guide the organisation. At the most, the auditor can do that the importance of finding out the matter, truths, correctness or verification may be indicated to the organisation as a value-adding activity or identifying the opportunity for improvement. He can also say where the knowledge is available and the reference. The auditor, by his experience and wisdom, must know the solutions, otherwise, he cannot rightly point out a gap or aberration (nonconformity)!

The statement of nonconformity is the first to be identified and documented, followed by objective evidence. This is because an internal auditor will observe situations that they 'feel' may be a potential nonconformity during an audit, even though they may not be sure about it at that point in time. The competent auditor will then document the audit evidence for the potential nonconformity in the audit notes before pursuing additional audit trails to confirm its nonconformity.

The next step the internal auditor will need to take is to identify and record the requirements that are not being met. Since nonconformity is a non-fulfilment of a requirement, and the requirements of an aerospace Quality Management System are given in the Standard, a situation which does not meet the Standard requirement should be stated as nonconformity. Requirements can be seen from many established sources. They may be from the clauses of AS 9100 series standards, the organisation's management system documentation, the applicable statutory and regulatory requirements, or the quality system imposed by the customer. If the auditor cannot identify a requirement, then the auditor cannot raise a nonconformity. Hence, the statement should always be a negation of the requirement. If there is no audit evidence, there is no nonconformity. If there is evidence, it must be documented as nonconformity instead of being softened with another classification as 'observations', 'opportunities for improvement', 'recommendations', suggestions, etc. Softer classifications do not help the organisation derive the benefit of auditing. There will be a tendency to prioritise these matters less than the nonconformity statement for corrective action.

Classification of nonconformity as major or minor depends on the impact of the nonconformity in the process or across other functions within the organisation and whether any product is likely to be affected or has been affected. The classifications of nonconformities are reiterated here in the context of internal/external audits.

Major nonconformity is defined as a non-fulfilment of a requirement, which is likely to result in the failure of the Quality Management System or reduce its ability to ensure controlled processes or compliant products/

services. That is likely to affect the system or product and is a major nonconformity. Any one or more of the following situations warrant a statement of the nonconformity to be graded as a major one:

a. The effect of the nonconformity is judged to be detrimental to the integrity of the product or service.

b. The requirement of the Standard (clauses) or procedure or customer requirement is not at all addressed or met, which means there is a total breakdown of the Quality Management System.

c. The nonconformity identified would affect the product delivered to the customer or have already affected the product.

d. The nonconformity could cause product failure or reveal a deficiency in quality or performance during use.

Minor nonconformity is defined as 'non-fulfilment of a requirement, which is not likely to result in the failure of the Quality Management System or reduce its ability to assure controlled processes or compliant products/services'. That which is not likely to affect the system or product is a minor nonconformity. But it is a gap, aberration mistake or human error. The situation warrants a statement of nonconformity to be graded as minor when there is a single system failure or lapse in meeting the requirement of the Standard (clauses) or procedure established by the organisation. The requirement was established and documented, but there was an identified lapse in compliance during the auditing. A number of minor nonconformities against one requirement can represent a total system breakdown and thus be considered a major nonconformity. Similar minor nonconformities identified in different processes/sections/departments/functions within a single location can be graded as a major nonconformity.

The initial result of the auditing process is the nonconformity statement. Part of documenting the nonconformity is the writing of a statement of nonconformity. The statement of nonconformity should be supported by objective evidence. Then, the NC statement will help the auditee understand the gap/mistake, enable containment/correction and establish corrective action after proper root cause analysis of the nonconformity. The best way of writing the nonconformity is just reproducing the requirement from the Standard or the organisation's documented information in quotes and saying not done, not evidenced, ineffective, etc. The NC record should identify the requirement or the reference document where the requirement is stated so that the auditee can clearly understand the nonconformity and refer to the Standard or documents. The reference can be the relevant clause number/statement of the Standard or reference to the specific internal documents of the organisation, statutory/regulatory/contract requirements and/or the customer QMS or process/product requirements. Thus, the first part of the nonconformity report will have three sub-parts:

a. Statement of the nonconformity,

b. Objective evidence and

c. Reference to the requirement clause number/document para number.

In other words, the nonconformity statement is a record stating what is wrong, what the evidence is and where it is correctly mentioned (as required/criteria). A sample Nonconformity report is given in Figure 11.2.

All audits add value, and the internal audits carried out by the organisation add value in the pursuit of continual improvement of the Quality Management System. Therefore, bringing out the existing gaps and aberrations in the practice of the Quality Management System during auditing is a value addition/help to the organisation. Some findings, which the auditor may report separately as observations or opportunities for improvement, are not nonconformities. In the opinion/judgment of the auditor, if the organisation takes action on the findings, it could bring improvements and better efficiency to the organisation. These findings are reported as 'Opportunities for Improvement' (OFI). The findings need to be studied in detail or investigated by the organisation and need to be acted upon by the organisation only when convinced that the solution is likely to bring benefits or improvements to the organisation's Quality Management System. It may be identified as strength in some processes, which can be used to improve weaker areas/processes. Auditors may feel that the Opportunities For Improvements (OFIs) add value to the auditing process. However, the organisation must evaluate the 'Opportunities For Improvements (OFIs)' and whether it can bring improvement.

'The following incident shows how the audit findings greatly benefit the management in controlling the organisation's activities. The 'Marketing Process' chief knows the whole business, determining customer satisfaction and receiving customer complaints, answering the auditor's questions in a well-run organisation in south India. What is being done to reduce the rejections/customer complaints?

"We can afford to have that rejection rate of 5%. We still earn a good profit. We don't do anything in particular."

Non-Confomity Report

Statement of Nonconformity (Ma) Clause 8.6 (Requirements: The urganization shall retain documented information on the release of products and services: The documented information shall include evidence of conformity with the acceptance criteria)

"Measurement requirements for product acceptance does not include required records of the measurement results"

Objective Evidence: "Fits and Clearances measurement record for Main Landing Gear 161W1xxx-42 Sl.No.MC0898P0xxx vide O/H Report No E_WML000131568xxx measures -

(a) ID as 0.375, OD is recorded as 'Nil, Replaced' (to ensure an assembly clearance of min 0.0005 and max 0.0015) for mating dimensions of one item and

(b) OD as 2.748 and ID as 'Nil and Outhoose repair' (to ensure an assembly clearance of min 0 0010 and max 0 0035) for another item. Measurement record accepted for assembly (next operation) by Authorized Inspector. (There is no evidence of any deviation caused in assembly)". (Non-conformity is likely to affect the performance of Main Landing Gear, hence rated as Major.)

Containment: A briefing was conducted to stress the importance of recording all required dimensions under fits & Clearance section of the CMM. (Briefing attendance attached as evidence.)

Root Cause:

1.The dimensions are available in the initial inspection records, staff did not transfer them into the Fits and Clearance table for assembly clearance to be recorded due to unawareness.

2. Staff wrongly assumed the replacement parts with relevant released certificates are restored to within the CMM limits because the parts were sent for complete restoration or NDT hence did not transfer the relevant dimensions to the fit & clearance table.

Corrective Action(s): Inspection Record shall indicate actual measurements to ensure fit & clearance and must be recorded. No remarks shall be indicated. Previous assemblies for the last three months to be verified for correct recording of fit & clearance

Corrective Action(s) verified: Evidences verified as below, satisfactory.

1. Completed NCR with the record of attendance for the briefing on the recording of fit & clearance verified.

2. The fit & clearance record completed afresh. It is a sample of new fit & clearance record format used for the current and future work.

Frevious assemblies for the last three months verified for correct recording of fit & clearance, disposed as satisfactory

Figure 11.2 Sample nonconformity report.

The human error system not being followed indicated the reasons for the rejection, but no corrections or corrective actions were taken, and no record of steps taken was kept.

"What is your annual business volume?" Two hundred fifty crores, the cost of poor quality is 12.5 crores. If you improve the quality, the losses can be reduced. What do you say?

The discussions were concluded with silence and a smile.

Can you arrange it? The question came all of a sudden while food was in the mouth.

"Yes, give an expert one crore, time of one year and the authority to implement and drive the system in the organisation. Rejections will be reduced from 5% to 1 % in a year, and an additional 11.5 crores you earn during the year and all the years that follow if you are seriously and consistently performing!" The food got digested, but the auditor had to report the matter to the management since the complaints and rejections from the customers were not acted upon, and there was no intention to act.

What happened later is unimportant to the audit process but is a lesson for all. The manager lost his job the next week, along with a coterie of about 50 people who collected their compensation for leaving the organisation. Not going into the reasons for the managerial decision, but the reasons for not acting on the customer complaints, the management was kind enough to explain what was happening behind the smoke screen of affording 5% rejection. The quality system was effective, and the cost of rejection was only on paper!'

The respective process owners take follow-up actions for any audit where the nonconformity has been reported. The steps to containment/corrections, systematic root cause analysis, implementing corrective actions, etc., have already been explained. Whether nonconformity is on quality system elements or the process or product/service, the steps leading to corrective action are the same. On completion of the corrective actions, the effectiveness of the corrective actions has to be verified. There should be no recurrence of the nonconformity if proper corrective action has been taken. That has to be verified in the subsequent product realisation process on the next day, next part, next project, and in parallel across all the processes/projects in the organisation. A collection of similar nonconformities across the organisation and the actions taken can be further analysed for effectiveness and overall quality system improvement. Identifying the mistakes/gaps and aberrations in the practice of the quality system on a timely basis, systematically analysing it and taking corrective actions effectively are ways to continue improvement.

An organisation remains excellent while dwelling at the peak for the process effectiveness. The longer the duration, the better. However, keeping the processes at the top level of effectiveness for a long time is not always possible due to the variations and huge efforts required for its control and sustenance. The audit helps identify the slippages, gaps and aberrations in the process, which happen occasionally, and correct them effectively for improvement whenever they are identified.

How the audit findings add value or improve the organisation's reputation or how the management demonstrates their good intentions for making the process effective for their business excellence is elucidated by the following incident/example, which occurred while auditing a leading MRO organisation in Southeast Asia.

> *'The auditor took note of the achievements, continual improvements and performance indicators of the organisation in the previous year before proceeding to the Maintenance Workshop where the Main Landing Gear assembly was being undertaken. While going through the final inspection and acceptance records of a few samples, the auditor noticed some alphabetic entries, where numeric characters were expected to be used to record the results of measurements after using measuring equipment. (It was written "Nil, Replaced" in one row and "Nil, Outhouse repair" in another row). Further examination showed that the measurement results were not recorded in numeric characters; some comments were written, and acceptance criteria were unavailable for direct correlation. The auditor explained the nonconformity and asked the process owner how the minimum and maximum assembly clearances were ensured without measuring the mating dimensions. (Please see the box below). Unable to give a reasonable explanation, the auditee welcomed the nonconformity. By the time the word reached the ears of the top management, senior managers had come down and verified the facts on the spot for first-hand information.*
>
> *The Whimper is a central NC on the core process of the reputed organisation! - was soon turned to a whisper, admitting a mistake/aberration of practice. The senior manager indicated his good intentions to contain the issue and show the improvement of the AQMS before the week-long audit ended. Admitting the mistake or fault has liberated the organisation's process, removing the aberration/indiscipline of writing words/explanations instead of recording the measurement results. The nonconformity identified provided a genuine opportunity for improvement. There is no merit in good intentions unless they are acted on'.*

Suppose the products meet customers' quality and on-time delivery requirements. In that case, customer confidence increases, customer satisfaction is enhanced and the organisation progresses. If the product does not meet requirements, customer confidence in the product and the organisation which provides the product or service diminishes, and the credibility of the certification process suffers. The mere display of a certificate on the wall does not benefit the organisation! One good external audit or customer audit is enough to expose the weakness. Instead, the organisation should periodically perform a practical gap analysis or internal audits to improve the system.

Some organisations get elated when they receive no nonconformity at the end of an external or customer audit. It is just out of the ordinary results. No auditor will take pride in writing nonconformities, and no one will hesitate to write a nonconformity if it is evidenced in the samples he has verified. No auditor is paid proportional to the number of nonconformities they write! No organisation is perfect, and even if their processes are effective for some time, the trend chart of business will never be a plot of uniform acceleration for a long time.

> *'However, some auditors retain their reputation and survive for a long just by their charming personality, writing one or two non-conformities on the control of documents or management of records - they write, "the retention period of records expired, and the records are not disposed of". Some auditors make statements of nonconformity without following the definition of nonconformity (non-fulfilment of a requirement), for example, "The process of identifying and managing the risks is inadequate," with evidence that there exists a practice of risk management, but the process owner has not identified all the risks (perhaps known to the auditor!). Deficiency, inadequacy, incompleteness, or partial compliance is not a non-fulfilment. Therefore, auditors should desist from writing such statements as a nonconformity. If improvement is intended, an observation or opportunity for improvement would have been sufficient. Once they divide their attention on trivial issues, potential value-adding findings do not come to their notice! They claim value addition to the client in their audit process and prepare their report while waiting at the airport. The management also nominates some team leads to look into customer complaints and process deficiencies, attending the audit completion party with a sigh of relief, saying 'cheers' that the escape was great, no one let the cat out of the bag as the auditor was in a hurry to catch the 9000 km flight back! Apart from being competent to do the most revered job, the auditors need to have the passion and consciousness to look for safety and quality for ASD products and services and ensure that clients achieve it through an effective system in the organisation.*
>
> *The calibration methodology for the different grades of aerospace auditors needs to be strengthened. It may include thorough domain knowledge, professional acumen, and auditing knowledge. Some CB*

auditors combine or compensate their auditing abilities with their marketing or management expertise with the same clients during the audit. Accreditation bodies need to vigorously enforce the rules uniformly throughout the world'.

Organisations can have a say in choosing credible, value-adding, experienced auditors from the CB while planning for external audits and requesting the removal of any auditor from the team. After the audit, every CB has a process for addressing disputes and appeals, which the clients can use if they have genuine concerns about the audit and audit findings.

After all, for the CBs, certification/registration is a business process; most of the CBs all over the world are doing it in a balanced manner, following the ethics and principles of audit and commercial interests. Accreditation bodies are there to oversee the audit process, and occasionally, they issue nonconformities to erring CBs for necessary corrective actions.

An organisation is excellent when it continually improves performance while trying to achieve its objectives. This is an ideal position where most organisations wish to dwell and others envy or try to emulate. A dedicated, systematic and planned effort is necessary to succeed. Sustaining it for long periods requires more and better effort. A self-evaluation is always better. The third-party certification bodies can never fully judge the excellence of an organisation!

An organisation practising a modern Quality Management System and endeavouring to leave a legacy behind could be synonymous with one aiming for similar and sublime life objectives and trying to achieve ultimate objectives. Trying to reach higher levels of excellence is more important than getting the fruits of the efforts. Therefore, the quality of life and ultimate objectives could be achieved by following the AQMS standards, like any excellent organisation.

It is compliance and continual improvement – implementing the requirements of the Quality Management System and taking actions for continual improvement of its effectiveness – that can bring improvements to human life and performance improvements to organisations which look forward to business excellence.

Annexure A: Bibliography

[1] ISBN: 978-1-4452-6743-2, Thomas Orchard Publishers – Safety Follows Where Quality Leads – K. T. Thomas (2010)
[2] ISBN: 978-81-924457-5-5, Thomas Orchard Publishers Ltd. – A Complete Guide to Implement Aerospace Standard AS 9100 C – K. T. Thomas (2012)
[3] ISO 9004, 'Managing for the sustained success of an organisation – A quality management approach.'
[4] ISO 10001, 'Quality management – Customer satisfaction – Guidelines for codes of conduct for organisations.'
[5] ISO 10002, 'Quality management – Customer satisfaction – Guidelines for complaints handling in organisations.'
[6] ISO 10003, 'Quality management – Customer satisfaction – Guidelines for dispute resolution external to organisations.'
[7] ISO 10004, 'Quality management – Customer satisfaction – Guidelines for monitoring and measuring.'
[8] ISO 10005, 'Quality Management Systems – Guidelines for quality plans.'
[9] ISO 10006, 'Quality Management Systems – Guidelines for project quality management.'
[10] ISO 10007, 'Quality Management Systems – Guidelines for configuration management.'
[11] ISO 10008, 'Quality management – Customer satisfaction – Guidelines for business-to-consumer electronic commerce transactions.'
[12] ISO 10012, 'Measurement Management Systems – Requirements for measurement processes and measuring equipment.'
[13] ISO/TR 10013, 'Guidelines for Quality Management System documentation.'
[14] ISO 10014, 'Quality management – Guidelines for realizing financial and economic benefits.'
[15] ISO 10015, 'Quality management – Guidelines for training.'
[16] ISO/TR 10017, 'Guidance on statistical techniques for ISO 9001:2000.'

[17] ISO 10018, 'Quality management – Guidelines on people involvement and competence.'

[18] [16] ISO 10019, 'Guidelines for selecting Quality Management System consultants and using their services.'

[19] ISO 14001, 'Environmental Management Systems – Requirements with guidance for use.'

[20] ISO 19011, 'Guidelines for auditing Management Systems.'

[21] ISO 31000, 'Risk management – Principles and guidelines' ISO 37500, 'Guidance on outsourcing.'

[22] ISO/IEC 90003, 'Software engineering – Guidelines for applying ISO 9001:2015 to computer software.'

[23] IEC 603001, 'Dependability management – Part 1: Guidance for management and application.'

[24] IEC 61160, 'Design review.'

[25] Quality management principles, ISO

[26] Selection and use of the ISO 9000 family of standards, ISO3

[27] S1000D (Issue No. 4.1) International specification for technical publications using a common source database – Aerospace and Defence Industries Association of Europe.

[28] ISO 9001 for Small Businesses – What to do, ISO3.

[29] Belinda Waldock – 'Being Agile in Business', Pearson Education Limited, UK.

[30] ASQ, Arun Hariharan- 'The Strategic Knowledge Management Handbook.'

Annexure B: IAQG Aviation, Space and Defence Standards

[1] IAQG* 910 Quality Management Systems Audit Requirements for Aviation, Space and Defence Organisations
[2] IAQG* 9102 Aerospace First Article Inspection Requirement
[3] IAQG* 9103 Variation Management of Key Characteristics
[4] IAQG* 9107 Direct Delivery Authorization Guidance for Aerospace Companies
[5] IAQG* 9110 Quality Management Systems – Requirements for Aviation Maintenance Organisations
[6] IAQG* 9114 Direct Ship Guidance for Aerospace Companies
[7] IAQG* 9115 Quality Management Systems – Requirements for Aviation, Space and Defence Organisations – Deliverable Software
[8] IAQG* 9116 Aerospace Series – Notice of Change (NOC) Requirements
[9] IAQG* 9120 Quality Management Systems – Requirements for Aviation, Space and Defence Distributors
[10] IAQG* 9131 Quality Management Systems – Aerospace – Non-conformance Documentation
[11] IAQG* 9132 Data Matrix Quality Requirements for Parts Marking
[12] IAQG* 9133 Qualification Procedure for Aerospace Standard Products
[13] IAQG* 9134 Supply Chain Risk Management Guideline
[14] IAQG* 9162 Aerospace Operator Self-Verification Programs
[15] ISO 9001 Quality Management Systems – Requirements
[16] www.iaqg.org IAQG Standards Support Material
[17] www.everyspec.com
[18] IAQG Supply Chain Management Handbook

Annexure C: Documented Information

Note: If you are part of a certified organisation, the existing (AS 9100 C) documentation (like quality manual and procedures, work instructions and records) is adequate for implementing and practising the AQMS. If you are not unhappy to maintain additional 'Documented Information', identify and customise it. Update the documentation for the additional requirements, stakeholder needs and clarifications in AS 9100 D/EN 9100:2018 (***Italics and bold***). Review periodically for effective implementation. Carry out a Strength, Weakness, Opportunities and Threats (SWOT) analysis or a Political, Economic, Social, Technological, Legal and Environmental (PESTLE) factors analysis. The information can be used in the 'Context of the Organisation'
Context of the Organisation
All 'documented information' available on well-established websites of reputed organisations indicates the context of that organisation. Establish a similar documentation. Give a brief of products and services offered by the organisation. Indicate the following information, which may be necessary for certification bodies, prospective customers and other stakeholders. List out: Name, Address, Contact details, and E-mail id: Owners, Partners, Board of Directors, Share Holders, Other Stakeholders, Interested Parties. Background information about how the organisation was established, corporate governance/management .... Products, like manufacturing precision components, aircraft structures and systems, aero-engines, rod-end bearings ... MRO of aircraft, industries we serve ... strategic relationships we have We provide services like business technology consulting, software engineering services, engineering analysis, detail engineering, mechanical systems, structures, product lifecycle solutions, out-sourcing, engineering process services, supply chain solutions, software testing, manufacturing engineering, obsolescence management, embedded systems and electronics. We comply with the following statutory/regulatory requirements: Our customers are: Our special processes (in-house) Our external providers and suppliers are: Other interested parties are customers, government and non-government organisations, QMS Certification Bodies, and our employees. (Give a brief description of all.)

Our facilities, equipment, machinery, special processes, employees, competency, special skills … Strengths, opportunities, mission, vision, values, quality policy, policies and guidelines, social and community initiatives ecological sustainability: conserve energy, conserve water and sustainably manage waste, preserve and enhance biodiversity; Business volume for all customers Business volume for ASD customers, market reputation, share values, quarterly financial results trends for the last two years External issues/context can include issues arising from the legal, technological, competitive, market, cultural, social and economic environments, whether international, national, regional or local. Internal issues/context can include issues related to values, culture, knowledge and performance of the organisation, indiscipline, unauthorised agitations, etc. Interested parties could include the organisation's shareholders, employees, customers, end users, suppliers, regulators, etc. Needs and expectations, requirements of interested parties may consist of specifications from raw materials manufacturers/ suppliers, concerns on employee feedback forms, contractual requirements from customers, corporate policies and procedures, legal requirements, etc.
SCOPE (Examples)
1. Design, development, verification and validation of software, hardware and mechanical engineering services for aerospace applications. 2. Fabrication of stamping tools and metal stamping for precision components. 3. Manufacture of precision machined components, sub-assemblies and assemblies of control valves, transducers/sensors and actuator systems for aerospace industries. 4. Manufacture of machined parts – such as rod-end bearings, spherical bearings, ball, sleeve and swash bearings for aircraft and other industries; semi-finished parts such as aircraft fittings, race bush, body, nut, roller and bolt.
Quality policy (Examples)
1. ABC is committed to exceeding customer expectations through implementing and continuously improving our Quality Management System. Absolute customer satisfaction is the expectation and will be achieved through supplying a superior product on time, at a competitive price. 2. 'To fully meet commitments to all our stakeholders by complying with the Quality Management System and deliver to or above their expectations through continual improvement'.

3. 'We are committed to complying with customer requirements and shall strive to continually improve the effectiveness of our Quality Management System to exceed customer expectations'.
The safety policy of ABC Engg Pvt Ltd. is to maintain the highest standards of safety with respect to personnel, products and services at all times. We shall ensure that products released from the organisation meet the safety requirements of all personnel who use them. Safety objectives: We at ABC Engg Pvt Ltd • recognise safety as the primary objective at all times for all personnel; • ensure that safety standards are never compromised for commercial imperatives; • apply human factors principles diligently; • maintain safety standards and practices throughout the organisation as specified in the maintenance data, associated technical literature and documents. • encourage open reporting of all safety-related issues and continually improve systems and procedures; • - impart safety awareness training to all personnel.

Functional Objectives for Key Processes/Key Performance Indicators (KPI)s

Objectives/key process/ performance indicators (KPI)s as (examples in some processes)	Method for determining process/performance results	The target for the period Jan to Dec, Year	Achievement for the period Jan to Dec, Year	State process is effective or not. If not effective, indicate the actions taken for improvement
Safety objective				
Recognise safety as the primary objective at all times for all personnel	Counting the number of incidents/injuries noticed and/or reported. Each is classified as Minor/Major	Nil injury/ incidents	1 with minor injury, 3 of a minor nature without injury	Incident investigated. The equipment responsible has been replaced. All other equipment/ tools/facilities verified – oil spillage, housekeeping improved to prevent more instances/safety hazards
Production/Engg Service				
On-time delivery (OTD), on-time in full 100%	On-time delivery of products to customers in total quantity percentage	95 (internal)/99 (customer)	100 (internal)/100 (customer)	Process is effective
OTIF (on time delivery in total quantity)	Percentage completion of required quantity as planned and delivered	92%	95%	Process is effective
Defects critical	Per hundred hours (DPHH) (internal/ customer)	0.2 (internal)/ 0.0046 (customer)	0.065 (internal)/ 0 (customer)	Process is effective
Defects functional	Per hundred hours (DPHH) (internal/ customer)	0.35 (internal)/ 0.0106 (customer)	0.16 (internal)/ 0 (customer)	Process is effective

Defects appearance	Per hundred hours (DPHH) (internal/ customer)	0.75 (internal)/ 0.0268 (customer)	0.16 (internal)/ 0 (customer)	Process is effective
Defect density	Percentage	0.5%	0.5	Process is effective
Effort variance	Percentage within the budgeted/estimated effort (man-hours)	5%	3.5%	Process is effective
Schedule variance	Percentage within the budgeted/estimated time (hours)	3%	3%	Process is effective
FAI planned and achieved/ FAI prove out	No FAIs of parts accepted by the customer during the period	12 per month/ 60 days per part	14 per month (average)/55 days per part	Process is effective
Production plan vs. achieved	Percentage completion of quantity	As planned 100%	90%	The process is ineffective due to the prioritisation of some jobs. Production schedules have been revised. With better planning, 100% plan/quantity completion is expected from next month
Productivity:	Measured as percentage	5% min	8.6%	Process is effective
Marketing/sales				
Customer satisfaction (C SAT)	Percentage	75	85	Process is effective

(Continued)

Customer satisfaction	On a weighted average scale, max 132. Survey to customers and customer ratings of process performance	100/132	96/132	The process is not practical; deficiency was in delivery performance. Staff augmented to achieve better results
RFQ win rate	Orders received vs. quotations sent	25%	30%	Process is effective
Planning				
Availability of material and batch file	At least 60 days in advance	60 days	57 days	Process is effective
On-time receipt of materials and tooling	For more than 90% of orders per month	90%	95.4%	Process is effective
Issue material and tooling as per the production plan	More than 95% per month	95%	100%	Process is effective
(Innovation engineering) Deviation on parts	Due to mistakes in process plans, TPI, tools, and fixtures as a percentage of processes planned	1%	0.92%	Process is effective
Cycle time reduction for parts	Achieved by effective process planning	hours saved per part x No of parts per year =1000	1105 part hours/year	Process is effective
Schedule adherence	Date of delivery agreed	100%	100%	Process is effective
Quality control				
Defect injection rate (internal)	Defects per 100 person hours of efforts	3.0 (Max)	1.91	Process is effective

Cost of quality:	Appraisal cost as a percentage of total effort	Max 20.0 %	18.4%	Process is effective
Quality escaped	Defects per 100 person hours	0.5 (max)	0	Process is effective
Rejection in first inspection (in-process)	Percentage of parts rejected	Less than 2%	0.89%	The first inspection resulted in less than 1% rejections on average throughout the year from April to March, effective
Rejection in final inspection	Percentage of parts rejected	3.0%	3.6%	Final inspection rejections did not meet the targets, marginally more than the desired level. This was analysed and found to be due to the large lot size, 100% check and time constraints. Management has addressed the issue and allowed adequate time and focus to detect defects through process improvements
Cycle time improvement	Through tooling, process changes and process parameter changes (10 operations/ week mandatory)	10	9	The target was almost met. Holidays during the week; hence, the target was not achieved. Acceptable to the management
Production percentage	Average production output vs. production plan	95%	103.2 %	Production exceeded the plan, effective

(Continued)

Roll through yield	No. of good parts produced within the month/total no. of parts produced within the month	99%	91%	Process not effective. The target was reduced to 91% based on trend capacity and capability
Capacity utilisation	Capacity utilisation of plant	More than 90%	91%	Target met effectively – further improvement in capacity utilisation possible through scrap reduction
Reduce overtime	Fewer overtime bookings planned	Less than 100%	92.8%	Overtime is allowed and is limited to being within the targets set. Achieved the plan and found an effective
Production	Minimum shipment of parts to plants at ABC/ DEF	More than 99%	98.24%	Not effective. The minimum shipment was marginally found less due to a shipping shortfall in May 2014, when the OTD was also less. Action taken to improve the delivery /shipment
Machine utilisation	Machine utilisation of more than 75% month wise average of all CNC machines	75%	69.70%	Waiting for more jobs, the orders will increase by next month, and the machines will be utilised by 75%
Quality: rejection/ rework	Rejection and rework not more than 5% of production	5%	(3.28) 2.13% Rejection Rework 1.15%	95% yield is good for the complex process. Rejections are less process-effective

Preventive maintenance	Adherence to a preventive maintenance schedule of not less than 95%	95%	97.60%	Process effective. Preventive maintenance is doing well
Zero recurrence (maintenance)	No repeat breakdowns on equipment after rectification	zero	zero	Well maintained. Process effective
Production	Assembly rejection for the transducer assembly	2%	1.89%	Production exceeded plan, process effective
Surface treatments quality	Defects in surface treatment processes	1%	0.80%	Production exceeded plan, process effective
Purchase				
Project plan OTD	Ensuring timely delivery of desired quality materials for each project (as per project plan)	98%	100%	Production exceeded plan, process effective
	Savings in purchase cost through negotiations	1% of estimated cost	1.08 %	Process is effective
Information technology				
Server uptime %	Actual available server up time *100/total available time	98%	98.59	Process effective

Annexure D: Positive Recall Procedure

The alert/escape notification may sometimes be documented as 'the positive recall procedure' by the organisation. There are three occasions where positive recall shall be exercised while implementing the Standard AS 9100 D/ EN 9100:2018. Requirements are indicated in italics and bold.

Requirement 1. ***When an externally provided product is released for production use pending completion of all required verification activities, it shall be identified and recorded to allow recall and replacement if it is subsequently found that the product does not meet requirements.*** (9100:2016, Clause 8.4.2) A positive recall procedure is normally avoided for parts with incomplete verifications during receipt. Such parts are quarantined.

Requirement 2. ***The organisation shall establish, implement, and maintain a process for the recall of monitoring and measuring equipment requiring calibration or verification.*** (9100:2016, Clause 7.1.5.2)

Requirement 3. ***The controlled conditions shall include, as applicable, the identification and recording of products released for subsequent production use pending completion of all required measuring and monitoring activities to allow recall and replacement if it is later found that the product does not meet requirements.*** (9100:2016, Clause 8.5.1 q)

(In MRO organisations, it includes Maintenance Error Data Analysis and Reporting to Customer (MEDA).) The product on which measurements or monitoring is incomplete but progressed for assembly/testing will not be released to the customer.

Annexure E: Nonconformity Report (NCR) Process Example

Name of the organisation			Organisation logo	
			Format No..........	
Process/product			NCR No	
Requirements per document/ criteria for acceptance (indicate Part No, Sl No, Assembly No, etc.)	Pt No. Sl No.		Issue date	
	Process specification:			
Statement of nonconformity				
Objective evidence				
Due date:	Originator name and signature	Name Sd/-	Process owner/rep (acknowledgement) name and signature	
Planned actions: containment, including correction			# Wherever necessary, attach additional pages	
1.			Planned completion date	Actual Completion Date
2.				
3.				

Root cause analysis:		Cause code		
Proposed corrective actions		Process owner/rep name:	Planned completion date	
		Signature		
Process owner/rep name and signature			Actual date of completion	
Note below the verification of correction, corrective actions/evidence for acceptance:			Date:	
Sd/-			Sign accepted/ rejected by:	Authority approved
Effectiveness of corrective actions completed (after sufficient time/evidence obtained with date)			Verified/approved by:	Next due date
Note: Root Cause Code (enter code in adjoining space on the right): 1. Means/Resources 2. Methods 3. Environment 4. Materials 5. Machines/Equipment 6. Training/skill 7. Measurement 8. Weak application 9. Freak 10. Design related 11. Procedure not followed, 12. Tooling, 13. Process deficiency/inadequacy 14. Workmanship 15 Handling/transit damage 16. Others.				
Repeat verification of 'effectiveness of corrective actions completed (after a year/no recurrence of the nonconformity) evidence obtained with date:		Verified/approved by:		Next due date

Annexure F: Management Review Agenda/Minutes (Example)

Organisation Logo Agenda/Decisions: Management Review SL No. ……Scheduled/held on Date: ……, Time: …………Venue……………		Presented by process owner, record points of discussions	Responsibility for action	Target date
Sl. No.	Agenda points			
1	Follow-up actions from previous management review			
2	Changes in external issues Changes in Internal Issues			
3	Performance and effectiveness of QMS processes			
4	Performance measures, KPIs of all processes Objectives: Targets and achievements. Review objectives/KPIs			
5	The output of the monthly production review meetings was held every month.			
6	Review all audits (internal audit, customer audit, regulatory authority audits, certification/surveillance audit by certification body and audit on suppliers by the organisation) and audit results for continual improvement. Is the QMS in our organisation effective?			
7	Number of nonconformities reported on deliverable products and services (internal defects), including in-house rejections of products, rework data/ concessions and production permits			

8	Status of corrective actions, product quality issues, customer complaints			
9	Customer satisfaction – trends, resolutions status			
10	On-time delivery (OTD) performance			
11	Internal and external audits and findings, status			
12	Premature withdrawals/defect reports on equipment/components; incident/ accident reports on aircraft			
13	Review of resource requirements, business plans, budgets, upcoming projects, status of ongoing projects, significant changes and resource needs			
14	Equipment repair and maintenance, downtime; information on calibration of measuring devices – data and trend analysis			
15	Recommendations for improvement, changes that could improve the Quality Management System, review continued suitability, adequacy and effectiveness			
16	Review employee efficiency and objectives, as well as training requirements.			
17	Adequacy of the policy and objectives, periodic measurement of quality objectives, including trend analysis, improvements needed in policy and objectives			
18	Supplier (external provider) performance: Additional controls needed before the introduction/outsourcing of a new process			
19	Improvements related to product, service and customer requirements			
20	Any additional resource needs?			
21	Review of risk register, including any additional risks identified			
22	Review the risk register for the effectiveness of mitigation actions. Did they achieve planned results?			

23	Are there any external and internal issues likely to impact business operations? Please include it in the risk register and review the current status			
24	Needs and expectations of interested parties, and requirements, which are likely to impact business operations? Any additions are to be managed in the risk register			
25	Any opportunity for improvements? System, products/services, logistics, etc.			
26.	Any other points with the permission of the Chair. (Decisions taken in production review meetings, customer visits, social responsibilities of the organisation, commitment for development, etc. could also be discussed, if any.)			
27	Plan for the following review: method of review, date, time, and venue, and identify if there are any special invitees for the review			
All of you are requested to attend and present your report/points/observations for discussion, with a copy of inputs to the undersigned.		Minutes of the meeting, approved by the chair, are circulated herewith for information and necessary action.		
MANAGEMENT REPRESENTATIVE		MANAGEMENT REPRESENTATIVE		
Name……………………….Signature…………..…….. Date……….. **Addressed to List of Members (MRM)**		Name……………………….Signature……….. Date………..		
		Addressed to List of Members (MRM) (present and could not be present)		
Note: Include presentation data, analysis reports, charts, etc., to show examples of objective measurements, nonconformity, CAR, quality and on-time delivery performance figures, trend charts, related actions, etc. In some organisations, MRM is known as the Committee of Management.				

Annexure G: AS9104/1A Audit duration calculation (Consistent with IAF MD 5) is available in the TABLE 2 – AUDIT DURATION REQUIREMENTS vide SAE standard AS 9104/1 (page 23 of 49)

Number of employees	9100/9110			9100/9110 Less Design (7.3)			9120		
	Initial	Annual surveillance	Recertification	Initial	Annual surveillance	Recertification	Initial	Annual surveillance	Recertification
1–5	2.0	1.0	2.0	2.0	1.0	1.5	2.0	1.0	1.5
6–10	2.5	1.0	2.0	2.5	1.0	2.0	2.0	1.0	1.5
11–15	3.0	1.5	2.5	2.5	1.0	2.0	2.5	1.0	2.0
16–25	3.5	1.5	3.0	3.0	1.5	2.5	3.0	1.5	2.0
26–45	5.0	2.0	4.0	4.5	2.0	3.5	4.0	2.0	3.0
46–65	6.0	2.5	4.5	5.0	2.0	4.0	4.5	2.0	3.5
66–85	7.0	3.0	5.5	6.0	2.5	4.5	5.5	2.5	4.0
86–100	8.0	3.0	6.0	7.0	3.0	5.0	6.0	2.5	4.5
101–125	8.5	3.5	6.5	7.5	3.0	5.5	6.5	3.0	5.0
126–175	9.5	4.0	7.0	8.0	3.5	6.0	7.5	3.0	5.5
176–275	10.5	4.0	8.0	9.0	3.5	6.5	8.0	3.5	6.0
276–425	12.0	5.0	9.0	10.0	4.5	7.5	9.0	4.0	7.0

426–625	13.0	5.5	9.5	11.0	4.5	8.0	10.0	4.5	7.5
626–875	14.0	5.5	10.5	12.0	5.0	8.5	10.5	4.5	8.0
876–1000	15.0	6.0	11.0	12.5	5.0	9.0	11.5	5.0	8.5
1001–1175	16.0	6.5	12.0	13.5	5.5	10.0	12.5	5.5	9.5
1176–1550	17.0	7.0	12.5	14.5	6.0	11.0	13.0	5.5	10.0
1551–2025	18.0	7.0	13.5	15.0	6.0	11.5	13.5	5.5	10.5
2026–2675	19.0	7.5	14.0	16.0	6.5	12.0	14.5	6.0	11.0
2676–3450	20.0	8.0	14.5	17.0	7.0	12.5	15.0	6.0	11.0
3451–4350	21.0	8.0	15.5	17.5	7.0	13.0	16.0	6.5	11.5
4351–5450	22.0	8.5	16.0	18.5	7.5	13.5	16.5	6.5	12.0
5451–6800	23.0	9.0	16.5	19.0	7.5	14.0	17.0	7.0	12.5
6801–8500	24.0	9.0	17.5	20.0	8.0	14.5	18.0	7.0	13.0
8501–10700	25.0	9.5	18.0	21.0	8.0	15.0	18.5	7.5	13.5
10,701–14,564	26.0	10.0	18.5	21.5	8.5	15.5	19.5	7.5	14.0
14,565–19,630	27.0	10.0	19.5	22.5	8.5	16.0	20.0	8.0	14.5
19,631–24,695	28.0	10.5	20.0	23.0	9.0	16.5	20.5	8.0	15.0
24,696–33,571	29.0	11.0	20.5	24.0	9.0	17.0	21.5	8.5	15.5
33,572–45,031	30.0	11.0	21.5	25.0	9.5	17.5	22.0	8.5	16.0
45,032–59,258	31.0	11.5	22.0	25.5	9.5	18.5	23.0	9.0	16.5
59,259–79,784	32.0	12.0	22.5	26.5	10.0	19.0	23.5	9.0	17.0
79,785–10,1635	33.0	12.0	23.5	27.0	10.0	19.5	24.0	9.0	17.5

Annexure H: Risk Register (Example)

	Organisation			Product/process			
	Team leader			Risk register			
	Members:			Scale 1–5	Scale 1–5	RPN	Visual risk
Risk item No.	Risk identified in processes – What can go wrong?	Risk description		Impact on	Probability on	Risk criticality	RED, YELLOW, GREEN
		Cause	Impact				
	PART I Organisational Level Risk and Opportunities (Refer para 6.1.2, and refer to para 9.3, Management Review)						
1	Quality	Not likely					
2	OTD	Not likely					
3	C Sat 100%	Not likely					
4	Int issues	Potential					
5	Ext issues	Potential					
6	Int parties	Needs difficult to meet					
7							
	PART II Process/Operational Level Risk Identification and Mitigation (Refer para 8.1.1 for All Processes)						
A	Supplier selection process (examples)						
1	Quality certifications (ISO, IS 13485, AS 9100 C etc.)	Not certified for the quality certifications	Customer requirement not met/customer dissatisfaction	2	2	4	Y
2	Manufacturer/ distributor	If the supplier is not a manufacturer	Non-availability of manufacturer test certificates	3	5	15	R

		PFMEA/FMEA number				Logo:		
			Date original: Date last reviewed:					
		Current controls available or not	(Review Frequency: every three months)					
Date reviewed	Risk owner	Mitigation action plan	Plan status – pending or date completed	Impact on a scale of 1–5	Probability on a scale of 1–5	Residual RPN= Risk criticality	R, Y, G visual	Next review
		Ensure the suppliers are certified by authorised certified bodies (or) new supplier to be identified and developed	Completed	2	2	4	G	
		Ensure that the distributor can provide the manufacturer's test certificate	Completed	1	3	3	G	

3	The capacity of the supplier to meet the delivery commitments	Shortfall in capacity	This leads to failure in customer commitment	3	5	15	R
4	Process capability of the supplier	Not capable of the process execution	Customer requirement not met/customer dissatisfaction	3	5	15	R
5	Customer approved supplier	Not a customer-approved source	Customer requirement not met/customer dissatisfaction	3	2	6	Y
B	Purchasing process						
1	a) Meeting the PPC requirement	Failure in commitment	Production schedule failure/failure of customer requirements	3	5	15	R
2	b) Delivery schedule of materials	Failure in delivery commitment	Production schedule failure/failure of customer requirements	2	2	4	Y
3	c) Pricing and delivery agreement	Increase in RM pricing and process cost	Failure to meet the internal part cost, customer delivery schedule	2	4	8	Y
4	d) Logistics requirements	Delay in reaching consignments to us, excess freight due to delayed supply, Possibility of theft/missing of consignment during transit	This leads to line stoppage in production. Increasing freight charges/decrease in profit margin decrease in profit margin/shortage at customer end	2	4	8	YELLOW

		Evaluation of the supplier capacity and project management	Completed	2	2	4	G	
		Evaluation of the supplier and process validation	Completed	2	2	4	G	
		It ensures that the supplier is an approved source for the customer	Completed	1	3	3	G	
		Follow up with the supplier and monitoring the delivery schedule, establishing alternate source	Completed	2	2	4	G	
		Always maintain some buffer stock to overcome this issue, establishing alternate source	Completed	2	2	4	G	
		Establishing pricing/delivery contract with the supplier	Completed	1	3	3	G	
		Buffer stock to be maintained, having alternate transporter, consignments should be insured to avoid loss during transit	Completed	2	2	4	G R E E N	

5	Natural calamities	Accidents during transit, such as fire/ earthquake/ storm/tsunami/ epidemic/ pandemic	Customer dissatisfaction. This leads to line stoppage at the customer's end. Reduction in share of business	2	4	8	Y
C	Sales/marketing process/contract review process						
1	Drawing specs not given in the PO						
2	Order quantity is not acceptable, considering the MOQ						
3	Price is not acceptable for the quantity						
4	The raw material is not available in stock						
5	The delivery schedule is not acceptable						
6	Machine time is not available						
7	The drawing/3D model from the customer has not been received						
8	Revision of file for repeat orders						
9	Material specs not available for procurement						
10	The ease of machining is not good						
D	Programmer, planning process						
1	The relevance of the process sheet is not available						
2	Tool list not available programs not validated fixtures not proven						
3	Deviation against the agreed lead time for calibration and delivery failure in adherence to calibration plan						
E	Mechanical design and development						
1	Customer inputs for modifications/design changes at later stages of the project under progress						
2	Appropriate database not maintained						
3	Drawings concerning assembly and final application were not assessed						

		Always maintain some buffer stock to overcome this issue. Insurance is necessary if any unexpected accidents occur – claim the cost	Completed	1	3	3	G	
		Check-in PO						
		Feasibility study						
		Feasibility study						
		Check and ensure it is not reserved for another file						
		Feasibility study						
		Discuss the machining time required with production and refer to the production schedule						
		Verify from the customer whether it is validated. If not, compare the drawing and model and conform before the file is launched						
		Check the reason for revision and update all the internal documents						
		Check from the customer the interchangeability of norms and approval status of the manufacturer						
		Plan sufficient tools for difficult-to-machine materials like Inconel, Titanium, etc.						
		Check whether it is required. If yes, generate.						
		Check the availability of tools, verify the programme, tools and fixtures, and validate based on the first part's production						
	Tech and commercial issues faced by the external provider, non-availability of masters for calibration. Instruments will be sent to other accredited calibration labs already identified and listed in the master list of external service providers							

4	If not checked for assembly requirements, part modelling used for machining leads to part rejection						
F	Electrical operations						
1	Data inputs on testing not correct, failure of tools/instruments general component stocks less, work share/job order details are not available in advance, electronic component procurement from unauthorised dealers						
G	Planning, customer processes						
1	Capturing the drawing and customer-specific requirements in PO	Inadequate review of drawing Inadequate brainstorming with the team	Product requirements in terms of material, manufacturing and special processes cannot be met, and the project timeline cannot be adhered to, customer dissatisfaction	4	3	2	R E D
2	Drawing/ material specifications to be met	Basic system requirements are not established or not adhered to by the supplier.	Product requirements cannot be met, the project timeline cannot be adhered to since RM needs to be imported and lead time is high, customer dissatisfaction	4	3	12	R E D
3	Shipping quantity to be ensured against PO quantity	Basic system requirements not established or not adhered to at the supplier less lead time or human error	The project timeline cannot be adhered to since RM needs to be imported and lead time is high, or there is insufficient supply to customer, customer dissatisfaction	4	3	12	R

		Complete a review of the drawing and clarify the details with the customer before the project execution. Documenting all the specific requirements	Completed	2	1	2	G	
		Material certificates received from the supplier and ensure that the requirements are met before the dispatch of the material from the supplier end	Completed	2	1	2	G R E E N	
		Invoice and packing list to be received from the supplier and ensure the requirements are met before the dispatch of the material from the supplier end	Completed	2	1	2	G	

H	Training, IT resources						
1	Failure to conduct the planned training program on time (in-house) due to the non-availability of faculty	Non-availability of the faculty or other critical organisational priority	Non-execution of the planned training program	4	3	12	R
2	Functional failure of essential IT facilities (projector, computer, malfunctioning of supporting software) for explanation and demonstration	Unexpected failures of IT systems	The training program will be interrupted	3	4	12	R

		A list of trainers covered the competence over various organisation processes. Their qualifications and skill levels are evaluated to understand their strengths in multiple sections. Alternate trainers will be organised based on the list of trainers as contingency and will be utilised.				
		Periodic verification of IT facilities by IT function is in place based on a check sheet. Verification of all required IT facilities before starting the training programme is done. As a contingency, the IT function holds critical items as alternate in case of failures. Completed			G R E E N	

Annexure I: Contract Review Checklist (Example)

By all concerned departments/functions (marketing, sales, customer services, design, production, quality, purchase, stores, etc. before responding to requests for quotation/enquiry and after receiving customer orders)					
ABC Engg Pvt Ltd	Checklist No:	Customer:	Enquiry/ Order No:… / Name Project Proposal No:	Contract Review No.	
				Date of review	Date of review
				Enquiry stage: Fill it out before sending the quotation. INDICATE REMARKS/ OBJECTIVE EVIDENCE	Order acceptance stage: Fill it out after receipt of the order._INDICATE REMARKS/ OBJECTIVE EVIDENCE
SL No	Requirements stated by the customer				
	PO No…………………………			WO No…………	
1	Drawing/specification/CADD data, Quality Plan:				
2	Verify input checklist for all requirements (Projects)			Completedn following:	
3	Part No. ……Part Issue……			Attach sheets if more than one part is listed	
4	Drawing No. Drawing Issue.....				
5	Conditions of Supply…			Ref:	
6	Identify Special Requirements/Critical Items/Key Characteristics.....				
7	Product details: Part Name/Brand name/ description/Specification …				
8	Is the drawing/specification a controlled copy issued by the customer?				
9	Does the drawing/specification No./issue match with the customer PO?				
10	Name and address of customer:				
11	Contact person with telephone no				
12	Delivery site address:				

13	Delivery date asked by customer	Agreed/Not agreed	
	Date scheduled:	Need to write to the customer: No/Yes	
14	Has the customer asked for approval of production processes before production?	Yes/No	
15	Is it possible to produce within the planned delivery date?	Yes/No Yes/No	
	Is any outsourcing to be considered?		
16	Are there any customer-supplied products/raw materials? If yes, specify the agreement pertaining to the a) method of receipt, b) inspection, c) verification, and d) payment		
17	Has the customer mentioned any prescribed methods for the following?		
	A) Storage, B) Identification, C) Maintenance		
18	Does the customer have any preferences regarding sources of procurement of materials and components to manufacture the product?	Yes/No If yes, specify the source:	
19	What are the methods of acceptance and acceptance criteria? Quality Plan: ...	Inspection by ABC/ customer inspection drawing dimensions/ notes	
20	Has the customer hired any inspection agency for third-party inspection?	Yes/No	
21	Specify whether you have informed the customer regarding the minimum area to accommodate the product if it is installed in an enclosed space.	Yes/No	
22	Has the customer asked for our measuring devices/test equipment calibration certificates?	Yes/No	
23	Has the customer given any prescriptions about reviewing and disposing of non-conforming products?	Yes/No	
24	Specify the prescriptions of the customer regarding 1. Handling, 2. Packing, 3. Storage, 4. Preservation, 5. Delivery, 6. Method of transportation?	1. 2. 3. 4. 5. 6.	

25	Does the customer scrutinise the quality records we maintain?	Yes/No	
26	Does the customer want to approve the product drawings, design calculations, test methods, etc.? Before manufacture?	Yes/No	
27	Has the customer prescribed identification and traceability methods on the final product before dispatch? Conformity certificate	Yes, Heat/Lot No./ Batch No.	
28	Is there any requirement for special processes for the product? How is it being done? Are any special requirements (such as a special process) indicated in the order/drawing/ specification?	Yes/No	
29	If special processes are yes, what are the customer's prescriptions to validate the special procedures? What reports are required by the customer?	Yes/No	
30	What are the inspection and testing methods prescribed by the customer during receiving of materials, in-process, and before dispatch to our premises?	Yes/No	
31	Are there any customer-specified hold points?	Yes/No	
32	Has the customer provided any inspection plans, test plans, or quality plans to be followed by us?	Yes/No	
33	Any other requirement the customer has specified in the contract/order (both technical and commercial)	Yes/No	
34	Are there any requirements not specified by the customer, but required for the intended use? Can we add any additional features/ characteristics? Any value addition? To the product/service or of the customer	Yes/No	
35	Are there any applicable statutory and regulatory requirements for the product? During production? During the intended use of the product?	Yes/No	

36	Review the risk management checklist before filling up	Explain whether risk mitigation was discussed or not	
	Risk evaluation with respect to Yes/No refer:		
	a. Is it a new technology?	Refer:	
	b. Is the delivery time short or adequate?		
	c. Do we have the capability to manufacture and supply?		
	d. Is the raw material available?		
	e. Is there any requirement for additional personnel training?		
	f. Can we use the customer material to validate the CNC programme/make the first part?		
	g. If the first part fails/deviates, are we ready to compensate for the cost of materials? What is the alternative? Is the risk acceptable?		
	h. Is commercial material available? How much difference in cost?		
	i. Obsolesces		
	j. Intricacies		
	k.		
	l.		
	m.		
37	Do we have the required tools/jigs/fixtures/templates?	Yes/No	
	Is there a requirement for new tooling?		
38	Is the price agreed as per the quotation/rate contract/purchase order?	Yes/No	
39	Is it a repeat order of the same product? As per PO, check and identify the Part No, issue/drawing number, and Issue/Rev str PO	Checked and found correct: Yes/No	
40	What were the issues during the supply of items last time/previous times? Are they resolved?	Yes/No/Nil	
41	Is the scope of work unambiguous?	Yes/No	
42	Is there any requirement in the order differing from our offer?	Yes/No	
43	If so, has this been taken up with the client and resolved?	Yes/No	

44	Are the responsibilities of the client and ours clearly defined?	Yes/No	
45	Does the organisation possess the resources/capabilities to comply with the customer's requirements, including special ones?	Yes/No	
46	Can the copies of technical specifications/drawings referred to in the order provided by the client and the requirements specified therein be met?	Yes/No	
47	Has the client provided all other appropriate reference documents (such as quality plans and design documents)?	Yes/No	
48	Are the documents provided by the client compatible? Adequate?	Yes/No	
49	Is the period stipulated for executing the enquiry/order practical/adequate? If not, has this been resolved with a client?	Yes/No	
50	Are the payment terms indicated in the order acceptable to us? If not, has this been resolved with a client?	Yes/No	
51	Are particular terms like fixed price, penalty clause, liquidated damages, insurance liability, etc., specified in the order? Are they acceptable to us? If not, have they been resolved with a client?	Yes/No	
52	Can we manufacture and supply the product on time?	Yes/No	
53	Deliverables as per Statement of Works (SOW)		
54	Detailed technical proposal with delivery schedules and bill of materials		
55	Customer TNST/CST no*	Indicate	
	(In case the customer fails to provide ST/ CST no's,		
	The company is not responsible for any check-post-related problems.)		
56	Transportation will be arranged by	Customer/By us	
57	Transit insurance	Required/Not required	
58	Transportation charges	To pay/Customer to pay	

59	Technical (Brief) a) Design capability/ activity		
	b) Development activity		
	c) Production/prototype activity		
	d) Quality assurance plan		
	e) Documentation		
	f) Service during warranty		
	g) Post delivery warranty		
	h) Detailed technical proposal with delivery schedules and BOM		
60	COMMERCIAL		
	a) Quantity b) Delivery date c) Unit price d) Total price e) Sales tax/Service tax/ Payment terms acceptable		
61	Resources: Infrastructure, human resources, software availability, training needs, test facilities, training needs, test facilities, etc. Is there any deficiency?		
62	Statutory/regulatory authority involvement?		
	Some more checkpoints:		
63	Is the product adequately defined (application requirements, etc.) to enable feasibility evaluation?		
64	Can the product be manufactured to the tolerances specified on the drawing?		
65	Can products be manufactured with Cpk's that meet requirements?		
66	Are Cpk's of our machines greater than 1.33		
67	Is there adequate capacity to produce the product?		
68	Can the product be manufactured without incurring any unusual costs for capital equipment?		
69	Is the cost of tooling affordable?		
70	Can any alternate manufacturing methods be used?		
71	Are all documentation, i.e., process flow, process control pan, PFMEA, inspection check sheets, configuration control plan, etc., accurate and clearly describe the process?		

72	Can the initial sample/FAI schedule be met?		
73	Are there any other points for consideration? Pl state:		
	SIMPLE SOFTWARE ENGINEERING SERVICES: CONTRACT REVIEW		
1	Does the contract correctly identify the parties?		
2	The scope of work is mentioned		
3	Customer requirements are understood and can be met		
	For product acceptance (e.g., quality, inspections and tests, verification, validation and any special monitoring)		
	Process design		
	Customer SOW/drawing/specifications		
	For delivery expectations		
	For post-delivery expectations		
4	Related standards have been reviewed and can be met		
	Statutory		
	Regulatory		
	Quality standards		
	Control Plan, QAP		
5	Unclear or ambiguous requirements are resolved		
6	Feasibility has been determined		
	We can meet order requirements		
	We have adequate resources		
	We have skilled personnel		
	We have the infrastructure		
7	Differences between the contract and proposal, if any, are resolved		
8	Methods of communicating with the customer are defined		
	Related to:		
	Product information		
	Enquires		
	Feedback		
	Concerns and complaints handling		

9	Requirements not stated by the customer are defined		
10	Does the contract provide payment at least 30 days after receipt of the proper invoice?		
11	Does the contract call for late payment penalties or finance charges?		
12	Is there any specific mention of the warranty period?		
13	Are there any confidentiality requirements?		
14	Is there any clause in the contract when invoked makes the organisation liable to pay more than the contract values?		
15	Is the contract duration specified?		
16	Is the termination period identified?		
	Order amendments, nature of amendments	Review	
	(Review for impact on technical features, delivery, price and risk factors and decide on acceptance and mitigations.)		
	1		
	2		
	(Make sure that the relevant documents and personnel are made aware of the changes indicated by the customer)		
	Follow-up actions:		

<table>
<tr><td></td><td colspan="3"></td><td colspan="4">INDICATE REMARKS and OBJECTIVE EVIDENCE in the above columns
Include functional reps for reviews</td></tr>
<tr><td>Signature</td><td colspan="2">Production Dept:</td><td>Customer Services</td><td colspan="2">Quality control</td><td>Decision of the manage-ment</td><td>MD/Director/CEO Date:</td></tr>
<tr><td></td><td colspan="2">Marketing</td><td>Sales</td><td colspan="2">Production planning and control (PPC)</td><td>MD/ Director/ CEO Date:</td><td></td></tr>
<tr><td colspan="2">Notes</td><td>Name</td><td colspan="4"></td><td></td></tr>
<tr><td colspan="2">Notes</td><td>Name</td><td colspan="4"></td><td></td></tr>
<tr><td colspan="2">Notes</td><td>Name</td><td colspan="4"></td><td></td></tr>
<tr><td colspan="3" rowspan="4">HOD marketing/sales will indicate the quotation and the reference of order acceptance sent to the customer

Ref.</td><td colspan="2">Ref.</td><td colspan="2" rowspan="4">To <u>PPC</u> for info and release of work order</td><td rowspan="4">Director/CEO date:</td></tr>
<tr><td></td><td></td></tr>
<tr><td></td><td></td></tr>
<tr><td></td><td></td></tr>
<tr><td colspan="3">Attach additional notes</td><td colspan="5"></td></tr>
<tr><td colspan="3">Special notes:</td><td colspan="5">Customise and reduce the table as per your business</td></tr>
</table>

Annexure J: Configuration Management Audit Checklist (Example)

Configuration management: Audit checklist		Format No. ....
Sl No	Please answer with objective evidence	Record objective evidence below
1	What is the part required by the customer as indicated in the PO?	Part No. .... Rev ....
2	Configuration planning and identification: (a) Did we plan to make the part as per the configuration matrix? (Refer .....)	
	(b) Are the correct tools/jigs/fixtures/CNC programmes identified and used for production?	
	(c) What is the part configuration planned for production?	
3	Is there any change in the configuration identified/ instructed by the customer after the issue of the PO?	
4	Was the configuration change identified as the customer instructed, and was the documented/route card updated?	
5	Was the tool/fixture/CNC programme validated before use?	
6	What is the part configuration produced during production as per the route card?	
7	Status accounting: What is the part configuration inspected and accepted for dispatch to the customer?	
8	What part configuration is dispatched to the customer per the final inspection report planned for production?	
9	Any other issues?	
10	An audit carried out by name (ensure impartiality and avoid conflict of interest)	Auditee: Rep marketing, planning /project management, production, tooling, quality
11	Any nonconformity reported? Indicate vide NCR FORMAT NO.	NCR No. MR/CM/2015/01.... Dated ....
	Name: Signature, date:	
Note: Audit plan: At least one part/project per month is to be completed before dispatch of that part/project deliverables.		

Annexure K: Planning for Changes/Change Management Form

Indicate change suggested/proposed	
Name and designation of the proposer	
Change in the QMS?	
Change in the customer drawing/production process/tooling/any other (specify)	
Refer document (pre-revised)	
Refer to process and its elements/procedure.	
Does change require a customer proposal or acceptance/concurrence? Specify	
Indicate the purpose of the changes and their potential consequences	
Does it affect the integrity of the quality management system? What is the impact of change?	
Necessary resources are available for the change?	
Does it require the allocation or reallocation of responsibilities and authorities?	
Review of the changes by the cross-functional team Use additional pages for record of review if required	Signature: Production Quality AGM
Change is considered, and the decision of the appropriate authority is made	Approved/Not approved Signature MR (QMS changes)/AGM (any other changes)
Where are the changes documented?	
Date of implementation with effect from?	
Responsible person to effect changes	
Changes implemented details	
	Signature MR (QMS changes)/ Engineering head (process or drawing changes)/AGM (any other changes)

Annexure L: Test Your Knowledge

Here is a list of typical nonconformities you may find in your process. What Standard AS 9100 D requirement is not met? Identify the clause number.

Sl No	Nonconformity	9100 Clause No
1	**Statement of nonconformity:** The internal audit process is not fully effective	
	Objective evidence: In the internal audit carried out in July 2017 (IA/03), top management and purchase processes were not audited. Specific requirements in EN 9100:2016, such as organisational knowledge, product safety and prevention of counterfeit parts, were not audited. The system was established, but there was an aberration in practice; hence, the nonconformity was classified as Minor.	
2	**Statement of nonconformity:** Management did not review the following points during the last management review: a. On-time delivery, b. External issues, c. Decisions and actions related to new risks, if any.	
	Objective evidence: Refer to (Report 04) the record of management review dated July 15, 2017. A system established, an aberration in practice, hence classified as Minor.	
3	**Statement of nonconformity:** The procedure for control of NC products does not include reviewing the effectiveness of any corrective actions taken.	
	Objective evidence: Documented procedure MP/01 issue 1 Rev 0 and NC report FIR-W-19 dated 29 July 2017. The system was established, but there was an aberration in practice, hence classified as Minor.	

(Continued)

Sl No	Nonconformity	9100 Clause No
4	**Statement of nonconformity:** Evaluation and approval of M/s xx (calibration agency) as a supplier of services is not evidenced. The supplier's name and details are not available in the register.	
	Objective evidence: Calibration services have been obtained from M/s xx vide. Calibration report No. RMTL/01/ P316107691-A1 dated 18 November 2016 and 19 October 2016. Approved supplier's register with scope of supply dated 3 March 2016, Format Q/7.4/02. The system was established, but there was an aberration in practice, hence classified as Minor. Conformity of services not affected.	
5	**Statement of nonconformity:** Performance evaluation of suppliers is not evidenced for 2016–17.	
	Objective evidence: PO No. X/KS/05403 May 2016 to M/s XX XX, for HT BAC 5619, Pt No 115A2776. From M/s xx xx xx, oil was purchased twice in 2016–17. The performance of these suppliers was not evaluated after receiving the services/products from them. The system was established, but there was an aberration in practice, hence classified as Minor.	
6	**Statement of nonconformity:** "Organisation shall maintain the identification of the configuration of the product to identify any differences between the actual configuration and the required configuration". This is not evidenced in the parts manufactured from customer-given raw materials.	
	Objective evidence: DC 16903907 dated 24 February 2017, Housing Bearing P 693A0042-00 Rev F. The customer indicated no original material traceability. No material origin traceability is followed throughout for manufacturing at xx xx. Final inspection and delivery of parts do not identify the configuration with regard to the traceability of the material. System established, aberration in practice, no product configuration is not likely to be affected as the customer issued material/accepted parts based on their work order number. The customer's work order cannot be considered adequate control of the part configuration at xx as the material supplied could be from different sources/traceability, and the traceability/configuration is a lifecycle-related information. It is rated as Minor since the traceability is a customer requirement and could not be verified during manufacturing. It was learned that the product goes into an aerospace customer after a particular process and inspection at the immediate customer. Corrective action may be done in coordination with the immediate customer xx.	
7	**Statement of nonconformity:** (a) Master list of measuring equipment does not indicate acceptance criteria and check method of verifications/calibrations.(b) Inspection reports do not indicate the ID of the instrument used for verification/ measurement.	

	Objective evidence: (a) Master list maintained in system QF/7.6/01. (b) Dimension inspection report Part P693A-0043-00 Rev F does not indicate the ID of the instrument used for measurement. Three Verners, each with a unique ID, are in the master list, 'Vernier calliper is used', as mentioned in the report. Which specific Vernier instrument was used is not identified. The system was established, but there is an aberration in practice; hence, it is classified as a Minor.	
8	**Statement of nonconformity:** Configuration management plan does not identify the customer's required part.	
	Objective evidence: The customer order is for Part Number 54-4409-501 Rev Nil, and the manufacturing process is described in drawing EH-544-4409 Rev G. Route cards for all the assemblies only indicate the drawing number and Rev status. Test sheet /results, however, indicate part number by a default test program, and the parts delivered are conforming. The final identification of the part manufactured and delivered was missing in the production documentation. (However, the manufacturing and delivery of customer parts was unaffected.)	
9	**Statement of nonconformity:** 'The organisation shall take actions to acquire the necessary competence and evaluate the effectiveness of the actions taken'. This has not happened regarding some training conducted in April 2017 and April 2018.	
	Objective evidence: (1) Training for quality policy was provided on 20 April 2017. No record of evaluation of the effectiveness of the training was evidenced. Effectiveness verification was planned after three months but has not been verified/recorded. (2) Training to avoid cross wiring dated 15 April 2018, attendance record. Effectiveness verification was planned after three months but has not been verified. Effectiveness not verified/recorded to date. (Minor system aberration and lapse in practice; hence classified as minor.)	
10	**Statement of nonconformity:** When nonconformity occurred, the quality process took action to dispose of the nonconformity, but no correction/containment was identified. The corrective action identified is not related to the root cause analysed. There is no evidence for implementing corrective action, including verification, to determine if similar nonconformities exist or could occur. Validation of the effectiveness of implemented corrective actions (for non-recurrence of the nonconformity) is not included in the NC record retained. Reqt: AS9100D 10.2.1 d When a nonconformity occurs, including any arising from complaints, the organisation shall review the effectiveness of any corrective action taken.	
	Objective evidence: NC Report No. B429, dated 9 January 2018. After RCA, no further action was recorded. Systemic aberration in practising NC/CA procedure. The product being delivered to the customer is not likely to be affected. Hence, it is classified as Minor.	

(*Continued*)

Sl No	Nonconformity	9100 Clause No
11	**Statement of nonconformity:** The organisation has not retained documented information on the review results. Contract review records were not retained.	
	Objective evidence: Identification of the requirements for products ordered by the customer is carried out partly with a preliminary enquiry response and order acceptance. Details of the requirements review are planned using a checklist with details vide checklist format no. FT-QMS-CD/F01 and the format for multi-functional review was not completed with objective evidence of the customer order and part details. Ref (a) Enquiry No. 35968, dated 14 November 2019, for spare parts, material 4340, HT-9 pcs. (b) Customer xx xx xx, Job Order 19/27405, dated 23 November 2019. (c) Quote No. 35700-R2, dated 25 August 2019, for bottle picking heads, Customer PO No 39897, dated 18 September 2019. (d) Mold for PCB Housing, Enquiry No. 35860 dated 8 October 2019, quotation sent in October 2019. The absence of the contract review records has not affected the parts manufactured and delivered to the customers, meeting all quality requirements. Hence, it is classified as Minor.	
12	**Statement of nonconformity:** 'The organisation shall identify and manage the risks associated with the external provision of processes, products, and services, as well as the selection and use of external providers'. This has not happened in the case of one approved service provider. The absence of risk evaluation for the calibration service provider has not affected or is not likely to affect the system; therefore, the NC is rated as minor, a minor aberration in practice.	
	Objective evidence: Refer to the record of supplier evaluation for service provider – M/s Calcon Calibration Services Ltd. Risk register external provider PCT/PUR/8010 dated 5 December 2017, Ver 3.0, Template in the procedure not followed. PO No. 19-1.02759-1 dated 7 July 2019, issued to the service provider, M/s Calcon Calibration Services.	
13	**Statement of nonconformity:** Shelf-life control is not established for consumables like spray paint.	
	Objective evidence: No evidence is available to confirm that the organisation is monitoring the shelf life of Abro Spray Paint, batch T 8030, Date of manufacture is in July 2015 and date of expiry is in July 2018. There are other hazardous or sensitive materials, like cleaning solvents, etc., available, for which there are no MSDS sheets for first aid and safety are not available.	

14	**Statement of nonconformity:** The organisation maintains a monitoring and measuring equipment register. The register includes the equipment type, the equipment type, unique identification, location and the calibration or verification method, frequency and acceptance criteria. However, the information about 'the calibration or verification method' and 'acceptance criteria' noted down is inappropriate.	
	Objective evidence: The M&M Equipment list maintained in the system with calibration due date for all equipment is indicated as December 2019.	
15	**Statement of nonconformity:** Customer complaints were resolved. Corrective actions and effectiveness verification of corrective actions recorded in the CAR are inappropriate.	
	Objective evidence: Corrective actions identified and implemented are only rectifications; scrap or replacement is done, which is recorded as the verification of the effectiveness of corrective action. Refer CAR 19-06-022, CAR -19-11-054, and CAR 19-08-033.	
16	The QMS documentation does not address customer, statutory and regulatory requirements or the identification of outsourced processes and their control.	
	Objective evidence: QMSM-Issue 02, Rev 00, dated 15 October 2017, and QMSP-01 to 21. Production processes are outsourced. Procedure QMSP 17 describes the outsourcing procedure. Methods and their controls are not indicated/ explained.	
17	**Statement of nonconformity:** The method of disposal of records is not indicated in the procedure for control of documents, QMSP 02. The list of records does not include records given/generated by suppliers.	
	Objective evidence: QMSM-Issue 02, Rev 00, dated 15 October 2011 and QMSP-02 for control of records.	
18	**Statement of nonconformity:** Although some functional level objectives are defined and measured, measurement metrics and measurements for the past months are unavailable for quality objectives identified under para 5.4.1 of the QMSM. There is no evidence of measurement of product conformity and on-time delivery performance of products/ services.	
	Objective evidence: Paras 5.2 and 5.4.1 of QMSM Issue 2 Rev 00 dated 15 October 2011. Functional/process-level objectives are defined and measured as per departmental manuals.	

(Continued)

Sl No	Nonconformity	9100 Clause No
19	**Statement of nonconformity:** The internal audit process is not practical. The schedule for the internal audit, conducted in June 2018, does not indicate the following: • The organisation's top management processes were not audited. • Statutory, regulatory and customer QMS and contract requirements were not identified and audited. • Undue delay was seen in closing of detected nonconformity. • Use of checklist/flow diagrams for the audit of the processes was not available for effective auditing. • NCR format does not include containment for the detected nonconformity and verification of the effectiveness of corrective actions.	
	Objective evidence: Ref QMSM para 8.2.2 and QMSP-14. • Did not consider the external audit results of December 2017 while planning the audits. • NCR issued in June (2017/xx/011) was closed on 18 and 20 January with undue delay. • NC report no. 2017/xx/011 Quality objective not measured in production control. Corrective action is not taken for all the functions in the organisation. (Refer NCR: xx/S2/03). One objective's measurement method is not indicated or measured from the two design process objectives.	
20	**Statement of nonconformity:** Control of external documents is not effective. External documents are not identified/ listed. The provisional status of printed drawings is not clearly defined in the QMSP-01.	
	Objective evidence: QMSP-01 does not identify external documents and their control. Engineering Standard, ES 004, does not indicate the status of control. Drawing No. 2000 0498 Rev PO1 Top Bracket R, printed from a soft copy held in the system and control is not identified.	
21	**Statement of nonconformity:** No specific recording of project xx's special requirements and key characteristics is evidenced. The review meeting minutes indicate the milestones/stages, but the design planning output is not updated periodically. However, the defence project has progressed to the second field trials stage.	
	Objective evidence: Feasibility study report of xx project and Technical proposal no. 50030-xx-GS-WE-11 dated 28 August 2008. Project xx does not see the chronological design milestones, project plan, or PERT/Gantt chart (planning output). Plan vs. actuals and periodic updating of the design and development plan is not evidenced for project xx. In the xx project, there is no identification of special requirements, key characteristics or critical items, for example, drawing no. 2000 0498 Rev PO1. Identification of crucial characteristics was seen in other projects.	

22	**Statement of nonconformity:** D&D changes are not controlled in accordance with the configuration management process.	
	Objective evidence: The Bill of Materials (QF-DEV-02) dated 28 February 2011 for connector assembly is as per drawing 106487, in which the issue status is 4, but the issue status is not indicated in the BOM. Part drawing no. 111844 is at issue 2 status. BOM does not indicate the Issue 2 status of the part.	
23	**Statement of nonconformity:** 'The organisation shall identify and manage the risks associated with the external provision of processes, products, and services, as well as the selection and use of external providers'. This is not evidenced.	
	Objective evidence: 1 . xyz Electronics and Industries Ltd. and xx Technologies have been selected and approved in early 2020. POs are regularly issued to them, e.g., PO No 16979 dated 26–20 and PO No 01710900 dated 8 January 2020 2. Xx xx Pvt Ltd. has been selected, and initial orders have been placed vide PO No 17215 dated 30 January 2020. However, there is no evidence that risks in respect of these three suppliers were determined and are being managed.	
24	**Statement of nonconformity:** Even though a format for contract review has been developed, records of the review results and actions arising from the review are not maintained.	
	Objective evidence: Supply Order No XYZ/APD/ASD/GSC/20-20/001 dated 27-11-20 received from XYZ. Business process 02 for enquiry handling does not include identification of special requirements; the format developed for contract review indicates identification of special requirements and risk evaluation. The format is not used for records of review.	
25	[12]**Statement of nonconformity:** The organisation has not communicated to external providers the requirements for the right of access by the organisation, their customer and regulatory authorities to the applicable areas of facilities and applicable documented information at any level of the supply chain, ensuring that persons are aware of their contribution to product or service conformity; their contribution to product safety; the importance of ethical behaviour.	
	[13]**Objective Evidence:** PO No SEFL 4400000131, dated 15 June 2018 • PO No xx xx 4200001404 dated 12 September 2018, to M/s xx xx, Vadodara • PO No 4200001076, dated 20 February 2018, to M/s xx Ltd. The NC is graded as Minor since no product/system is affected by the absence of the info (terms and conditions) in POs.	
26	**Statement of nonconformity:** Risk management is not effective.	
	Objective evidence: Initial assessment on 8 June 2018 shows no high risks, mitigation or residual risk. No review was evidenced after that date. Even after high risk from suppliers was noticed by delay in supply, resulting in a high impact of the first article submission to the customer, no evidence of risk review was seen. Ref M/s Thales PO No 1180858. Though the business risk is reviewed every CRR, acceptance of the remaining risk is not indicated.	

(*Continued*)

Sl No	Nonconformity	9100 Clause No
27	**Process/Area/Department:** Warehouse (incoming inspection, storage, kitting) Classification (Ma/Mi): Mi **Statement of nonconformity:** Finer dust was found accumulated in the bins used to store the raw material in stores.	
	Objective evidence: The accepted material stored after inspection was not appropriately preserved in the material storage area, and finer dust accumulated in one rack. Ex. MIL-W-83420 cable wire 7*19 strands.	
28	**Statement of nonconformity:** Purchasing information does not fully describe the product/service being purchased, including d) the identification and revision status of specifications, process requirements and other relevant technical data. This is not evident in the pending delivery order issued to ACP and the conformity certificate issued by ACP.	
	Objective evidence: Certificate of compliance issued by ACP Metal Finishing vide Ref AN80817 PCC No. C60483 for yellow Chromate as per MIL – C-5541 Class 3 and Ref AN 80343 PCC No. C 59957, for Black, anodised as per MIL-A-8625 Type II Class 2. Mil Std Rev status not indicated. This is against the pending delivery order by XXXXXX to M/s ACP, attaching drawing/specification. The delivery order is required to indicate the rev status of the specification. COC is also not indicating the rev status of the Mil Std followed. The parts met the coating requirements to the latest specifications and were verified and accepted by XXXX. Hence, the product is not likely to be affected/not affected. Hence, the classification is Minor.	
29	[11]**Statement of nonconformity:** The organisation has not evaluated the effectiveness of the actions taken (training/skill to achieve the necessary competence), and appropriate records are not maintained /not shown/incomplete.	
	[12] **Objective evidence:** Training Plan 2021 Micro milling to V Robert Clive, planned in February 2021. No record is available to confirm training was provided. • Training provided a record for employee 1130 Vanathaiyan Robert Clive CNC Milling Training on 16 April 2017. The evaluation column indicates Not evaluated. Evaluation column is signed by someone, but what was not evaluated, is not recorded. • Refer Training/Skill Training attendance record PAR#003-14 MM-105051-01 dated 9 May 2020, four persons attended, Trainee -Kumaran. How effectiveness was evaluated on the same day of the Training (Internal) is not clear. The effectiveness evaluation method (discussions/issues raised during/after the session) is not suitable to determine the effectiveness recorded on the same day, as turning/milling are skills developed over time and not in a few hours within a day. • Training/skill training Record (TRA-04-A) Cheng Zhe (Employee No. 1174) CNC Milling training given on 20 February 2021, by Sami, Evaluated Yes, Evaluated by 75% recorded, No other details are noted regarding what was evaluated. (Internal)	

	• All training records from 2016 to date are in the same manner. • Training (AS 9100 System Procedure) AS-004 Para C 3.2 indicates that training effectiveness verification is not required internally if the external trainer conducts certificates or tests to ensure trainees' competency to perform job functions. The standard requirement does not allow any such organisational freedom or discretion with respect to the maintenance of records of training (external or internal).	
30	**Statement of nonconformity:** 'The organisation shall retain documented information on releasing products and services. The documented information shall include Evidence of conformity with the acceptance criteria'. This is not evidenced.	
	Objective evidence: Unique identification of measuring equipment is not indicated in (a) inspection report dated 30 June 2018, for PT No 9AB200077-88 B1, used Pin gauge and VMS, etc., (b) report for PT No. RM1201857 Rev D, dated 24 June 2018, where there is space to write the instrument and unique identification, but not filled up. (c) Template validation SF/GR/FX report dated 11 January 2018.	

Answers

Note: TRY ANSWERING THE QUESTIONS INITIALLY WITHOUT REFERRING TO THE STANDARD.
VERIFY YOUR ANSWERS LATER WITH THE STANDARD **AS 9100D/EN9100:2018**
YOU GAIN CONFIDENCE.

Sl No.	9100 Clause No.	Sl No	9100 Clause No.	Sl No.	9100 Clause No.	Sl No.	9100 Clause No.	Sl No.	9100 Clause No.	Sl No.	9100 Clause No.
1	9.2	6	8.5.2	11	8.2.3	16	4.4.1	21	8.3.3	26	8.1.1
2	9.3.2	7	7.1.5/8.6	12	8.4.1	17	7.5.3.2	22	8.3.6	27	8.5.5
3	8.7/10.2.1	8	8.1.3	13	8.5.4	18	6.2/9.3.2	23	8.4.1	28	8.4.2
4	8.4.1.1	9	7.2	14	7.1.5.2	19	9.2.2	24	8.2.3.2	29	7.2
5	8.4.2	10	10.2.1	15	8.7/10.2.1	20	7.5.3	25	8.4.3	30	8.6

Index

D

E

About the Author

Thomas Kannattumadom is a leading quality professional and certified Lead Auditor for ISO 9001 and AS 9100 D (IAQG/PROBITAS approved/IAQG-OASIS) registered (OIN 6144704664) Aerospace Industry Experienced Lead Auditor, AIEA (Cert. No. 1619032761, valid until January 2022). Over the last 14 years, he has conducted over 900 man-days of audits across India, Sri Lanka, UAE, Singapore, Malaysia, Thailand, Saudi Arabia, Japan and the UK for EN9100:2018 certification. Thomas Kannattumadom was recognised as a SAMAR ASSESSOR by QCI in 2023. He is a Visiting Faculty at the HAL Management Academy, Bangalore. He has delivered nearly 300 training days to DRDO labs, HAL divisions throughout India and many other public sector and private sector aviation, space and defence organisations.

Thomas Kannattumadom retired on 28 February 2006, superannuating as Deputy Director/Principal Scientific Officer in DGAQA (Directorate General of Aeronautical Quality Assurance, Ministry of Defence) with nearly 40 years of experience. This includes over 11 years of teaching experience in mechanical engineering subjects in the College of Military Engineering, Pune, and 28 years of quality assurance experience with developing, producing and overhauling aircraft, aero-engines and airborne equipment of the Indian Air Force from trainer to frontline fighters. He was a technical member in several defect and accident investigations of aircraft/aero-engines, systems/components and special study/audit teams of the IAF and Ministry of Defence. His contributions to the quality and safety of Indian military aviation are significant. He has presented technical papers at several national/international conferences and seminars.

Career history:

- 2010–2021: Aerospace Industry Experienced Lead Auditor for AS 9100 C & D with various Certification Boards: Bureau Veritas, Intertek, TUV

Rh, NQA, ABS, DQS-UL and ROS; served clients globally across UK, Vietnam, Japan, Singapore, Malaysia, etc.

- 2006–2010: QMS-AS Auditor, KEMA, Bangalore, India.
- 1981–2006: Dy. Director/Principal Scientific Officer, DGAQA, Ministry of Defence, Bangalore.
- 1979–1981: Foreman, Quality Assurance (Eng. Equipment), DGQA, DGI, Min of Defence, Pune.
- 1966–1979: Lecturer, College of Military Engineering, Pune.

Relevant awards and certifications:

- Commendation from Director, DTD&P (Air)/DGAQA for modification to Jaguar Aircraft (Special Task), (1990); for modification to Ejection Seat (1988); for modification to defence Equipment (1981).
- 2017 IAQG-SANCTIONED 9100 D/9100:2016 REVISION TRAINING (Cert. No. Q9100-3U4-088155-300549), approved by AAB Probitas, status updated in OASIS, valid until January 2022.
- 2016 Exam about the ISO 9001:2015 training, TUV Rheinland.
- 2014 IAQG-SANCTIONED 9101E/9101:2014 REVISION TRAINING.
- 2012 IAQG PEAR and NCR Commentary (IAQG Aerospace Auditor Transition Training).
- 2010 IAQG-Sanctioned Aerospace Auditor Transition Training – 9100.
- Plexus/RABQSA approved.
- 2007 Aerospace Industry Experienced Lead Auditor, AS 9100 C and ISO 9001:2008, ANAB/RABQSA approved/IAQG-OASIS Registered, AO 1049010.
- 2005 Lead Auditor QMS ISO 9001–2000, Zandig, Bangalore (QCI approved).

For Product Safety Concerns and Information please contact our EU representative GPSR@taylorandfrancis.com Taylor & Francis Verlag GmbH, Kaufingerstraße 24, 80331 München, Germany